PROTOPLASMA
Supplementum 1

M. Tazawa (ed.)

Cell Dynamics

Volume One

Cytoplasmic Streaming
Cell Movement—Contraction and Migration
Cell and Organelle Division
Phototaxis of Cell and Cell Organelle

Springer-Verlag Wien New York

Prof. Masashi Tazawa, Ph. D.
Department of Botany, Faculty of Science,
University of Tokyo, Tokyo, Japan

Printed in Austria by A. Holzhausens Nfg., A-1070 Wien

With 147 Figures

ISSN 0934-8727
ISBN 3-211-82094-9 Springer-Verlag Wien-New York
ISBN 0-387-82094-9 Springer-Verlag New York-Wien

To a Pioneer who Introduced
a New Way of Thinking and Novel Methodologies
into Cell Biology

Prof. Dr. NOBURO KAMIYA

on the Occasion of His 75th Birthday

Contents

Cytoplasmic Streaming

Cell Movement—Contraction and Migration

Cell and Organelle Division

Phototaxis of Cell and Cell Organelle

Cytoplasmic Streaming

Protoplasma (1988) [Suppl 1]: 3–9

Cytoplasmic Streaming Regulated by Adenine Nucleotides and Inorganic Phosphates in *Characeae*

T. SHIMMEN*

Department of Botany, Faculty of Science, University of Tokyo, Hongo, Tokyo

Received August 21, 1987
Accepted November 18, 1987

Dedicated to Professor Dr. NOBURO KAMIYA on the occasion of his 75th birthday

Summary

The effect of adenine nucleotides and inorganic phosphates on cytoplasmic streaming was studied in *Nitellopsis* cells whose tonoplast had been removed by intracellular perfusion. The cytoplasmic streaming was energized by ATP and the half maximal concentration was about 0.09 mM, which was consistent with the values reported in *Chara* and *Nitella* (SHIMMEN 1978). ADP strongly inhibited the streaming, but AMP did not at physiological concentrations. Orthophosphate and pyrophosphate weakly inhibited the streaming. Based on these results, regulation of the cytoplasmic streaming by physiological concentrations of these compounds was discussed.

Keywords: Actin; ADP; AMP; ATP; *Characeae*; Cytoplasmic streaming; Myosin; *Nitellopsis*; Orthophosphate; Pyrophosphate.

Abbreviations: ADP adenosine 5′-diphosphate; AMP adenosine 5′-monophosphate; APW artificial pond water, ATP adenosine 5′-triphosphate; ATPase adenosine 5′-triphosphatase; CCCP carbonyl cyanide *m*-chlorophenylhydrazone; DAPP P^1,P^5-di- (adenosine-5′)pentaphosphate; DCCD dicyclohexylcarbodiimide; EDTA ethylenediamine tetraacetic acid; EGTA ethyleneglycol- bis-(β-aminoethyl ether)N, N, N′, N′-tetraacetaic acid; HEPES N-2-hydroxyethylpiperazine-N′-2-ethanesulfonic acid; MES 2-(N-morpholino)ethnesulfonic acid; PEP phosphoenol pyruvate; P_i orthophosphate; PIPES piperazine-N, N′-bis(2-ethanesulfonic acid); PP_i pyrophosphate; Tris tris(hydroxymethyl)aminoethane.

1. Introduction

Since the pioneering finding of KAMIYA and KURODA (1956) that the motive force of cytoplasmic streaming in *Characeae* is generated at the sol-gel interface where active sliding of the flowing endoplasm along the stagnant cortical gel layer occurs, attention has been focused on the molecular mechanism of the sliding. Extensive studies done so far by many workers clarified that the motive force is generated by interaction between bundles of actin filaments affixed to the inner surface of the stagnant chloroplast layer and a putative myosin in the flowing endoplasm (*cf.* KAMIYA 1986). Although biochemical studies on motility are scanty in this material, two types of cell models, tonoplast-free cell (WILLIAMSON 1975, TAZAWA *et al.* 1976) and plasmalemma-permeabilized cell (SHIMMEN and TAZAWA 1982 b, 1983) have contributed much to elucidating mechanism of cytoplasmic streaming. Direct demonstration of ATP as an energy source of the cytoplasmic streaming of Characeae was first carried out in tonoplast-free cells (WILLIAMSON 1975, TAZAWA *et al.* 1976) and then in plasmalemma-permeabilized cells (SHIMMEN and TAZAWA 1982 b, 1983). The Ca^{2+} control of cytoplasmic streaming has been throughly studies also taking advantage of the two cell models (cf. WILLIAMSON 1975, HAYAMA *et al.* 1979, TOMINAGA *et al.* 1983).

The quantitative relationship between the ATP concentration and the velocity of the cytoplasmic streaming was first studied in tonoplast-free cells (SHIMMEN 1978). The velocity of the streaming showed saturation kinetics in relation to ATP concentration and the half maximal concentration was 0.07 mM in *Nitella* and 0.09 mM in *Chara* (SHIMMEN 1978). Later, REID and WALKER (1983) measured the streaming velocity and the ATP concentration during application of energy metabolism inhibitors. The relation between ATP con-

* Correspondence and Reprints: Department of Botany, Faculty of Science, University of Tokyo, Hongo, Tokyo 113, Japan.

centration and the streaming velocity was linear. Such a relation is not possible in the kinetics of enzyme action. They postulated the inhibition of streaming by ADP, which increased upon the inhibition of energy metabolism. The present study is aimed at checking the inhibitory action of adenine nucleotides and inorganic phosphates on cytoplasmic streaming of tonoplast-free *Nitellopsis* cells.

2. Material and Methods

Nitellopsis obtusa was cultured as reported previously (SHIMMEN and TAZAWA 1982a). Internodal cells were isolated from neighboring cells and calcium depositions on the cell surface were removed by incubating cells in APW (0.1 mM K_2SO_4, 0.1 mM $CaCl_2$ and 1 mM NaCl) the pH of which was adjusted to 5.0 with 10 mM MES-Tris. Cells recovered in APW at least overnight before use.

Removal of the tonoplast was carried out according to TAZAWA *et al.* (1976). After both cell ends were removed by cutting with scissors, the natural cell sap was replaced with Mg medium (Table 1) by vacuolar perfusion. After the perfusion, the cell was kept on the perfusion bench, which was kept in a moist chamber to prevent cells from drying. After 20 min, during which the tonoplast was disintegrated, the interior of the tonoplast-free cell was perfused with Mg medium supplemented with adenine nucleotides and/or inorganic phosphates. After the second perfusion, both cell ends were closed by ligation with strips of polyester thread and the streaming velocity or ATP concentration was measured. By the second perfusion after disintegration of the tonoplast, the chemical composition in the cell is almost identical to that of the perfusion medium (SHIMMEN 1978). Intracellular ATP concentration after the second perfusion was sometimes buffered by adding 2 mM phosphoenolpyruvate (PEP) and 0.5 mg/ml pyruvate kinase (EC 2.7.1.40) to the perfusion medium. ATP concentration is fixed at a given level according to the following reaction.

$$\text{ADP} + \text{PEP} \rightarrow \text{ATP} + \text{pyruvate} \qquad (1)$$

After the second perfusion, most of the endoplasm effused out but some remained attached to the gel layer and moved in the presence of ATP.

ATP content was measured by luciferin-luciferase method according to MIMURA *et al.* (1983). Cells were frozen for 10–20 sec with liquid nitrogen and were boiled in the medium containing 10 mM EDTA, 25 mM HEPES-KOH and 0.3% H_2O_2 (pH 7.4) for 5 min. ATP concentration was measured in the medium containing 50 mM $MgSO_4$, 10 mM K_2SO_4 and 50 mM HEPES (pH 7.4) using the ATP photometer (Chemglow Photometer J 4-7411, American Instrument Company, Silver Spring, MD, U.S.A.).

Table 1. *Composition of Mg medium (mM)*

EGTA	5
$MgCl_2$	6
PIPES	30
KOH	67
Sorbitol	200
pH	7.0

Measurement of the streaming velocity were carried out at room temperature (22–25 °C) with an eyepiece micrometer and a stop watch. Approximately 0.5–1 min elapsed between the end of perfusion and the onset of microscopic observation. Since the velocity decreased with time, velocity measurements were finished within 3 min. Three cells were used for one experiment. Organelles or organelle aggregates having the highest velocity in the microscope field were selected for the measurment. Six measurements were done for each cell. Average velocities of 18 organelles or organelle aggregates are shown with the SD.

3. Results

In the first part of the experiments, adenylate kinase (EC 2.7.4.3.) activity was studied. Adenylate kinase affects the $(ATP)(AMP)/(ADP)^2$ ratio according to the following reaction.

$$2\,\text{ADP} \rightleftharpoons \text{ATP} + \text{AMP} \qquad (2)$$

After the tonoplast disintegrated by the perfusion with Mg medium, the tonoplast-free cell was perfused with Mg medium supplemented with 1 mM ADP but lacking ATP. At the beginning, the streaming velocity was very low but increased with time (Fig. 1) as reported previously (SHIMMEN 1978). This gradual increase of the streaming velocity may be caused by the action of adenylate kinase. Adenylate kinase activity in tonoplast-free cells was evaluated by measuring ATP production from ADP. After disintegration of the tonoplast by the first perfusion with Mg medium, the cell interior was perfused again with Mg medium lacking adenine nucleotide and incubated in APW for 4 min. The cells were frozen, boiled and subjected to ATP measurement. The ATP concentration after the second perfusion with Mg medium was 2 μM (Table 2, Exp. 1). When tonoplast-free cells were perfused with Mg medium supplemented with 1 mM ADP and incubated for 4 min in APW, 0.22 mM ATP was synthesized (Table 2, Exp. 2). By adding 0.1 mM DAPP, an inhibitor of adenylate kinase (LIENHARD and SECEMSKI 1973) to the second perfusion medium containing 1 mM ADP, the increase in the streaming velocity (Fig. 1) and the ATP synthesis (Table 2, Exp. 3) were significantly inhibited. DAPP at 0.01 mM slightly inhibited the increase in the velocity (Fig. 1). Next, the streaming was reactivated by Mg medium supplemented with 0.025 mM ATP, 2 mM PEP and 0.5 mg/ml pyruvate kinase and the effect of DAPP on the streaming was studied (Fig. 2). The streaming velocity was slightly inhibited by 0.1–0.2 mM DAPP and inhibition was increased with increasing DAPP concentration. DAPP at 0.1 mM scarcely inhibited the streaming reactivated with 0.05 or 0.1 mM ATP (not shown). In the following exper-

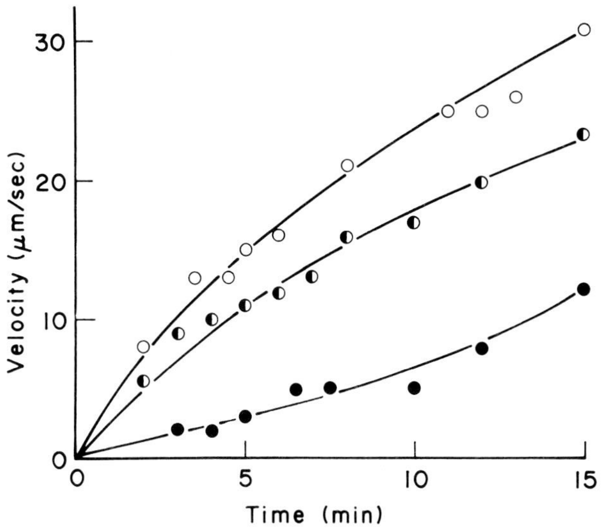

Fig. 1

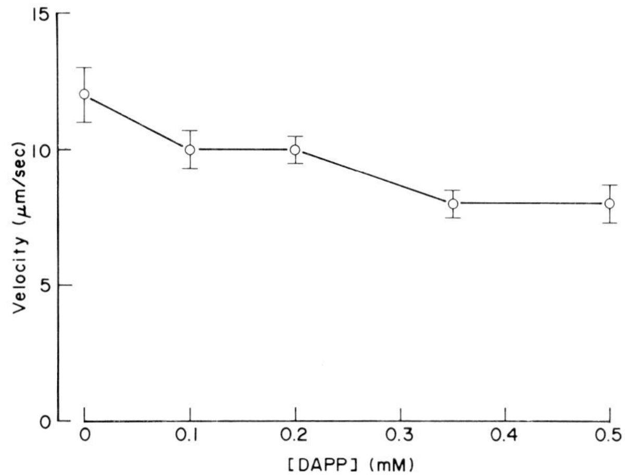

Fig. 2

Fig. 1. Recovery of cytoplasmic streaming after application of 1 mM ADP. Mg medium supplemented with 1 mM ADP was applied into tonoplast-free cells in the absence or presence of DAPP at time zero. Zero (○), 0.01 mM (◐), and 0.1 mM (●) DAPP

Fig. 2. Effect of DAPP on cytoplasmic streaming. Tonoplast-free cells were perfused with Mg medium supplemented with 0.025 mM ATP, 2 mM PEP, 0.5 mg/ml pyruvate kinase and DAPP at various concentrations. Streaming velocity was measured 1–4 min after the perfusion

Table 2. *Synthesis of ATP in tonoplast-free cells after the second perfusion with Mg medium supplemented with ADP*

Exp.	Added (mM)		Time after perfusion (min)	ATP (mM)
	ADP	DAPP		
1	0	0	4	0.002 ± 0.0003
2	1	0	4	0.220 ± 0.038
3	1	0.1	4	0.071 ± 0.005

ATP buffering system was not added
Three cells were used for each experiment

iments, DAPP was used at 0.1 mM to inhibit adenylate kinase activity in tonoplast-free cells.

The effect of ADP on ATP-reactivated streaming was studied (Fig. 3). In this experiment, the ATP buffering system can not be used, since all ADP is converted to ATP according to the reaction of Eq. (1). Tonoplast-free cells were perfused with Mg medium supplemented with 0.1 mM DAPP and ATP at various concentrations and the streaming velocity was measured after 1–4 min. The relationship between ATP concentration and streaming velocity showed a saturation curve. When 1 mM ADP was added, the streaming velocity was not affected at lower ATP concentration but drastically inhibited at higher concentrations.

Since the ATP concentration was not buffered, a decrease in the ATP concentration after the second perfusion due to ATPase (EC 3.6.1.3) action is supposed (SHIMMEN 1978). In addition, incomplete inhibition of adenylate kinase activity by 0.1 mM DAPP (Table 2) should change the ATP concentration in the presence of 1 mM ADP. To assess the result in Fig. 3, the ATP

Table 3. *Changes in the ATP concentration of tonoplast-free cells after a second perfusion in the absence or presence of ADP*

Added (mM)		Measured ATP (mM)
ATP	ADP	
0.025	0	0.010 ± 0.002
0.025	1	0.095 ± 0.012
0.050	0	0.026 ± 0.013
0.050	1	0.105 ± 0.026
0.100	0	0.059 ± 0.021
0.100	1	0.138 ± 0.029
0.200	0	0.157 ± 0.026
0.200	1	0.270 ± 0.059
0.500	0	0.329 ± 0.109
0.500	1	0.464 ± 0.044
1.000	0	0.794 ± 0.129
1.000	1	0.974 ± 0.110

DAPP at 0.1 mM was added but not ATP buffering system
ATP was measured 4 min after perfusion
Three cells were used for each experiment

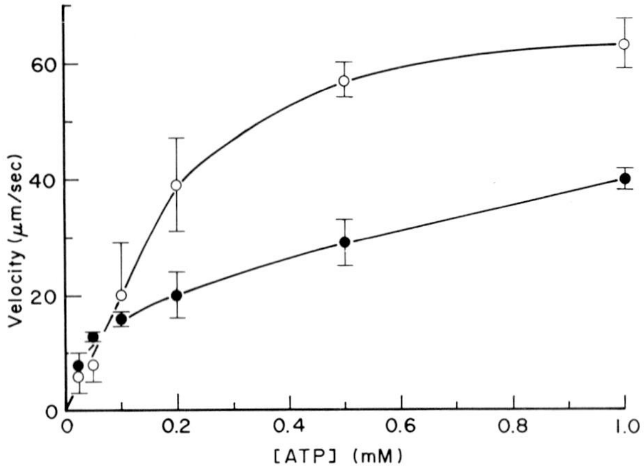

Fig. 3

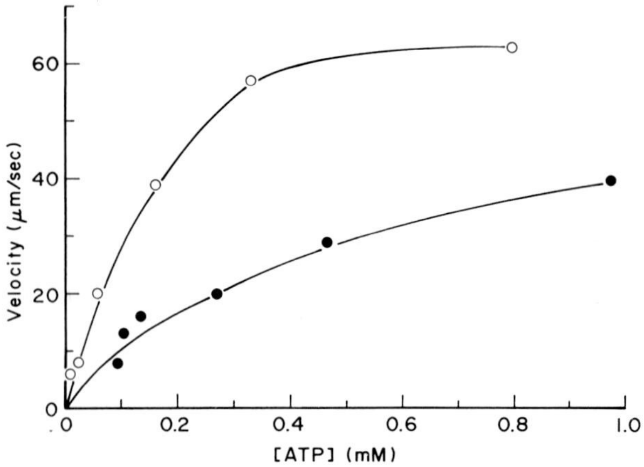

Fig. 4

Fig. 3. Effect of ADP on cytoplasmic streaming. Tonoplast-free cells were perfused with Mg medium supplemented with 0.1 mM DAPP, ATP of various concentrations, and either 0 mM (○) or 1 mM (●) ADP. Streaming velocity was measured 1–4 min after the perfusion. ATP concentration was not buffered. Abscissa represents ATP concentration added to the perfusion medium

Fig. 4. Effect of ADP on cytoplasmic streaming. Result of Fig. 3 was replotted using ATP concentration from Table 3. Abscissa represents ATP concentration measured 4 min after the perfusion. ○ Without ADP, ● with ADP

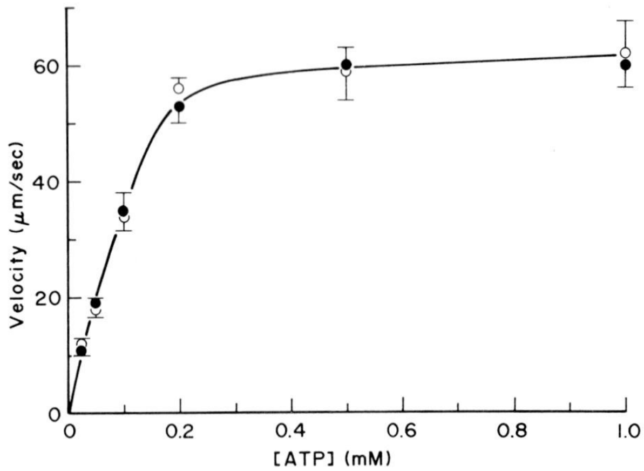

Fig. 5

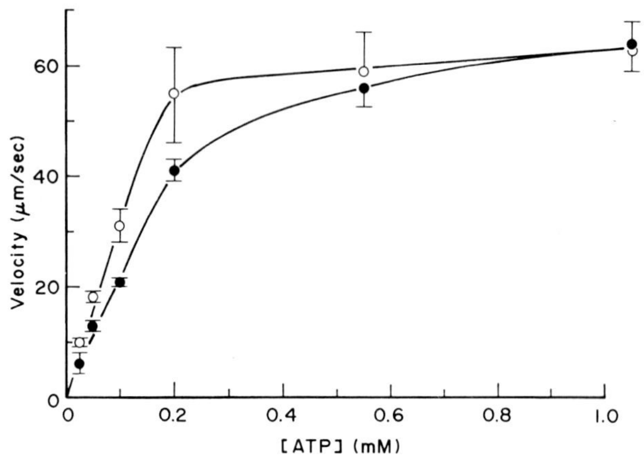

Fig. 6

Fig. 5. Effect of AMP on cytoplasmic streaming. Tonoplast-free cells were perfused with Mg medium supplemented with 0.1 mM DAPP, ATP of various concentrations, 2 mM PEP, 0.5 mg/ml pyruvate kinase, and either 0 mM (○) or 1 mM (●) AMP. Streaming velocity was measured 1–4 min after the perfusion

Fig. 6. Effect of P_i on cytoplasmic streaming. Tonoplast-free cells were perfused with Mg medium supplemented with 0.1 mM DAPP, ATP at various concentrations, 2 mM PEP, 0.5 mg/ml pyruvate kinase, and either 0 mM (○) or 10 mM (●) P_i. Streaming velocity was measured 1–4 min after the perfusion

concentration was measured 4 min after the second perfusion with Mg medium containing ATP at various concentrations and 0.1 mM DAPP with or without 1 mM ADP (Table 3). In the absence of ADP, ATP decreased after the second perfusion at all concentrations. In the presence of 1 mM ADP, however, ATP increased at lower concentrations (0.025–0.2 mM). At higher ATP concentrations (0.5 and 1 mM), ATP slightly decreased. Result of Fig. 3 was replotted using the ATP concentrations measured (Fig. 4). The cytoplasmic streaming was significantly inhibited by ADP at all ATP concentrations.

The relation between the ATP concentration and the streaming velocity was studied with Mg medium supplemented with 0.1 mM DAPP and an ATP buffering system (PEP and pyruvate kinase) (Fig. 5). The relation

Table 4. *Streaming velocity after a second perfusion with Mg medium supplemented with adenine nucleotides and inorganic phosphates*

Exp.	Added (mM)					Velocity (μ/sec)
	ATP	ADP	AMP	P_i	PP_i	
1	1	0	0	0	0	63 ± 3
2	1	0	5	0	0	63 ± 5
3	1.6	0	0	0	0	66 ± 4
4	1.6	0.6	0.8	14.7	0.2	61 ± 4

DAPP at 0.1 mM was added but not ATP buffering system
Three cells were used for each experiment

showed a saturating curve with half maximal concentration of about 0.08 mM ATP. AMP at 1 mM scarcely affected the streaming velocity at all ATP concentrations tested (Fig. 5). In the presence of AMP and ATP, the backward reaction of Eq. (2) should proceed to synthesize ADP, which inhibits the cytoplasmic streaming. At lower ATP concentrations, however, the reaction should be slow, since the concentration of one of the substrate (ATP) is low. At higher ATP concentration, certain amounts of ADP should be synthesized. However, inhibition by synthesized ADP may be insignificant due to higher ATP/ADP ratio. These predictions were verified by the complete absence of inhibition after the addition of 1 mM AMP at all ATP concentrations (Fig. 5). The streaming reactivated by 1 mM ATP was not affected by 5 mM AMP (Table 4, Exp. 2).

The effect of 10 mM P_i was studied in the presence of an ATP buffering system and 0.1 mM DAPP (Fig. 6). At higher ATP concentrations, the streaming was not inhibited but at lower ATP concentrations, the streaming was significantly inhibited.

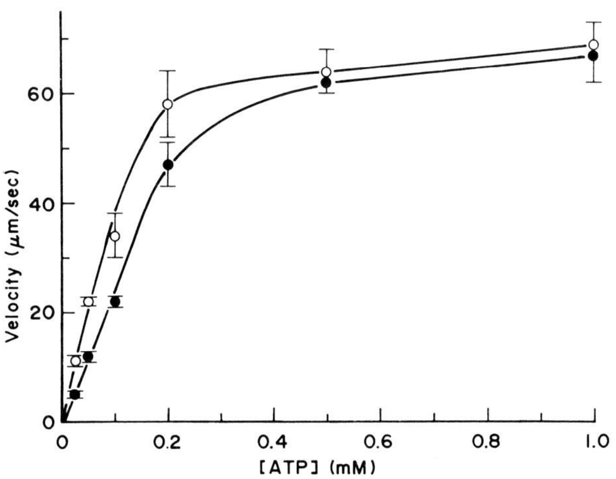

Fig. 7. Effect of PP_i on cytoplasmic streaming. Tonoplast-free cells were perfused with Mg medium supplemented with 0.1 mM DAPP, ATP at various concentrations, 2 mM PEP, 0.5 mg/ml pyruvate kinase, and either 0 mM (○) or 1 mM (●) PP_i

The effect of 1 mM PP_i was studied in the presence of ATP buffering system and 0.1 mM DAPP (Fig. 7). At higher ATP concentrations, the streaming was not affected by 1 mM PP_i but was significantly inhibited at lower ATP concentrations.

The streaming velocity at physiological concentrations of adenine nucleotides and inorganic phosphates was studied, 1.6 mM ATP, 0.6 mM ADP, 0.8 mM AMP (MIMURA *et al.* 1984) and 14.7 mM P_i (MIMURA and KIRINO 1984). Since PP_i concentration in *Chara* cytoplasm is 0.21 mM (TAKESHIGE and TAZAWA, personal communication), 0.2 mM PP_i was used. In this experiment, an ATP buffering system was not added but 0.1 mM DAPP was added. The streaming velocity was slightly increased by increasing the ATP concentration from 1 mM (Table 4, Exp. 1) to 1.6 mM (Table 4, Exp. 3). By adding ADP, AMP, P_i and PP_i, a little inhibition occurred (Table 4, Exp. 4).

4. Discussion

The relationship between the ATP concentration and the streaming velocity showed a saturation curve as reported previously in *Nitella and Chara* (SHIMMEN 1978). The result did not follow Lineweaver and Burk's equation (not shown). This is presumably due to the fact that the streaming velocity is dependent on both the motive force and the resistance including viscosity. Therefore, results were analyzed by fitting plots by eye. The half maximal ATP concentration was 0.09 mM (average value of Figs. 5–7), which is consistent with values reported for *Nitella axillaris* (0.07 mM) and *Chara corallina* (0.09 mM) (SHIMMEN 1978).

The effect of ADP, AMP and PP_i on reactivating the

cytoplasmic streaming in tonoplast-free *Chara* cells was first studied by WILLIAMSON (1975). ADP partially reactivated the streaming presumably due to adenylate kinase action, while AMP and PP_i could not support the streaming. The present study first demonstrated the inhibitory action of ADP, P_i and PP_i on the cytoplasmic streaming activated by ATP. Since the results can not be analyzed by the Lineweaver-Burk's equation, it is difficult to unequivocally determine whether mode of inhibition is competitive or noncompetitive. Since the extent of inhibition by ADP, P_i and PP_i decreased with increase in ATP concentration, it is suggested that inhibition by these phosphorous compounds is competitive.

REID and WALKER (1983) studied the relationship between ATP concentration and the streaming velocity during the application of CCCP, and uncoupler of oxidative phosphorylation. The relation did not show a saturation curve but a straight line. The authors postulated the inhibition of the cytoplasmic streaming by ADP, which increased as a consequence of energy metabolism inhibition. It must be noticed that the energy metabolism inhibition might cause not only a change in the adenine nucleotide level but also a change in the concentration of H^+ and/or Ca^{2+}, which significantly affect the streaming velocity (TAZAWA and SHIMMEN 1982, TOMINAGA *et al.* 1983). In the present study, we could control levels of adenine nucleotides and inorganic phosphates without changing other factors by taking advantage of the intracellular perfusion method. Inhibition by ADP, P_i and PP_i was unequivocally demonstrated. REID and WALKER (1983) observed a transient small increase of streaming velocity at the beginning of inhibition by DCCD, an inhibitor of proton ATPase. They attributed the velocity increase to a transient increase in ATP/ADP ratio and argued that under normal conditions the streaming process does not operate at maximum velocity. In the present study (Table 4), we found that the velocity in the medium of physiological compositions (Table 4, Exp. 4) was 7.5% lower than that in the medium of ATP alone (Table 4, Exp. 3).

A high activity of adenylate kinase was observed in the tonoplast-free cells whose endoplasm was drastically effused out by the second perfusion. Therefore, it is suggested that adenylate kinase activity is associated with the thin ectoplasm and chloroplasts. Most of the activity was lost after removing chloroplasts from tonoplast-free cells by strong centrifugation (unpublished data). Association of adenylate kinase with chloroplasts has been reported in spinach (MURAKAMI and STROTMANN 1978). Adenylate kinase of muscle was completely inhibited by 0.01 mM DAPP (LIENHARD and SECEMSKI 1973). In *Nitellopsis*, however, the activity was only partially inhibited by 0.1 mM DAPP. The disturbance of ADP experiments (Fig. 3) by adenylate kinase was overcome by directly measuring ATP (Fig. 4).

The effect of adenine nucleotides and inorganic phosphates on motility has been extensively studied in muscle. ADP inhibited myosin ATPase (CHAPLAIN *et al.* 1965, CHAPLAIN 1966). P_i decreased isometric force in glycerinated muscle (COOKE and PATE 1985). PP_i inhibited myosin ATPase (DANKER 1983) and decreased force under isometric contraction or velocity under isotonic contraction (PATE and COOKE 1985). In the study of skeletal muscle contraction, the effects of adenine nucleotides and inorganic phosphate had been discussed in relation to formation of an intermediate in the ATPase reaction, myosin-ADP-P_i (cf. DANCKER 1983, HIBBERT *et al.* 1985). The present study first demonstrated inhibition of cytoplasmic streaming in plants by ADP, P_i and PP_i, and suggested the possible formation of an intermediate in the *Characeae* "myosin" ATPase reaction.

Acknowledgements

I express my cordial thanks to Dr. M. TAZAWA and Dr. R. WAYNE for critical reading of this manuscript. This work was partly supported by a Grant in Aid for Scientific Research from the Ministry of Education, Science and Culture of Japan.

References

CHAPLAIN RA (1966) Indication for an allosteric effect of adenosine diphosphate in actomyosin gels from insect fibrillar flight muscle. Arch Biochem Biophys 115: 450–461

— ABBOTT RH, WHITE DCS (1965) Indication for an allosteric effect of ADP on actomyosin gels and glycerinated fibres from insect fibrillar flight muscle. Biochem Biophys Res Comm 21: 89–93

COOKE R, PATE E (1985) The effects of ADP and phosphate on the contraction of muscle fibers. Biophys J 48: 789–798

DANCKER P (1983) The competition between adenosine triphosphate and inorganic pyrophosphate for myosin and its suppression by substoichiometric actin concentrations. Biochim Biophys Acta 749: 296–301

HAYAMA T, SHIMMEN T, TAZAWA M (1979) Participation of Ca^{2+} in cessation of cytoplasmic streaming induced by membrane excitation in Characeae internodal cells. Protoplasma 99: 305–321

HIBBERD MG, DANTZIG JA, TRENTHAM DR, GOLDMAN YE (1985) Phosphate release and force generation in skeletal muscle fibers. Science 228: 1317–1319

KAMIYA N (1986) Cytoplasmic streaming in giant algal cells: a historical survey of experimental approaches. Bot Mag Tokyo 99: 441–467

— KURODA K (1956) Velocity distribution of the protoplasmic streaming in *Nitella* cells. Bot Mag Tokyo 69: 544–554

LIENHARD GE, SECEMSKI II (1973) P^1,P^5-di(adenosine-5′)pentaphosphate, a potent multisubstrate inhibitor of adenylate kinase. J Biol Chem 248: 1121–1123

MIMURA T, KIRINO Y (1984) Changes in cytoplasmic pH measured by ^{31}P-NMR in cells of *Nitellopsis obtusa*. Plant Cell Physiol 25: 813–820

— SHIMMEN T, TAZAWA M (1984) Adenine-nucleotide levels and metabolism-dependent membrane potential in cells of *Nitellopsis obtusa* Groves. Planta 162: 77–84

MURAKAMI S, STROTMANN H (1978) Adenylate kinase bound to the envelope membranes of spinach chloroplasts. Arch Biochem Biophys 185: 30–38

PATE E, COOKE R (1985) The inhibition of muscle contraction by adenosine 5′(β, γ-imido) triphosphate and by pyrophosphate. Biophys J 47: 773–780

REID RJ, WALKER NA (1983) Adenylate concentrations in *Chara*: Variability, effects of inhibitors and relationship to protoplasmic streaming. Aust J Plant Physiol 10: 373–383

SHIMMEN T (1978) Dependency of cytoplasmic streaming on intracellular ATP and Mg^{2+} concentrations. Cell Struct Funct 3: 113–121

— TAZAWA M (1982a) Effects of intracellular vanadate on electrogenesis, excitability and cytoplasmic streaming in *Nitellopsis obtusa*. Plant Cell Physiol 23: 669–677

— — (1982b) Cytoplasmic streaming in the cell model of *Nitella*. Protoplasma 112: 101–106

— — (1983) Control of cytoplasmic streaming by ATP, Mg^{2+} and cytochalasin B in permeabilized Characeae cell. Protoplasma 115: 18–24

TAZAWA M, KIKUYAMA M, SHIMMEN T (1976) Electric characteristics and cytoplasmic streaming of Characeae cells lacking tonoplast. Cell Struct Funct 1: 165–176

— SHIMMEN T (1982) Artificial control of cytoplasmic pH and its bearing on cytoplasmic streaming, electrogenesis and excitability of Characeae cells. Bot Mag Tokyo 95: 147–154

TOMINAGA Y, SHIMMEN T, TAZAWA M (1983) Control of cytoplasmic streaming by extracellular Ca^{2+} in permeabilized *Nitella* cells. Protoplasma 116: 75–77

WILLIAMSON RE (1975) Cytoplasmic streaming in *Chara*: A cell model activated by ATP and inhibited by cytochalasin B. J Cell Sci 17: 655–668

Protoplasma (1988) [Suppl 1]: 10–14

Apparent Viscosity of the Endoplasm of Characean Internodal Cells Measured by the Centrifuge Method

E. KAMITSUBO[1,*], M. KIKUYAMA[2], and I. KANEDA[3,**]

[1] Biol. Lab., Hitotsubashi University, Kunitachi, Tokyo, [2] Faculty of Pharmacology, Niigata College of Pharmacy, Kamishin-ei-cho, Niigata, [3] Biol. Lab., Tokyo Institute of Technology, Meguro, Tokyo

Received October 29, 1987
Accepted December 4, 1987

Dedicated to Professor Dr. NOBURO KAMIYA on the occasion of his 75th birthday

Summary

The apparent viscosity of *Nitella* endoplasm was determined *in vivo* by measuring the movement of oil droplets through the endoplasm under various centrifugal accelerations (100–500 × **g**). It was shown that the endoplasm is highly viscous when a shearing force is absent and is very fluidic at fairly high rates of shear.

Keywords: Apparent viscosity; *Nitella* endoplasm; Centrifuge microscope.

1. Introduction

Cytoplasmic streaming in characean cells is caused by the interaction between the stationary subcortical fibrils (KAMITSUBO 1966 b, 1972 a, b, 1980) and the endoplasm (KAMITSUBO 1981). The physical properties of the endoplasm contiguous to the subcortical fibrils are of crucial importance for studying the mechanism of streaming in characean cells.

When cytoplasmic streaming is going on in the cells of *Characeae*, the viscosity of the endoplasm contiguous to the subcortical fibrils seems to be very low when observed with Nomarski optics at the highest resolution. In fact, a thin endoplasmic layer, 1–2 µm thick, was found at the boundary between the subcortical fibrils and the flowing bulk endoplasm by optically sectioning the cells with Nomarski optics. The rate of shear of this thin layer was estimated to be around 50–70 s^{-1} at 20 °C (KAMITSUBO 1980, 1986 a). By contrast, the viscosity of the bulk endoplasm seems to be very high, because it is flowing as a whole, or is sliding against the cortex as if it were a mass of gel (KAMIYA and KURODA 1956, KAMITSUBO 1979). In other words, there is no appreciable rate of shear within the flowing bulk endoplasm. In order to clarify the mechanical properties of the endoplasm contiguous to the subcortical fibrils and the bulk endoplasm subjected to various conditions, we measured the viscosity of the endoplasm *in vivo* with a new centrifuge microscope (CMS) of the stroboscope type.

2. Materials and Methods

2.1. Plant Material

Internodal cells of *Nitella axilliformis*, ca. 2.5 cm in length and 300–500 µm in diameter, were used for the present study. The algae were cultured in the laboratory as previously described (KAMITSUBO 1981). The internodal cells isolated from adjacent cells were kept in artificial pond water (APW; containing 0.1 mM each of KCl, NaCl and $CaCl_2$) for at least one day before use. Prior to the measurement of the endoplasmic viscosity, a sufficient amount of the endoplasm was collected at one end of the cell by centrifugation. During this treatment, chloroplasts were detached from the cell cortex resulting in an improvement of the optical condition for observing the movement of oil droplets through the accumulated endoplasm. This treatment was carried out using a simple CMS of the Brown type (BROWN 1940, HAYASHI 1957, KAMITSUBO 1966 a, 1972 a) constructed at our laboratory. The appropriate strength and duration of centrifugation for the above purpose were around 1000 × **g** for 5–10 min. Figure 1 schematically shows a part of an internodal cell after the treatment. At the centrifugal end of the cell, chloroplasts (*p*) detached from the

* Correspondence and Reprints: Biol. Lab., Hitotsubashi University, Kunitachi, Tokyo 186, Japan.

** Present address: Pharmacology Section, Teijin Institute for Biomedical Research, Hino, Tokyo 191, Japan.

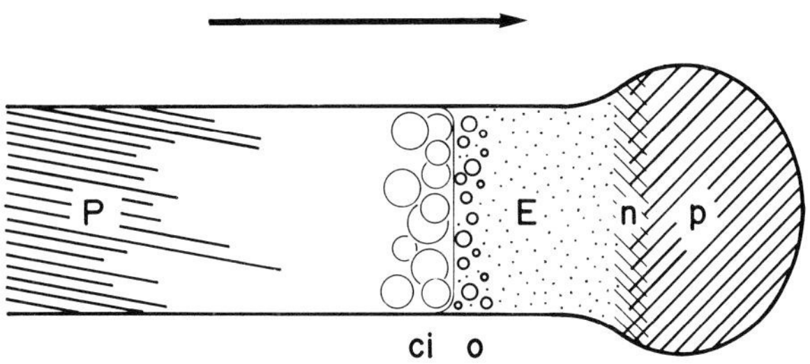

Fig. 1. Schematic view of a part of a centrifuged *Nitella* internodal cell. The arrow indicates the direction of centrifugation. *p* Chloroplasts are detached from the cell cortex and are accumulated at the centrifugal end of the cell, *n* accumulated nuclei, *E* accumulated endoplasmic mass, *o* large oil drops accumulated during centrifugation with a Brown CMS, *ci* cell sap inclusions collected at the centrifugal end of the central vacuole, *P* chloroplasts remain attached to the cortex

cell cortex, nuclei (*n*) and endoplasm (*E*) were accumulated and stratified. Cell sap inclusions (*ci*) were accumulated at the centrifugal end of the central vacuole. Cytoplasmic streaming did not take place for several hours at the chloroplasts-detached area and therefore the interference of cytoplasmic streaming on the movement of oil droplets through the accumulated endoplasm was eliminated. The viscosity of the endoplasm was determined by measuring the movement of oil droplets.

2.2. *Measurement of Viscosity of the Endoplasm*

The internodal cells prepared as described above were further centrifuged under observation with the new CMS of the stroboscope type. As a consequence of centrifugation, oil droplets (ca. 10 µm in diameter), escaped one after the other from the endoplasm trapped within the accumulated chloroplasts and nuclei, and moved in the centripetal direction through the accumulated endoplasm (Fig. 2). The velocity of the oil droplets was measured at different centrifugal forces (100–500 × **g**), *i.e.*, at different shear stresses since the endoplasm (cytoplasm) might be a non-Newtonian fluid. From the values of the velocities and the diameters of the oil droplets, which were measured using a video tape recording system with a computer (PC-6001mkII, NEC, Tokyo), the viscosity of the endoplasm (η) was calculated by the Stokes' formula;

$$\eta = \frac{a^2\alpha(D - D')}{9v} \quad (\text{poise; dyne cm}^{-2}\,\text{s}) \qquad (1)$$

where a is the radius of the oil droplet, α is the acceleration of the centrifugal force, D and D′ are respectively the density of the endoplasm and the oil droplet and v is the velocity of the oil droplet. The shear stress (σ) and the rate of shear ($\dot{\gamma}$) at the equatorial region of the oil droplet are expressed as follows;

$$\sigma = \frac{a\alpha(D - D')}{3} \quad (\text{dyne cm}^{-2}) \qquad (2)$$

$$\dot{\gamma} = \frac{3v}{2a} \quad (\text{s}^{-1}) \qquad (3)$$

The density of the oil droplets (D′) was estimated from their velocities in distilled water at a certain centrifugal force measured with the CMS.

2.3. *CMS of the Stroboscope Type*

A new CMS of the stroboscope type recently constructed at our laboratory in collaboration with Nikko Engineering Co. (Yoko-

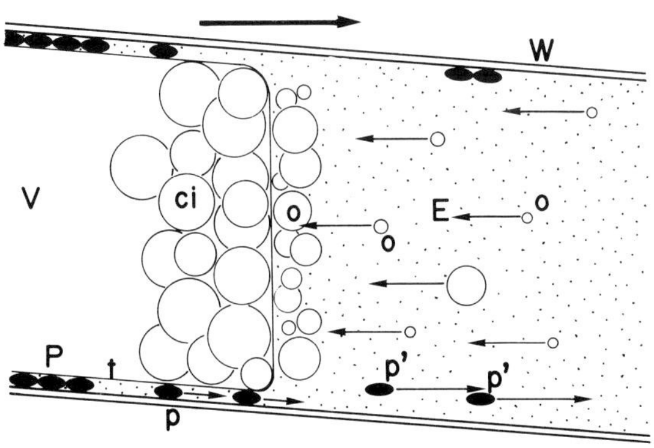

Fig. 2. Schematic view of a part of the profile of centrifuged *Nitella* internode. A long arrow indicates the directions of centrifugation. Short arrows indicate the direction of movement of the cell organelles. Velocity of each organelle is roughly shown by the length of each arrow. *W* cell wall, *P* chloroplasts remain *in situ*, *p* chloroplast detached from the cortex by the centrifugal force and sliding against the cortical gel, *p′* chloroplast detached from the cortical gel and gliding through the endoplasmic mass, *o* oil droplet, *t* tonoplast, *V* central vacuole, *ci* cell sap inclusions. The cell membrane and the cortical gel lining the cell membrane are not illustrated

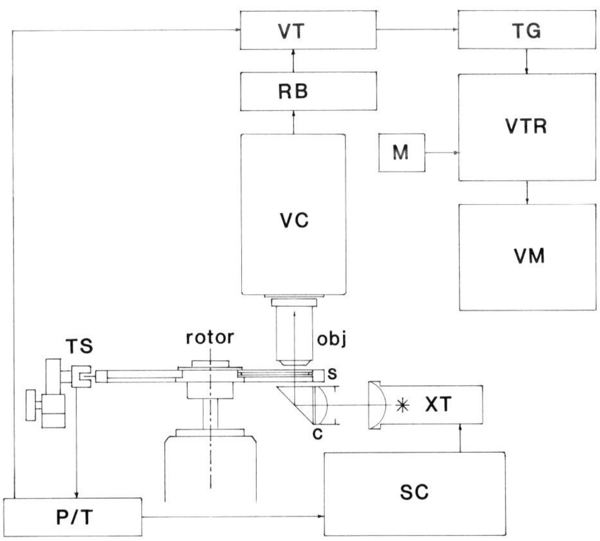

Fig. 3. Diagram of new centrifuge microscope of the stroboscope type. *s* Specimen in APW in a centrifuge cuvette made from an acrylic sheet with a inner dimension of 30 mm × 4 mm × 1 mm (with an external size of 33 mm × 10 mm × 3 mm), *TS* timing sensor detects the position of the specimen, and sends an electrical signal to *P/T*, *P/T* tachometer serving as a pulse generator which triggers a stroboscope controller (*SC*), *XT* xenon tube flashes 180 nsec/pulse, *c* condenser lens, *obj* objective lens, *VC* video camera, *RB* record booster, *VT* video timer, *TG* title generator for displaying the rotor-speed (*rpm*), *VTR* video tape recorder, *VM* video monitor, *M* microphone

hama, Japan), was used for measuring the velocity of the oil droplets under centrifugal acceleration. The CMS is composed of a video-enhanced contrast microscope combined with a stroboscopic light source, a mechanism for centrifugation of a specimen, and a control panel for triggering the light source synchronously with the rotation of the specimen (Fig. 3). A Nikon CF DIC 20 × (NA. 0.40) objective lens was mainly used with a Nikon ELWD condenser lens (NA. 0.3). The optical image was taken by a video camera equipped with a 2/3 inch Newvicon camera tube (WV-1550, Matsushita, Yokohama) and was contrast-enhanced by a record booster (RB-10R, For.A, Tokyo). A 1/2 inch VTR (NV-8030, Matsushita, Osaka) was used for recording the images. The speed of the rotor in revolutions per min (rpm) was always superimposed digitally on the video monitor (PM-171, Ikegami Tsushinki, Tokyo). The date and time were also displayed on the monitor.
All experiments were carried out at ca. 20 °C.

3. Results

To obtain the apparent viscosity (η) of the endoplasm from the velocity (v) and the diameter (a) of the oil droplets moving in the *Nitella* endoplasm, a knowledge of the density of the oil droplets (D′) and that of the endoplasm (D) are required. To determine D′, the diameters and the velocities of the oil droplets in distilled water under various centrifugal accelerations were measured using the new CMS (data not shown). Assuming that the density and the viscosity of distilled water are respectively 0.998 g cm^{-3} and 1.002×10^{-2} poise (dyne cm^{-2} s) at 20 °C, the density of the oil droplets was estimated to be 0.96 g cm^{-3}. For the density of the endoplasm (D), the value of 1.014–1.015 g cm^{-3} (KAMIYA and KURODA 1957) was employed.
The velocity and the diameter of the oil droplets forced to move or to float through the accumulated endoplasm were measured at different centrifugal forces between 100–500 × **g** (ca. 1500–3300 rpm), in other words, at different shear stresses using the new CMS of the stroboscope type.
The measurements were carried out on 9 different oil droplets under different centrifugal accelerations, and the shear rate ($\dot{\gamma}$), the shear stress (σ) and the apparent viscosity (η) were calculated from each value with formulae (3), (2), and (1) and are summarized in Table 1. Figure 4 shows the relationship between the shear stress and the rate of shear of the endoplasm in *N. axilliformis*. Although there is a slight scatter in the data, it is clear that the endoplasm shows a non-Newtonian character having a yield value of about 5 dyn cm^{-1}. Figure 5 shows the relationship between the rate of shear and the viscosity of the endoplasm of *N. axilliformis*. Clearly, the endoplasm is highly viscous if a shearing force is absent, and is very fluidic at fairly high rates of shear.

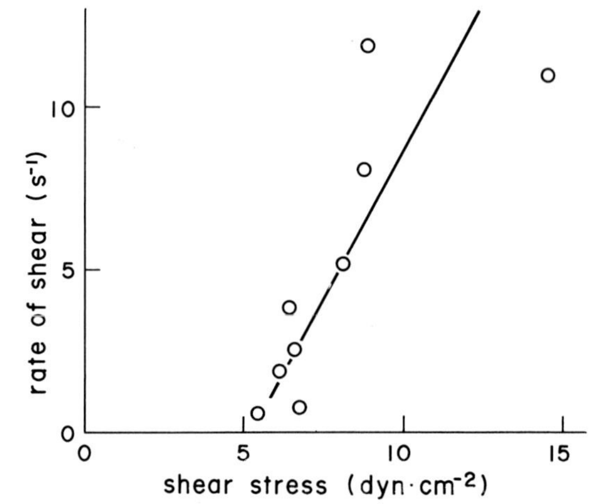

Fig. 4. The relationship between the shear stress and the rate of shear in the endoplasm of *N. axilliformis* measured *in vivo* by the centrifuge method

Table 1. *The rate of shear ($\dot{\gamma}$), the shear stress (σ), and the viscosity (η) of the endoplasm in internodal cells of Nitella axilliformis measured in vivo by the centrifuge method*

No. of oil droplet	$\dot{\gamma}$ (s^{-1})	σ (dyn cm^{-2})	η (poise)
1	0.69	5.45	7.90
2	0.79	6.73	8.52
3	1.88	6.09	3.24
4	2.56	6.60	2.58
5	3.85	6.44	1.67
6	5.18	8.05	1.55
7	8.09	8.77	1.08
8	11.0	14.5	1.32
9	11.9	8.87	0.75

During the centrifugation of *Nitella* internodes with a Brown CMS, it was frequently observed that chloroplasts detached from the cortical gel and moved or were sliding at a certain speed in the centrifugal direction in contact with the cortical gel (*p* in Fig. 2). When an internode was set obliquely to the direction of the centrifugal force in the centrifuge cuvette, or when an internode was bending slightly at the part near the centrifugal end of the cell in the cuvette, a chloroplast sliding in contact with one side of the cortical gel departed from the gel at the centripetal level of the accumulated endoplasmic mass. Once the chloroplast departed from the cell cortex and came into the accumulated endoplasm, it glided through the endoplasm in the direction parallel to that of centrifugal force with a speed several times higher than the previous one (*p′* in Fig. 2). The measurement of the endoplasmic vis-

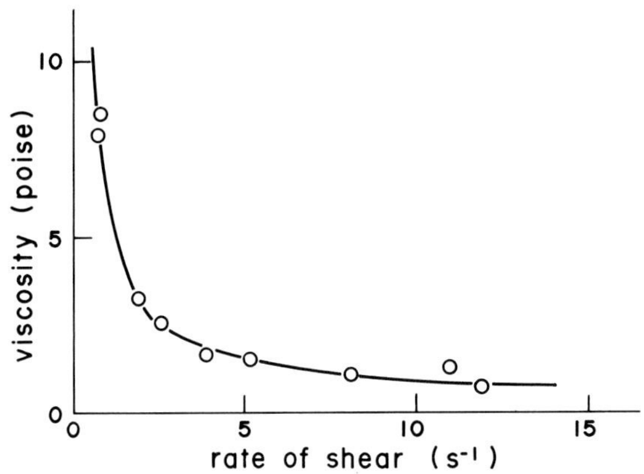

Fig. 5. The relationship between the rate of shear and the viscosity of the endoplasm of *N. axilliformis* internodes

cosity from the velocity of the chloroplasts gliding in the endoplasm was not performed for two reasons. First, the shape of the chloroplast is too irregular to apply the Stokes' formula. Secondly, the starch content in the chloroplast differs from one to another, and therefore, the measurement of the density of this organelle is thought to be considerably difficult.

4. Discussion

KAMIYA and KURODA (1965, 1973) studied the viscosity of "living" endoplasm isolated from internodal cells of *Nitella megacarpa* and *N. expansa*. Using the agar capillary method they analyzed *in vitro* the shear stress/shear rate relationship of the "living" endoplasm and revealed the non-Newtonian properties of the endoplasm, *i.e.*, the higher the shear force applied the lower was the apparent viscosity. For instance, the apparent viscosities η in *N. megacarpa* were 10 and 2.5 poise when the rates of shear were 0.2 and $1.7\,s^{-1}$, respectively. There were two differences of the endoplasmic properties between *N. megacarpa* and *N. expansa*. First, the absolute value of the apparent viscosity of the endoplasm of *N. expansa* was considerably lower than that of *N. megacarpa* at about the same temperature (24–25 °C). Secondly, the shear stress/shear rate diagram of *N. megacarpa* showed the presence of a yield value (ca. $1.5\,dyn\,cm^{-2}$), while that of *N. expansa* did not. The reason for the difference is still obscure. The endoplasmic viscosity of *N. axilliformis* obtained in the present study coincides well with that of *N. megacarpa* obtained by KAMIYA and KURODA (1965).

KIKUYAMA and TAZAWA (1972) estimated the viscosity of *Nitela flexilis* endoplasm *in vivo* by measuring the swimming rate of *Tetrahymena* in the endoplasm. They introduced *Tetrahymena*, which had been adapted to the artifical cell sap (ACS) (KURODA and KAMIYA 1968), into the vacuolar space of *Nitella* by means of vacuolar perfusion (TAZAWA 1964). After closing both cell ends by ligation, the *Tetrahymena* were forced into the endoplasm by centrifuging *Nitella* in a longitudinal direction. Soon after the centrifugation, the cell was ligated at a site near the centrifugal end to obtain a small cell fragment filled with endoplasm containing the ciliate. In the cell fragment, *Tetrahymena* were swimming in contact with the endoplasm. The viscosity of endoplasm was then estimated using the relation between the swimming rate of *Tetrahymena* in ACS containing various concentrations of methyl cellulose and the viscosity of the solution. The viscosity of endoplasm was found to be 0.4–9 poise, which coincides with the present results (Table 1).

Using the new CMS of the stroboscope type, the present study revealed *in vivo* the mechanical properties of the endoplasm in *N. axilliformis* under various conditions, *i.e.*, under different shear stresses and rates of shear ranging from 0.69 to $11.9\,s^{-1}$. The viscosity of the thin endoplasmic layer contiguous to the subcortical fibrils seems to be very low as pointed out in the Introduction. The rate of shear of this thin endoplasmic layer was estimated to be around $50–70\,s^{-1}$ at 20 °C (KAMITSUBO 1980, 1986a). This value of the rate of shear leads to the conclusion that the viscosity of the endoplasm contiguous to the subcortical fibrils is fairly low (about 0.5 poise or less) when analyzed in light of the results of the present study (Fig. 5).

By contrast, the viscosity of the flowing bulk endoplasm must be extremely high, because no appreciable rate of shear is present in it. This has been determined by studying the velocity profiles (KAMIYA and KURODA 1956, KAMITSUBO 1979). Furthermore when an internodal cell is centrifuged longitudinally with a moderate acceleration force, the endoplasm is forced to move or to slip down toward the centrifugal direction as a whole (KAMIYA and KURODA 1958). The endoplasm slips down holding oil droplets and other cell organelles in it. This is further evidence showing that the bulk endoplasm is highly viscous under low rates of shear.

It was observed that the movement of the oil droplets in the accumulated endoplasm was sometimes not straight but irregular. This phenomenon suggests the presence of some kind of structure(s) inhibiting the movement of oil droplets, or it simply shows the heterogeneity of the endoplasm.

ALLEN (1974) estimated *in vivo* the endoplasmic viscosity of *Nitella translucens* to be around 0.06 poise at 25 °C from an analysis of the displacement of cytoplasmic particles after streaming cessation by a mechanical stimulus. However the viscosity of endoplasm contiguous to the subcortical fibrils increases greatly at the time of streaming cessation by an electrical stimulus (KAMITSUBO 1986 b). The viscosity of the bulk endoplasm is also extremely high because no shear rate is present within it whether it is flowing or is at a standstill. The reason why the viscosity measured by ALLEN is so low is still unknown.

Acknowledgements

The work was supported partly by Grant-in-Aid for Developmental Scientific Research from the Japanese Ministry of Education, Science and Culture (56840038, 57840036). We thank Drs. N. KAMIYA (M.J.A., National Institute for Basic Biology), T. HAYASHI (Ibaraki Univ.), and Y. MASUDA (Osaka City Univ.) for their encouragement and Dr. R. WAYNE (Cornell Univ.) for his critical reading of the manuscript. We should like to express our obligation to Messrs. H. NAKAGAWA (Sanki Engineering K.K.), Y. OHASHI (Hamamatsu Photonics K.K.), and M. NISHIKO (Quantel K.K.) for their valuable advice and technical support on constructing the CMS.

References

ALLEN NS (1974) Endoplasmic filaments generate the motive force for rotational streaming in *Nitella*. J Cell Biol 63: 270–287

BROWN RH (1940) The protoplasmic viscosity of *Paramecium*. J exptl Biol 17: 317–324

HAYASHI T (1957) Some dynamic properties of the protoplasmic streaming in *Chara*. Bot Mag (Tokyo) 70: 168–174

KAMITSUBO E (1966 a) Motile protoplasmic fibrils in cells of *Characeae*. I. Movement of fibrillar loops. Proc Japan Acad 42: 507–511

— (1966 b) Motile protoplasmic fibrils in cells of *Characeae*. II. Linear fibrillar structure and its bearing on protoplasmic streaming. Proc Japan Acad 42: 640–643

— (1972 a) Motile protoplasmic fibrils in cells of *Characeae*. Protoplasma 74: 55–70

— (1972 b) A "window technique" for detailed observation of characean cytoplasmic streaming. Exp Cell Res 74: 613–616

— (1979) Rotation of cytoplasmic protrusions in the centrifuged internodal cells of *Nitella*. Univ. Tokyo Press, Tokyo (Cell motility: molecules and organization, pp 241–246)

— (1980) Cytoplasmic streaming in characean cells: role of subcortical fibrils. Can J Bot 58: 760–765

— (1981) Effect of supraoptimal temperatures on the function of the subcortical fibrils and an endoplasmic factor in *Nitella* internodes. Protoplasma 109: 3–12

— (1986 a) Velocity distribution of cytoplasmic streaming at the vicinity of subcortical fibrils in characean internodal cells. J Muscle Res Cell Motil 7: 287

— (1986 b) Flowing endoplasm contiguous to the cortex gelates when streaming is stopped by an electric stimulation. J Muscle Res Cell Motil 7: 287

KAMIYA N, KURODA K (1956) Velocity distribution of the protoplasmic streaming in *Nitella* cell. Bot Mag (Tokyo) 69: 544–554

— — (1957) Cell operation in *Nitella*. III. Specific gravity of the cell sap and endoplasm. Proc Japan Acad 33: 403–406

— — (1958) Measurement of the motive force of the protoplasmic rotation in *Nitella*. Protoplasma 50: 144–148

— — (1965) Rotational protoplasmic streaming in *Nitella* and some physical properties of the endoplasm. John Wiley, New York (Proc IVth intern congr rheology, part 4, symp biorheol, pp 157–171)

— — (1973) Dynamics of cytoplasmic streaming in plant cells. Biorheology 10: 179–187

KIKUYAMA M, TAZAWA M (1972) An attempt to introduce various unicellular organisms into the cytoplasm of *Nitella flexilis*. (Proc 37th ann meet Bot Soc Japan, p 161) (in Japanese)

KURODA K, KAMIYA N (1968) Cell operation in *Nitella*. V. Implantation of foreign bodies into an internodal cell. Proc Japan Acad 44: 823–828

TAZAWA M (1964) Studies on *Nitella* having artificial cell sap. I. Replacements of the cell sap with artificial solutions. Plant Cell Physiol 5: 33–43

Protoplasma (1988) [Suppl 1]: 15–26

Screw-mechanical Models Related to Cytoplasmic Streaming

R. Jarosch*

Institut für Pflanzenphysiologie der Universität Salzburg

Received November 21, 1987
Accepted April 19, 1988

Dedicated to Professor Dr. Noburo Kamiya on the occasion of his 75th birthday

Summary

The mechanical characteristics and the motile behaviour of helical protein filaments are illustrated by experiments with macroscopic helices. These model experiments demonstrate
1) the mechanochemistry of the α-helix: The transformation of ionic interactions to torsional motion;
2) selfintertwining of twisted coiled-coil proteins like myosin;
3) the translocation of particles and cross-bridges along actin filaments and microtubules;
4) the translocational polymerization, treadmilling, and depolymerization of actin filaments in migrating cells;
5) Periodical formation and sliding of actin filaments in plasmodia.
Rotation, winding and unwinding of the associated proteins are suggested as the basic events that determine the behaviour of cytoskeletal filaments.

Keywords: α-Helix; Mechanochemistry; Myosin; Actin filaments; Microtubules; Cytoplasmic streaming; Translocation; Cross-bridges; Plasmodia.

1. Introduction

Cell motility is closely related to cellular filaments. Many details of structure and chemistry have been elucidated in the last decades, but how the filaments move is still a matter of ignorance and speculation. The reason for this is obviously the unknown motion mechanics of the filaments which is not comprehensible by the current methods of cell biology and biochemistry. Since the cellular filaments are elastic helical bodies they are subject to the special mechanics ("screw-mechanics") of helical bodies that can be studied by using suitable macroscopic helices (Jarosch 1987 a, b). A simple example may demonstrate what screw-mechanics means: The sliding motion of the thin filaments against the thick myosin filaments and their cross-bridges is well established for a vertebrate muscle sarcomere. Since the known shape of the thin filaments is a double helix it follows by simple screw-mechanics that a motion of such a helical body against the attached cross-bridges is connected with rotations—as the motion of a cork-screw in the cork-stopper is connected with rotations. These overlooked rotations of the actin filaments during their sliding motion has important implications for the structure and function of muscle (Jarosch 1986 a, 1987 a).

The experiments with helical models simulate many phenomena of cell motility in a striking manner. Some experiments in relation to cytoplasmic streaming are described in this paper. Such experiments are simplifications but they demonstrate a hitherto overlooked mechanical aspect of the molecular events.

2. Models Demonstrating the Mechanochemistry of Protein Molecules

The direct conversion of chemical into mechanical energy is a basic ability of life (compare *e.g.*, Katchalsky and Oplatka 1971, Kedem 1983) which is performed by mechanoenzymes. A good general model for cell motility should therefore be founded on the structure and function of these proteins. That foundation is missing with the power- or beating-stroke model of cross-bridges originally proposed for the myosin cross-bridges of muscle and later adopted for cytoplasmic streaming and other kinds of cell motility. Some workers tried to describe the cross-bridge kinetics by ther-

* Correspondence and Reprints: Institut für Pflanzenphysiologie, Universität Salzburg, Hellbrunnerstrasse 34, A-5020 Salzburg.

modynamic formalism (*e.g.*, HILL 1977) but nobody revealed the molecular mechanism how a myosin-, dynein- or kinesin-cross-bridge should "beat" actively or how it should "walk" along an actin filament or a microtubule. These and similar postulated capabilities of enzymes overestimate the possibilities of proteins and are reminiscent of the omnipotent "vital forces" of the past.

It has been ignored that a basic protein configuration—the α-helix—converts chemical in mechanical energy directly (JAROSCH 1963, 1964, 1969, 1986 b, 1987 a, b). This is demonstrated by the models of Figs. 1 to 7. A stretched helical spring shows a mechanical relation between pitch-height and twist: The counterclockwise rotation (arrow cc) of the freely movable end of a stretched helical spring after releasing is demonstrated in Fig. 1. The generated torsional force is strongly increased if ribbonlike helices are used. A tubular helix (ribbon plane parallel to the helix axis) rotates likewise counterclockwise (arrow *cc* in Fig. 2) after releasing. But a helix with the ribbon plane perpendicular to the helix axis (Fig. 3) rotates in opposite clockwise direction (arrow *c*, all helices are right-handed).

The α-helix (Fig. 4) was described with a slightly variable torque (PAULING and COREY 1951). Cross sections of two configurations are drawn in Fig. 5. The α-helix corresponds with a tubular helix as shown by the model (Fig. 6) of FRASER and MACRAE (1973). Little shortening of all hydrogen bonds of a molecule model generates counterclockwise torsion (arrow *cc* in Fig. 7 *b*) as already demonstrated with the model of Fig. 2.

Work on synthetic polypeptides *e.g.*, polyglutamic acid (APPLEQUIST and DOTY 1962) has shown that every change of the electrical side chain potential of an α-helix changes the electrostatic repelling forces which are indicated by double arrows between the charged side chains in Fig. 4. The repelling forces are so strong at high pH that this α-helix is not stable. However, protein α-helices contain different side chains and a decrease of the pH value can only decrease the repelling forces between the negatively charged side chains since they are discharged by the addition of protons. This causes a cooperative shortening of the intrahelical hydrogen bonds and results in a pitch decrease and consequently in torsional rotations. The accumulated torsional rotation of a long α-helix is impressively demonstrated if a tubular helix (Fig. 2) a few meters long is used. However, a charged α-helix is not very stable in an aqueous environment and may easily be destroyed by charge modifications (SCHELLMAN 1955). It is the great advantage of the coiled-coil proteins to be very resistant *e.g.*, against pH changes, Ca^{2+} and ATP^{4-} oscillations that are common inside of a cell. The presupposition for the formation of a coiled-coil are apolar residues at the positions 1 and 4 of each group, if the aminoacid sequence is devided in groups of seven residues. This allows the face to face connection of two intertwined α-helices which are stabilized by the strong hydrophobic interaction of the side chains 1-1' and 4-4'. Salt-bridges may also be important (for details see *e.g.*, MCLACHLAN and STEWART 1975, JAROSCH 1979, 1985).

3. Selfintertwining of Twisted Coiled-Coil Proteins

An important mechanical effect of torsion in a filament is selfintertwining. This phenomenon is frequently shown on electron micrographs by isolated R(D)NA filaments and coiled-coil proteins (Figs. 8–10). Isolated myosin molecules appear "folded" under low salt conditions and in the presence of ATP and Mg^{2+} (TRYBUS *et al.* 1982, ONISHI and WAKABAYASHI 1982, CRAIG *et al.* 1987). These configurations can be easily produced by twisting a rubber string (Fig. 8). Increased salt concentration, ATP-removal, and light chain phosphorylation result in tail-extension (CRAIG *et al.* 1987). The folding at a definite region of the myosin tail (ONISHI and WAKABAYASHI 1982) can be explained by a special aminoacid sequence at that region which may weaken the stiffness of the α-helices as it is well known *e.g.*, for proline. The two mirror-image models proposed by ONISHI *et al.* (1983) for the "looped" myosin molecules correspond here with the models of Fig. 8 *b* and *c*. The L-form (Fig. 8 *b*) leads to a right-handed double helix (Fig. 11 *c, d*) the D-form (Fig. 8 *c*) to a left-handed double helix (Fig. 11 *a*) if longer molecules are present (compare the oligomers of Fig. 10). The torsion generated by the coiled-coil proteins may strongly accumulate when polymers (minifilaments) are formed. It seems that a network of minifilaments (cytomatrix) connects all cytoskeletal elements of a cell. A strongly twisted molecular filament can easily adopt the shape of a globular aggregate if it is not stabilized by rope formation (*e.g.*, within intermediate filaments) or by winding about a cytoskeletal element. Thus, the torsional force of polymeric α-helical minifilaments that can generate many hundreds or thousands of rotations if stabilized by a fibril, may be dissipated in many counterdirected motions of selfintertwining with special structural and enzymatic effects.

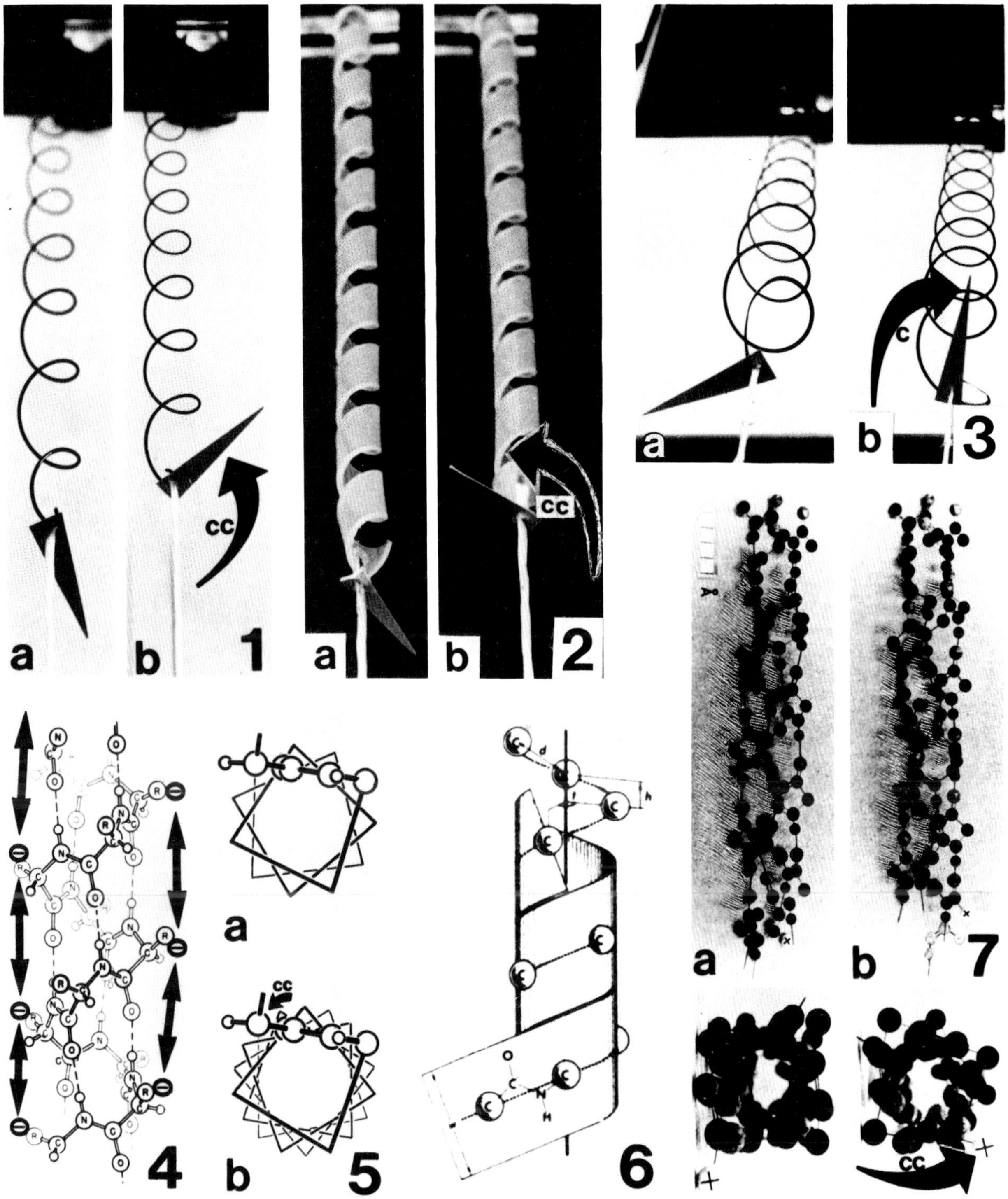

Figs. 1–7. The mechanochemical behaviour of the alpha-helix

Fig. 1. *a* A stretched helical spring; *b* the helix end rotates in counterclockwise direction (arrow *cc*) during release from tension

Fig. 2. Similar behaviour of a tubular plastic helix but a much stronger force is generated

Fig. 3. The end of a ribbon-like helix rotates in clockwise direction (arrow *c*) during release

Fig. 4. The alpha-helix from Pauling and Corey. Electrostatic repelling forces between negatively charged side-chains that stretch the hydrogen bonds are indicated by double arrows

Fig. 5. Cross-section of *a*: the 11-residue 3 turn helix (3,67 residues per turn), *b*: the 18-residue 5 turn helix (3,60 residues per turn), changed from Pauling and Corey (1951). The conformation change from *a* to *b* means H-bond shortening and torsional rotation (arrow *cc*)

Fig. 6. The tubular alpha-helix model of Fraser and Macrae (1973).

Fig. 7. The torsional rotation (arrow *cc*) after shortening of all H-bonds is demonstrated with an elastic molecule model (from Jarosch 1969)

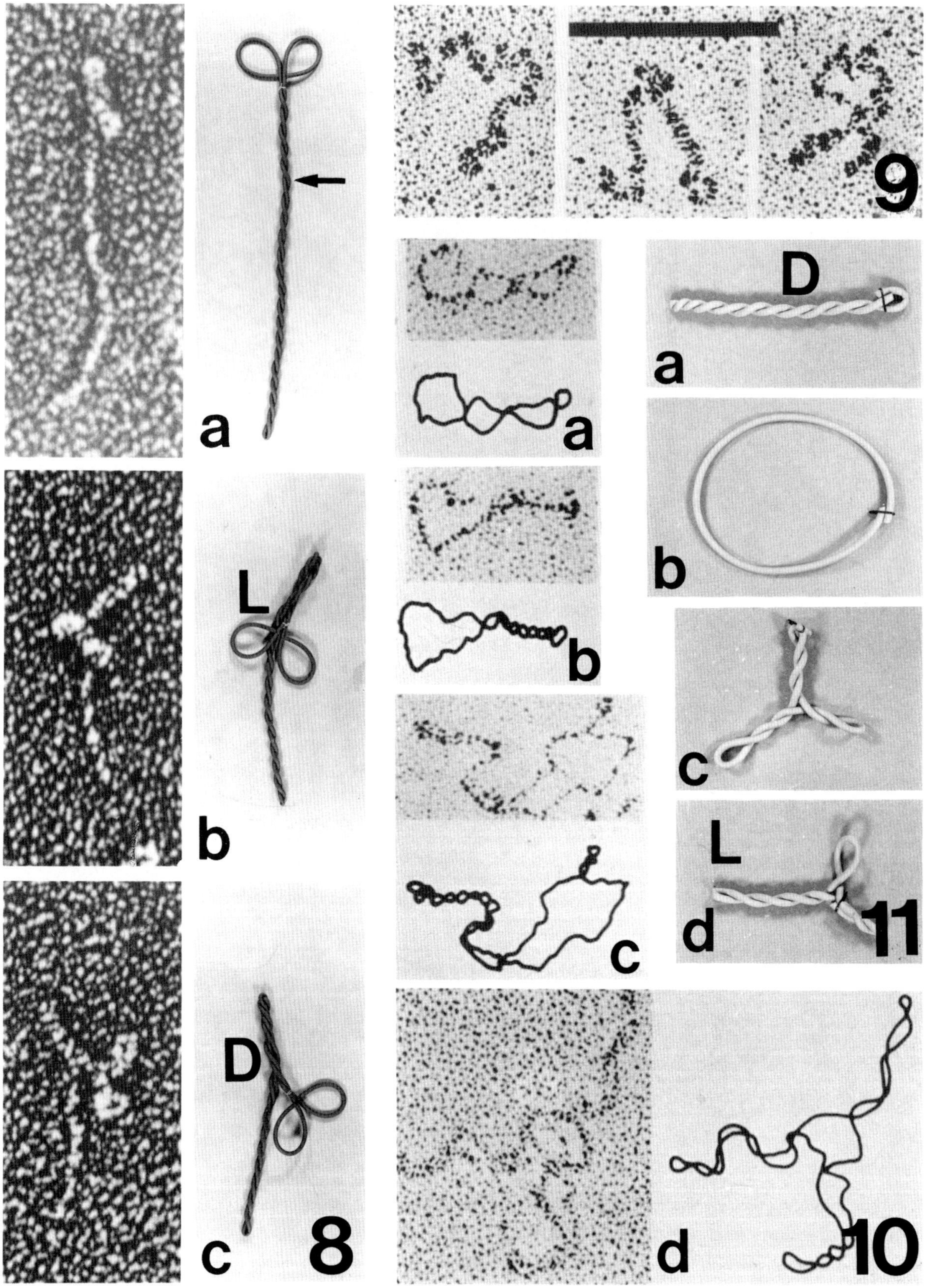

Figs. 8–11. Twisting behaviour of coiled-coil molecules and its simulation by a rubber string

Fig. 8. Myosin, *a* non twisted, *b*, *c* twisted (electron micrographs from Elliot and Offer 1978)

Fig. 9. Actin-binding protein (electron micrographs from Hartwig and Stossel 1981)

Fig. 10. Some conformations of spectrin (electron micrographs from Morrow and Marchesi 1981)

Fig. 11. Torsion of a ringlike rubber string. The torsion produces left-handed (*D*-form) and right-handed (*L*-form) double helices

4. Models for the Translocation of Particles (Cytoplasmic Streaming) and Cross-Bridges

It was early discovered that the ability for active movement is supported by cytoplasmic fibrils (JAROSCH 1956 a, b, 1957, 1960) and that the motive force for the fast cytoplasmic streaming in the characean cell is generated by the cortical layer (KAMIYA and KURODA 1956, for reviews see KAMIYA 1959, 1962) that contains thousands of actin filaments (NAGAI and REBHUN 1966, PICKETT-HEAPS 1967, KAMITSUBO 1972, WILLIAMSON 1974, PALEVITZ *et al.* 1974, PALEVITZ and HEPLER 1975). Besides, a second endoplasmic factor indispensable for the streaming has been supposed to be myosin (CHEN and KAMIYA 1975) and has been described later in electron micrographs as helically coiled minifilaments also associated with cytoplasmic membranes (NAGAI and HAYAMA 1979, WILLIAMSON 1979, ALLEN 1980, for review see KAMIYA 1981). Since these filaments are at least a few microns long they may consist of myosin-like oligomers or polymers. Many recent papers demonstrated either the sliding of myosin along F-actin bundles or the sliding of F-actin on myosin coats (HIGASHI-FUJIME 1982, SHEETZ and SPUDICH 1983, SHIMMEN and YANO 1984. YANAGIDA *et al.* 1987). Many authors interpreted their observations by the power stroke cross-bridge model and did not consider the molecular difficulties connected with that model (but see also YANAGIDA 1984, YANAGIDA *et al.* 1987). Simpler and consistent with many motion phenomena of the highly dynamic actin fibrils (*e.g.*, the waving rings and polygons, dynamic networks) is the model of a fast filament rotation (JAROSCH 1976, FOISSNER and JAROSCH 1981, JAROSCH and FOISSNER 1982, 1983, FOISSNER 1985, JAROSCH 1986 b, 1987 a, b). This model assumes that many coiled-coil filaments (myosin and other mechanoenzymes) which generate the torsional rotation contact a cytoskeletal filament, turn it (arrow *C*), and are wound at the same time as it is shown with the helix of Fig. 12 *a, b*. Here a double helix is used which is simultaneously untwisted. The oscillative untwisting and twisting motion of the molecular helix may also take place if no actin filament is present. That oscillation may represent the behaviour of myosin which becomes much more effective if it contacts an actin filament. Only now a helix-attached particle can be translocated (arrow) along the rotating actin filament or can perform a saltatoric motion if the helix is short. A lateral or helical motion of the particle around the rotating actin filament is prevented here by the other actin filaments of a bundle and also by the strong resistance of the medium against the lateral motion of the particle. It is possible that the endoplasmic minifilaments do not contact the actin filaments by their ends but somewhat laterally as shown by the model of Fig. 12 *c, d*. So the free remaining end may be unwound without changing the rotational direction. If projecting free ends of minifilaments will contact a particle it is translocated by unwinding.

In this model the speed of the particle translocation depends on the number of filament rotations per second and on the pitch-height of the winding helix. One pitch-height of a helical filament progresses one time per helix rotation, and the coils of a rotating helix move as apparent helical waves. F-actin forms a double helix with a rather variable pitch. If we assume 100 nm as an average value for the pitch and 20 μm (= 20,000 nm)/s for protoplasmic streaming velocity, 20.000/100 results in 200 rot/s. The particle velocity in direct contact with the actin cables is never higher than 15–20 μm/s. The quicker velocity of the characean endoplasmic particles (60–100 μm/s) is probably a summation effect (JAROSCH and FOISSNER 1983, JAROSCH 1986 b).

The pitch of the F-actin double helix may correspond with the pitch of the associated superhelix as shown by the superhelix of tropomyosin in the grooves of muscular actin filaments. This needs not always to be the case *e.g.*, when superhelices with different pitches are wound. The variability of the F-actin pitch (EGELMAN *et al.* 1982, TAYLOR *et al.* 1984, EGELMAN 1985) may indicate strong torsional stress. Besides, a regular apposition and winding, *e.g.*, of myosin, can increase the smoothness of the F-actin double helix and thus its ability to screw.

Translocation of particles and cytoplasmic streaming is also performed by microtubules (*e.g.*, KURODA and MANABE 1983, TRAVIS *et al.* 1983, SCHLIWA 1986). The similarity of the particle motion is striking and a relation as found for myosin and F-actin has been described for dynein or kinesin and microtubules. However, particles can move here simultaneously in both directions along a single microtubule (ALLEN *et al.* 1985, KOONCE and SCHLIWA 1985, MILLER and LASEK 1985, VALE *et al.* 1985). The same motion mechanics as with actin filaments is applicable if single protofilaments of a microtubule are rotated by winding and unwinding of mechanoenzymes like kinesin or dynein. At least parts of these giant protein combinations may be filamentous and may appear folded on electron micrographs. Unfolding during their action is expected as it is shown by the twisted double helix of Fig. 12.

The model of Fig. 13 *a, b* proposes a microtubule as a bundle of rotating profilaments that can wind smaller associated filaments (MAPs) which may appear as cross-bridges on electron micrographs. Right (R)- and left (L)-handed helices wind here in opposite directions and move their associated particles in opposite directions (arrows). The translocation of particles may be expected to follow a steep helix if the protofilaments are not quite parallel to the axis of the microtubule but show a small supertwist. A further consequence of such a supertwist is a slow rotation of the whole microtubule. Filament rotation generated by the active torsional rotations of mechanoenzymes should be powered by ATP-splitting. However, passive winding and unwinding along rotating filaments is also possible as demonstrated by the compound cross-bridge of Fig. 14 which moves to the left (arrow). One site of a helix is translocated here from A at the lower filament (Fig. 14 *a*) to B at the upper filament (Fig. 14 *b*). Another site is translocated from C to D. Small particles which are fixed at the helix undergo strong mechanical stress and can be set free when they pass the cross-bridge. They may correspond to substrate molecules that are metabolically changed when they are carried from filament to filament. The passively wound carrier helices should correspond to enzymes (*e.g.*, kinases) or their subunits whose metabolic work is thus energetically coupled with ATP-splitting.

5. Winding and Unwinding of Associated Proteins as a Basis for Addition and Removal of Subunits During Controlled Polymerization and Depolymerization

Actin filaments polymerize at their membrane-associated ends. This leads to a progressive growth *e.g.*, of acrosomal processes (TILNEY *et al.* 1981, TILNEY and INOUÉ 1982), microvilli (MOOSEKER *et al.* 1982), microspikes (HÖGLUND *et al.* 1980, SMALL and LANGANGER 1981) and platelets (KARLSON *et al.* 1984). LINDBERG *et al.* (1981) proposed a detailed molecular model for the translocational polymerization (arrows *Pr*) of actin filaments in the leading lamella of migrating cells which is shown in Figs. 15 and 17. He assumes a transmembrane protein that adds G-actin from profilactin to the tips of the actin filament which rotates. The resulting motion is either a progression (Fig. 17 *b*) or a screwing in the opposite direction (Fig. 17 *c*). The model of Fig. 17 is fully consistent with the screw-mechanics of a rotating double helix like F-actin that gets new subunits by winding of associated helices which carry the subunits. Not only the site of winding progresses here as indicated by the arrow *Pr* in Fig. 16 but also the helical filament can screw (slide) or treadmill in the opposite direction (arrow *Tr*). The same motion mechanics is also demonstrated with a ring-shaped helix model (Fig. 18) which winds on a rod at the plus (+) site and unwinds at the minus (—) site (compare with Fig. 17). This model simulates (1) the progressive motion of the leading lamella, (2) the lateral motions (small arrows in Fig. 15) of contacting filaments in the interspike region described as "hammock" by HÖGLUND *et al.* (1980). These lateral motions may be generated by the waves (arrows *w* in Fig. 16 *b*) and by the mutual rolling motions of the contacting filaments. (3) F-actin treadmilling as proposed by WEGNER (1967) and observed by WANG (1985), (4) F-actin polymerization in the front and (5) F-actin depolymerization in the tail of a migrating cell. Unwinding of associated proteins with bound G-actin subunits means depolymerization and collapse or degeneration of the actin filaments (see below). It is not shown by the model of Fig. 18 that the actin filament rotations (arrow *c*) can be driven everywhere along the filaments by endoplasmic myosin.

6. Periodical Winding and Unwinding of Filaments

Selfintertwound double helices (E_1) are wound (a → b) and unwound (b → a) on rods in the model of Fig. 19. The helices may represent endoplasmic myosin or other associated proteins respectively. The rods may represent actin filaments. The torsional force generated in

Figs. 12–14. Translocation of particles and cross-bridges

Fig. 12. *a, b* Translocation (large arrow to the right) of a particle by winding of a particle-associated right-handed (*R*) helix on a filament like F-actin. The helix has been twisted at first and forms therefore a double helix like the coiled-coil proteins of Figs. 8–11. Untwisting the double helix (arrow *c*) turns the actin filament (arrow *C*) by winding the single R-helix. *c, d* The particle translocation is performed by winding and immediate unwinding of the helix

Fig. 13. Bidirectional translocation (arrows) of particles that pass each other on a filament bundle that may represent a microtubule. The particle-associated helices show opposite coiling direction and move along single "protofilaments" which rotate in the same direction

Fig. 14. Translocation (arrow to the left) of a compound cross-bridge along two rods by unwinding and winding of its two helices. For more details see the text

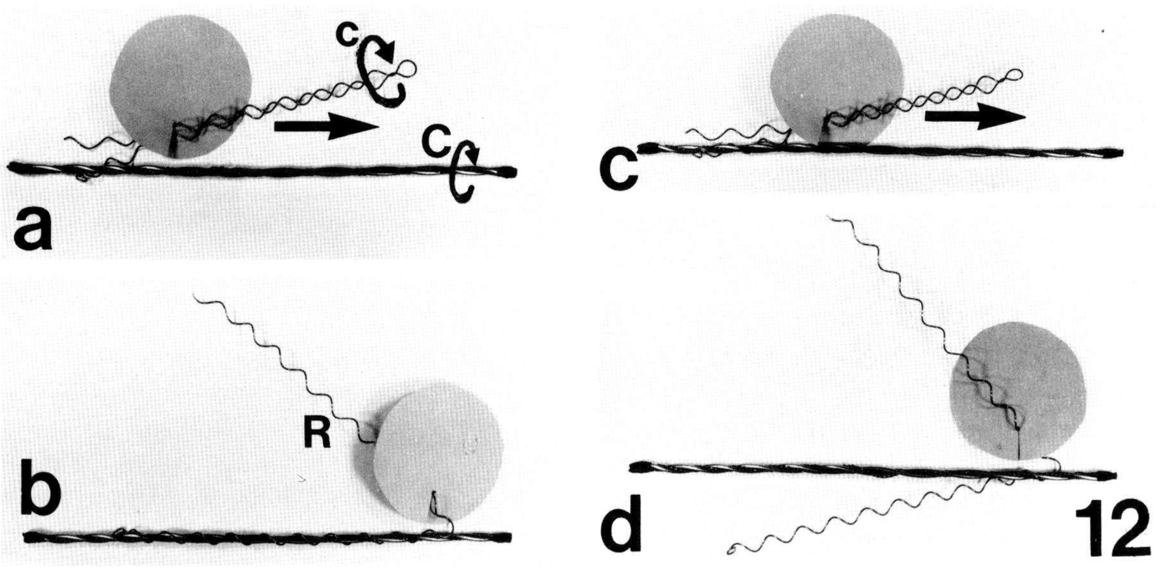

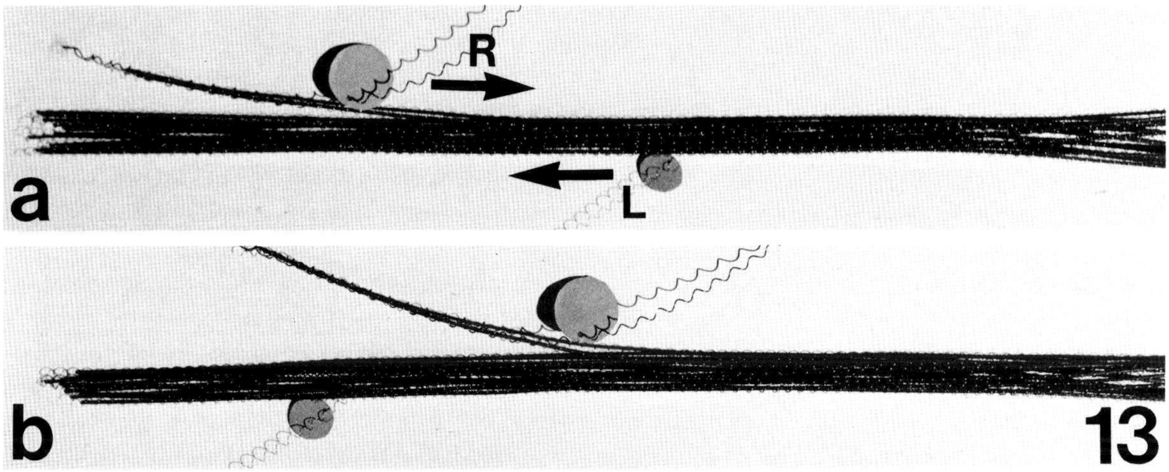

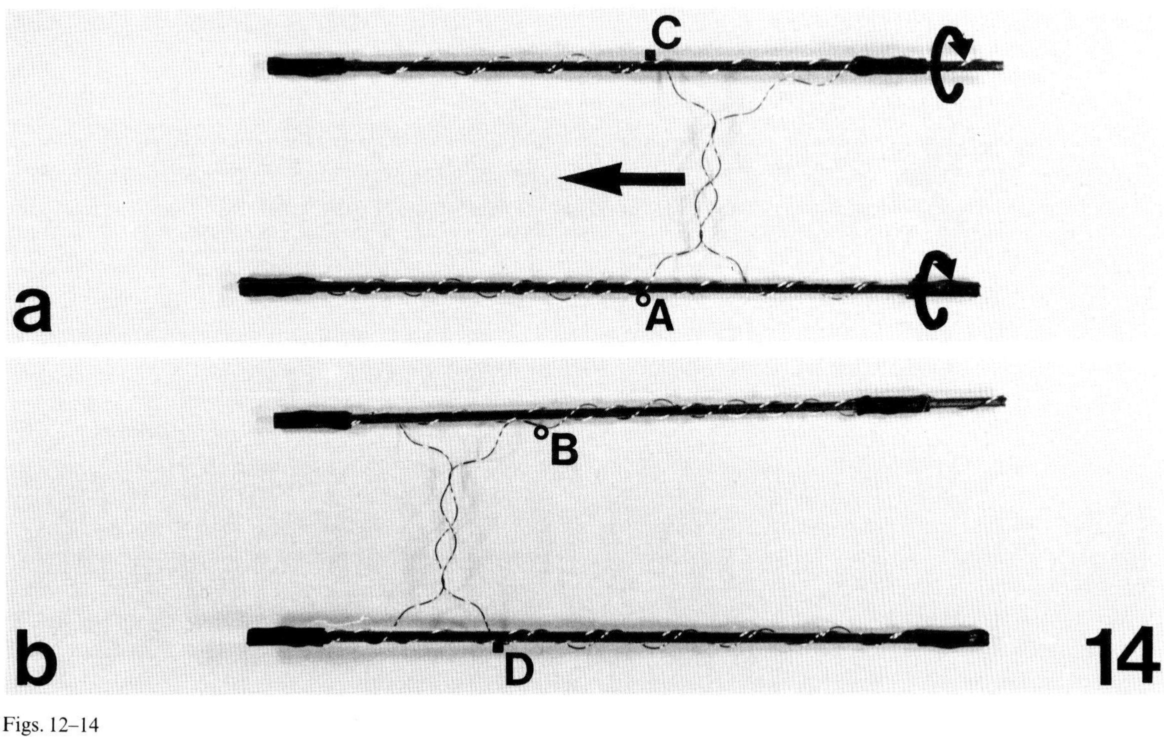

Figs. 12–14

the compound myosin helices E_1 by a binding of cations and $ATPMg^{2-}$ unwinds E_1. The single coiled-coils wind on the actin filaments forming actomyosin E_2 where the binding conditions for ions have been changed. Therefore, some ions are displaced as indicated by e in the model. This causes the opposite torsional force in the myosin coiled-coils and leads again to unwinding. Thus a feedback controlled oscillation may take place which can be described $E_1 > E_2 \rightleftarrows E_1 < E_2$.

A molecule indicated by *8* in Fig. 19 *a* if bound by both helices of E_1 can be split by unwinding of E_1. One part (o) is displaced by winding, the other can remain as shown in Fig. 19 *b*. Molecules bound at E_2 by myosin and F-actin can similarly be split by unwinding. This is a suggestion for a possible mechanism of ATP-splitting (JAROSCH and FOISSNER 1983, JAROSCH 1985).

If the selfintertwining or selfwinding process of molecular filaments described in the Figs. 8–12 occurs with proteins located in membranes they should transport ions through membranes as it is well known for the ion pumps. Models for such ATPases and translocases from the literature assume manyfold penetrations of the membrane by α-helical parts of the molecule and loops (!) on the inner membrane side. The ions may dissociate more easily *e.g.*, from the double helical part (= twisted loop) of the molecule.

The motions performed by the rods of Fig. 16 and Fig. 19 occur simultaneously in the same direction since they are rather parallel. If actin filaments are more randomly distributed in the cytoplasm their growing motions may occur in different directions as shown by the model of Fig. 20. Periodical sliding motions of actin filaments intermediated by myosin were already supposed by KAMIYA (1968) for the periodic volume changes responsible for the shuttle streaming in plasmodia. A periodical transformation of F-actin bundles to a network corresponding to the contraction cycle was described by NAGAI *et al.* (1975, 1978). On the other hand cyclic polymerization and depolymerization of the actin filaments was proposed for the plasmodial periodicity by FLEISCHER and WOHLFARTH-BOTTERMANN (1975), WOHLFARTH-BOTTERMANN and FLEISCHER (1976) and ISENBERG and WOHLFARTH-BOTTERMANN (1976). Both interpretations are consistent with the winding and unwinding model (JAROSCH and FOISSNER 1983). It is possible that the kinky filaments described by NAGAI *et al.* (1975, 1978) for the relaxation phase is the appearance of the degenerating actin filaments when the associated proteins are unwound. The rods bare of helices are dotted in Fig. 20. The straight actin filaments which show progressive elongation (arrows *Pr* in Fig. 20 *a*) or screwing motion (arrows *Tr*, compare Fig. 16) during their non-parallel sliding motion should form a more parallel arrangement in the contraction phase and during stretch as indicated in Fig. 20 *b*. This model could describe in a simplified manner the behaviour of the polygonal pattern, ramification points or nodes of actin filaments observed in plasmodia (NAIB-MAJANI *et al.* 1984, ISHIGAMI 1986, ISHIGAMI *et al.* 1987) and in some tissue culture cells (references in NAIB-MAJANI *et al.* 1984).

Figs. 15–20. Filament motions connected with polymerization and depolymerization

Fig. 15. Progressive motion (arrows *Pr*) of microspikes and the leading lamella by the translocational polymerization of actin filaments of a migrating glia cell (from LINDBERG *et al.* 1981). The small arrows indicate the putative rolling motions of contacting single actin filaments.

Fig. 16. Model for the translocational polymerization of three actin filaments by winding (white arrows) of associated helices that are thought to carry actin subunits. The site of winding (+) progresses (arrow *Pr*) with the helical waves (arrows *w*) to the right because the screwing or treadmilling motion (arrow *Tr*) to the left is small with the rotating F-actin double helices. The rods not yet wrapped by the helices are dotted

Fig. 17. The addition of G-actin subunits by a rotating transmembrane protein (*a*) causes either the progressive motion of the membrane (*b*) or the screwing motion of the actin filament in the opposite direction (*c*), from LINDBERG *et al.* (1981)

Fig. 18. Screw-mechanical model for a migrating cell that is consistent with LINDBERG's model. A ringlike helix is wound at the +site and unwound at the −site (white arrows) in order to simulate actin polymerization and depolymerization. The rotational direction of the actin filaments is always the same. For details compare Figs. 16 and 17

Fig. 19. Periodic winding and unwinding of helices on rods by changing the rotational direction (arrow *c* and *cc*). The single helices form either double helices (E_1) or are wound on the rods (E_2). This model may simulate the feedback-controlled oscillation of actin filaments by a periodical change of configuration and torsional force within associated filaments like myosin. It is also a model for an ion pump. For more details see the text

Fig. 20. The same model as shown in Fig. 19 but with non parallel arrangement and sliding of the rods. This model may simulate the periodical generation and degeneration of the straight actin filaments in plasmodia by winding (compare Fig. 16) and unwinding of helices that should represent associated proteins. For more details see the text

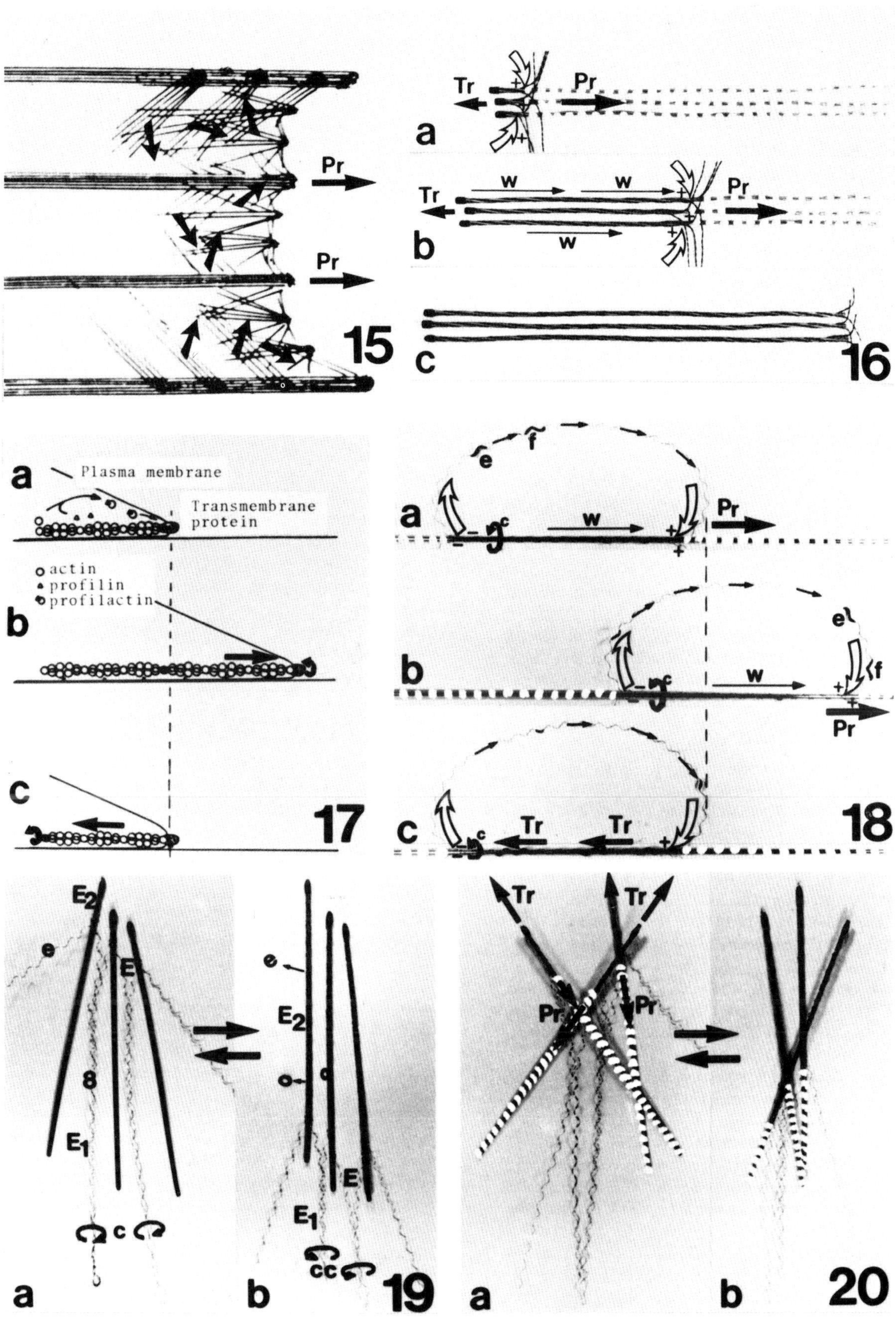
Pr
Pr
15
a
Tr
Pr
b
Tr
w
w
w
Pr
c
16
a
Plasma membrane
Transmembrane protein
o actin
profilin
profilactin
b
c
17
a
e
f
w
Pr
b
e
f
w
Pr
c
Tr
Tr
18
E2
e
E
8
E1
c
a
e
E2
E1
E
cc
b
19
Tr
Tr
Pr
Pr
a
b
20

Figs. 15–20

The demanded displacements of ions by winding and unwinding may correspond with the synchroneous oscillation of ATP, Ca^{2+} and H^+ described for plasmodia (Yoshimoto *et al.* 1981 a, b, Kamiya *et al.* 1982, Nakamura and Kamiya 1985 a, b). However, the increased tension after a stretch of a plasmodial strand is not expressed by these oscillations. Therefore no passive rotations of actin filaments as proposed for the stretch activation of muscle (Jarosch 1986 a, 1987 a) may occur when a plasmodium is stretched. The "stress relaxation phase" observed immediately after the stretch, the new generation of endoplasmic actin filaments and the disappearance of these filaments under isotonic conditions (Wohlfarth-Bottermann and Fleischer 1976, Wohlfarth-Bottermann 1987) indicate that the filament mechanics after a stretch is here much more complex as compared with a vertebrate muscle.

References

Allen NS (1980) Cytoplasmic streaming and transport in the characean alga *Nitella*. Can J Bot 58: 786–796

Allen RD, Weiss DG, Hayden JH, Brown DT, Fujiwake H, Simpson M (1985) Gliding movement of and bidirectional transport along single native microtubules from squid axoplasm: Evidence for an active role of microtubules in cytoplasmic transport. J Cell Biol 100: 1736–1752

Applequist J, Doty P (1962) In: Stahmann MA (ed) Polyamino acids, polypeptides and proteins. University of Wisconsin Press, Madison, Wisconsin, p 161

Chen JCW, Kamiya N (1975) Localization of myosin in the internodical cell of *Nitella* as suggested by differential treatment with N-ethylmaleimide. Cell Structure Function 1: 1–9

Craig R, Padron R, Kendrick-Jones J (1987) Structural changes accompanying phosphorylation of *Tarantula* muscle myosin filaments. J Cell Biol 105: 1319–1327

Egelman EH (1985) The structure of F-actin. J Muscle Res Cell Motility 6: 129–151

— Francis N, Derosier DJ (1982) F-actin is a helix with a random variable twist. Nature 298: 131–135

Elliot A, Offer G (1978) Shape and flexibility of the myosin molecule. J Mol Biol 123: 505–519

Fleischer M, Wohlfarth-Bottermann K-E (1975) Correlation between tension force generation, fibrillogenesis and ultrastructure of cytoplasmic actomyosin during isometric and isotonic contractions of protoplasmic strands. Cytobiologie 10: 339–365

Foissner I (1985) The rotation model as a basis for *Nitella* filament-dynamics. In: Alia EE, Arena N, Russo MA (eds) Contractile proteins in muscle and non-muscle cell systems. Praeger, New York, pp 213–218

— Jarosch R (1981) The motion mechanics of *Nitella* filaments (cytoplasmic streaming): their imitation in detail by screw-mechanical models. Cell Motility 1: 371–385

Fraser RDB, Macrae TP (1973) Conformation in fibrous proteins. Academic Press, London

Hartwig JH, Stossel Th P (1981) Structure of macrophage actin-binding protein molecules in solution and interacting with actin filaments. J Mol Biol 145: 563–581

Higashi-Fujime S (1982) Active movement of bundles of actin and myosin filaments from muscle: A simple model for cell motility. Cold Spring Harbor Symp Quant Biol 46: 69–75

Hill TL (1977) Free energy transduction in biology. The steady-state kinetic and thermodynamic formalism. Academic Press, New York, 299 pp.

Höglund A-ST, Karlsson R, Arro E, Fredriksson B-A, Lindberg U (1980) Visualisation of the peripheral weave of microfilaments in glia cells. J Muscle Res Cell Motil 1: 127–146

Isenberg G, Wohlfarth-Bottermann K-E (1976) Transformation of cytoplasmic actin. Cell Tissue Res 173: 495–528

Ishigami M (1986) Dynamic aspects of the contractile system in *Physarum* plasmodium: I. Changes in spatial organization of the cytoplasmic fibrils according to the contraction-relaxation cycle. Cell Motil Cytoskeleton 6: 439–447

— Kuroda K, Hatano S (1987) Dynamic aspects of the contractile system in *Physarum* plasmodium. III. Cyclic contraction-relaxation of the plasmodial fragment in accordance with the generation-degeneration of cytoplasmic actomyosin fibrils. J Cell Biol 105: 381–386

Jarosch R (1956 a) Aktiv bewegungsfähige Plasmaelemente und Chloroplastenrotation bei Characeen. Anzeiger Math-nat Kl Österr Akad Wiss 6: 58–60

— (1956 b) Plasmaströmung und Chloroplastenrotation bei Characeen. Phyton (Argentina) 6: 87–107

— (1957) Zur Mechanik der Protoplasmafibrillenbewegung. Biochim Biophys Acta 25: 204–205

— (1960) Die Dynamik im Characeen-Protoplasma. Phyton (Argentina) 15: 43–66

— (1963) Grundlagen einer Schrauben-Mechanik des Protoplasmas. Protoplasma 57: 448–500

— (1964) Screw-mechanical basis of protoplasmic movement. In: Allen RD, Kamiya N (eds) Primitive motile systems in cell biology. Academic Press, New York, pp 599–622

— (1969) Das sterische Verhalten der Alpha-Helix. Z Naturforsch Teil 24 B: 672–680

— (1976) Dynamisches Verhalten der Actinfibrillen von *Nitella* auf Grund schneller Filament-Rotation. Biochem Physiol Pflanzen 170: 111–131

— (1979) The torsional movement of tropomyosin and the molecular mechanism of the thin filament motion. In: Hatano S, Ishikawa H, Sato H (eds) Cell motility: Molecules and organization. University of Tokyo Press, Tokyo, pp 291–319

— (1985) The filament rotation model: Molecular and screw-mechanical details. In: Alia EE, Arena N, Russo MA (eds) Contractile proteins in muscle and non-muscle cell systems. Praeger, New York, pp 239–251

— (1986 a) The thin filament rotation model of muscle contraction. In: Frey M, Freilinger G (eds) 2nd Vienna muscle symposium proceedings. Facultas Universitätsverlag, Wien, pp 1–7

— (1986 b) A model for the molecular basis of contractility and sliding as demonstrated by helix models. Cell Motil Cytoskeleton 6: 229–236

— (1987 a) Screw-mechanical models for the action of actin- and tubulin-containing filaments. In: Wohlfarth-Bottermann KE (ed) Nature and function of cytoskeletal proteins in motility and transport. Gustav Fischer, Stuttgart New York (Progress in zoology, vol 34, pp 231–249)

— (1987 b) Mechanics and hydrodynamics of rotating filaments. In: BEREITER-HAHN J, ANDERSON OR, REIF W-E (eds) Cytomechanics. Springer, Berlin Heidelberg New York Tokyo, pp 54–75

— FOISSNER I (1982) A rotation model for microtubule and filament sliding. Eur J Cell Biol 26: 295–302

— — (1983) The rotation model for filament sliding as applied to the cytoplasmic streaming. In: EARNSHAW JC, STEER MW (eds) The application of laser light scattering to the study of biological motion. Plenum, New York, pp 545–558

KAMITSUBO E (1972) Motile protoplasmic fibrils in cells of Characeae. Protoplasma 74: 53–70

KAMIYA N (1959) Protoplasmic streaming. In: HEILBRUNN LV, WEBER F *et al* (eds) Protoplasmatologia, vol 8. Springer, Wien, part 3 a

— (1962) Protoplasmic streaming. In: RUHLAND W (ed) Handbuch der Pflanzenphysiologie XVII/2. Springer, Berlin Heidelberg New York, pp 979–1035

— (1968) The mechanism of cytoplasmic movement in myxomycete plasmodium. In: Aspects of cell motility. Symp Soc Exp Biol 22: 199–214

— (1981) Physical and chemical basis of cytoplasmic streaming. Ann Rev Plant Physiol 32: 205–236

— KURODA K (1956) Velocity distribution of the protoplasmic streaming in *Nitella* cells. Bot Mag Tokyo 69: 544–554

— YOSHIMOTO Y, MATSUMURA F (1982) Contraction-relaxation cycle of *Physarum* cytoplasm: concomitant changes in intraplasmodial ATP and Ca^{++} concentrations. Cold Spring Harbor Symp Quant Biol 46: 77–84

KARLSSON R, LASSING I, HÖGLUND A-ST, LINDBERG U (1984) The organization of microfilaments in spreading platelets: A comparison with fibroblasts and glial cells. J Cell Physiol 121: 96–113

KATCHALSKY A, OPLATKA A (1971) Mechano-chemical conversion. In: LOEWENSTEIN WR (ed) Handbook of sensory physiology, vol 1, principles of receptor physiology, Springer, Berlin Heidelberg New York, pp 1–17

KEDEM O (1983) Mechano-chemical conversion. In: OPLATKA A, BALABAN ISS (eds) Biological structures and coupled flows. Academic Press, New York, Balaban Publishers, Philadelphia, Rehovot, pp 27–32

KOONCE MP, SCHLIWA M (1985) Bidirectional organelle transport can occur in cell processes that contain single microtubules. J Cell Biol 100: 322–326

KURODA K, MANABE E (1983) Microtubule-associated cytoplasmic streaming in *Caulerpa*. Proc Japan Acad 59 B: 131–134

LINDBERG U, HÖGLUND AS, KARLSSON R (1981) On the ultrastructural organization of the microfilament system and the possible role of profilactin. Biochimie (Paris) 63: 307–323

MCLACHLAN AD, STEWART M (1975) Tropomyosin coiled-coil interactions: evidence for an unstaggered structure. J Mol Biol 98: 293–304

MILLER RH, LASEK RJ (1985) Cross-bridges mediate anterograde vesicle transport along microtubules in squid axoplasm. J Cell Biol 101: 2181–2193

MOOSEKER MS, POLLARD TD, WHARTON KA (1982) Nucleated polymerization of actin from the membrane-associated ends of microvillar filaments in the intestinal brush border. J Cell Biol 95: 223–233

MORROW JS, MARCHESI VT (1981) Self-assembly of spectrin oligomers in vitro: A basis for a dynamic cytoskeleton. J Cell Biol 88: 463–468

NAGAI R, HAYAMA T (1979) Ultrastructure of the endoplasmic factor responsible for cytoplasmic streaming in *Chara* internodial cells. J Cell Sci 36: 121–136

— REBHUN LI (1966) Cytoplasmic microfilaments in streaming *Nitella* cells. J Ultrastruct Res 14: 571–589

— YOSHIMOTO Y, KAMIYA N (1975) Changes in fibrillar structures in the plasmodial strand in relation to phase of contraction-relaxation cycle. Proc Japan Acad 51: 38–43

— — — (1978) Cyclic production of tension force in the plasmodial strand of *Physarum polycephalum* and its relation to microfilament morphology. J Cell Sci 33: 205–225

NAIB-MAJANI W, ACHENBACH Fr, WEBER K, WOHLFARTH-BOTTERMANN K-E, STOCKEM W (1984) Immunocytochemistry of the acellular slime mold *Physarum polycephalum*. IV. Differentiation and dynamics of the polygonal actomyosin system. Differentiation 26: 11–22

NAKAMURA S, KAMIYA N (1985 a) Oscillation of cytoplasmic pH of *Physarum* plasmodium in relation to motility. Cell Struct Funct 10: 133–141

— — (1985 b) Regional difference in oscillatory characteristics of *Physarum* plasmodium as revealed by surface pH. Cell Struct Funct 10: 173–176

ONISHI H, WAKABAYASHI T (1982) Electron microscopic studies of myosin molecules from chicken gizzard muscle. I. The formation of the intramolecular loop in the myosin tail. J Biochem (Tokyo) 92: 871–879

— — KAMATA T, WATANABE S (1983) Electron microscopic studies of myosin molecules from chicken gizzard muscle. II. The effect of thiophosphorylation of the 20 k-Dalton light chain on the ATP-induced change in the conformation of myosin monomers. J Biochem (Tokyo) 94: 1147–1154

PALEVITZ BA, HEPLER PK (1975) Identification of actin in situ at the ectoplasm-endoplasm interface of *Nitella*. J Cell Biol 65: 29–38

— ASH JF, HEPLER PK (1974) Actin in the green alga *Nitella*. Proc Natl Acad Sci USA 71: 363–366

PAULING L, COREY RB (1951) The structure of synthetic polypeptides. Proc Natl Acad Sci USA 37: 241–250

PICKETT-HEAPS JD (1967) Ultrastructure and differentiation in *Chara* sp. I. Vegetative cells. Austr J Biol Sci 20: 539–551

SCHELLMAN JA (1955) The stability of hydrogen-bonded peptide structures in aqueous solutions. Compt Rend Trav Lab Carlsberg Ser Chim 29: 230–259

SCHLIWA M (1986) The cytoskeleton.Springer, Wien New York (Cell biology monographs, vol 13)

SHEETZ MP, SPUDICH JA (1983) Movement of myosin-coated fluorescent beads on actin cables *in vitro*. Nature 303: 31–35

SHIMMEN T, YANO M (1984) Active sliding movement of latex beads coated with skeletal muscle myosin on *Chara* actin bundles. Protoplasma 121: 132–137

SMALL JV, LANGANGER G (1981) Organization of actin in the leading edge of cultured cells: Influence of osmium tetroxide and dehydration on the ultrastructure of actin meshworks. J Cell Biol 91: 695–705

TAYLOR KA, REEDY MC, CÓRDOVA L, REEDY MK (1984) Three-dimensional reconstruction of rigor insect flight muscle from tilted thin sections. Nature 310: 285–291

TILNEY LG, INOUÉ S (1982) Acrosomal reaction of *Thyone* sperm. II. The kinetics and possible mechanism of acrosomal process elongation. J Cell Biol 93: 820–827

— BONDER EM, DEROSIER DJ (1981) Actin filaments elongate from their membrane associated ends. J Cell Biol 90: 485–495

TRAVIS JL, KENEALY JFX, ALLEN RD (1983) Studies on the motility of the foraminifera. II. The dynamic microtubular cytoskeleton of the reticulopodial network of *Allogromia laticollaris*. J Cell Biol 97: 1668–1676

TRYBUS KM, HUIATT TW, LOWEY S (1982) A bent monomeric conformation of myosin from smooth muscle. Proc Natl Acad Sci USA 79: 6151–6155

VALE RD, REESE TS, SHEETZ MP (1985) Identification of a novel force-generating protein, kinesin, involved in microtubule-based motility. Cell 52: 39–50

WANG Y-L (1985) Exchange of actin subunits at the leading edge of living fibroblasts: possible role of treadmilling. J Cell Biol 101: 597–602

WEGNER A (1976) Head to tail polymerization of actin. J Mol Biol 108: 139–150

WILLIAMSON RE (1974) Actin in the alga, *Chara corallina*. Nature 248: 801–802

— (1979) Filaments associated with the endoplasmic reticulum in the streaming cytoplasm of *Chara corallina*. Eur J Cell Biol 20: 177–183

WOHLFARTH-BOTTERMANN K-E (1987) Dynamic organization and force production in cytoplasmic strands. In: BEREITER-HAHN J, ANDERSON OR, REIF W-E (eds) Cytomechanics. Springer, Berlin Heidelberg New York Tokyo, pp 154–166

— FLEISCHER M (1976) Cycling aggregation patterns of cytoplasmic F-actin coordinated with oscillating tension force generation. Cell Tissue Res 165: 327–344

YANAGIDA T (1984) Angles of fluorescently labelled myosin heads and actin monomers in contracting and rigor stained muscle fiber. In: POLLACK GH, SUGI H (eds) Contractile mechanisms in muscle. Plenum Press, New York, pp 397–411

— HARADA Y, KISHINO A, PROCHNIEWICZ-NAKAYAMA (1987) Direct observations of molecular motility by light microscopy. Symposium in honor of Robert Day Allen: Optical approaches to the dynamics of cellular motility, October 5–8, 1987, Marine Biol Lab Woods Hole, Mass. Abstract 13

YOSHIMOTO Y, SAKAI T, KAMIYA N (1981 a) ATP Oscillation in *Physarum* plasmodium. Protoplasma 109: 159–168

— MATSUMURA F, KAMIYA N (1981 b) Simultaneous oscillations of Ca^{2+} efflux and tension generation in the permealized plasmodial strand of *Physarum*. Cell Motil 1: 433–443

Protoplasma (1988) [Suppl 1]: 27–34

Inhibitory Effects of K^+ and Cl^- on Cytoplasmic Streaming in Characean Cells

Y. Tominaga[1] and M. Tazawa[2,*]

[1] St. Agnes' Junior College, Nampeidai Takatsuki, and [2] Department of Botany, Faculty of Science, University of Tokyo

Received December 28, 1987
Accepted May 2, 1988

Dedicated to Professor Dr. Noburo Kamiya on the occasion of his 75th birthday

Summary

The effects of vacuolar osmolality on cytoplasmic streaming were studied in relation to the ionic concentrations of the cytoplasm in *Chara australis*. Ionic concentrations of the cytoplasm were changed in a wide range using tonoplast-free and plasmalemma-permeabilized *Chara* cells.

Three-fold increase in vacuolar osmolality almost stopped the streaming of vacuole-perfused *Chara* cells with intact tonoplast, which is consistent with the data on *Nitella* (Kuroda 1956, Tominaga and Tazawa 1981 b).

In order to see whether the osmolality effect can be accounted for by the ionic effects, we modified the ionic concentrations in the cytoplasm. The streaming of tonoplast-free cells that contain much endoplasm was stopped by adding 110 mM KCl to the perfusion medium containing 64 mM K^+ and 12 mM Cl^-. Superfusion of the plasmalemma-permeabilized cells with the same medium supplemented with 110 mM KCl also stopped the streaming. This inhibition was completely reversed by removing the added KCl.

The streaming inhibition caused by a 1.2-fold increase in the osmotic pressure of the vacuole in cells having artificial cell sap was mostly explained by the increase of K^+. Above this range, the streaming became more sensitive to Cl^- than to K^+. Thus it is concluded that the inhibition of streaming caused by an increase in the vacuolar osmolality in cells having artificial cell sap is accounted for by the ionic effects on the sliding resistance at the sol-gel interface.

The streaming of tonoplast-free cells whose endoplasm is dispersed becomes insensitive to KCl (Tominaga and Tazawa 1981 b). Some endoplasmic factors, which sensitize the motile actomyosin system to KCl, are assumed to exist in the endoplasm.

Keywords: *Characeae*; Ca^{2+}; Cytoplasmic streaming; KCl; Permeabilized cell; Tonoplast-free cell; Trifluoperazine.

Abbreviations: ACS artificial cell sap; APW artificial pond water; ATP adenosin-5′-triphosphoric acid; EGTA ethyleneglycol-bis-(β-aminoethyl ether)N,N,N′,N′-tetraacetic acid; KMS K-methanesulfonate; PIPES piperazine-N,N′-bis(2-ethanesulfonic acid); TFP trifluoperazine.

* Correspondence and Reprints: Department of Botany, Faculty of Science, The University of Tokyo, Hongo, Tokyo 113, Japan.

1. Introduction

Kuroda (1956) and Tazawa (1968) reported that an increase in the osmotic pressure of the vacuolar sap decreases the rate of cytoplasmic streaming in characean internodal cells. Tazawa (1968) and Tominaga and Tazawa (1981 b) suggested that an increase in the osmotic pressure causes an increase in the sliding resistance at the sol-gel interface where the motive force for streaming is generated (Kamiya and Kuroda 1956). The streaming almost stops when the vacuolar osmolality is increased up to 4-fold the normal one by means of vacuolar perfusion. It is clear that a dehydration of the cytoplasm caused by an increase in the vacuolar osmotic pressure brings forth an increase in the ionic concentrations of the cytoplasm. When the tonoplast is removed by application of EGTA into the vacuole (Tazawa *et al.* 1976), the streaming rate is markedly affected by increasing the intracellular osmolality with K^+ salts but not with sorbitol.

There is a discrepancy in the concentration of KCl needed to inhibit streaming in cells having artificial cell sap and tonoplast-free cells. When the osmolality of the vacuolar sap in cells having ACS is increased to 900 mOsm (3-fold hypertonic), the streaming rate decreases to 13% of that of the normal cells (300 mOsm). In this case the K^+ concentration in the cytoplasm is estimated to rise from 78 mM (Tazawa *et al.* 1974) to 230 mM according to van't Hoff's law. However, when tonoplast-free cells are intracellularly perfused with a medium containing 322 mM K^+, the streaming rate decreases to only 38% of the normal rate. The discrepancy between 13 and 38% was interpreted by Tominaga and Tazawa (1981 b) to be a consequence of an increase in the cytoplasmic viscosity in cells having artificial cell sap, which causes an increase in the sliding resistance at the sol-gel interface where the motive force for the cytoplasmic streaming is generated.

In the present paper the authors again tackled the direct effects of K^+ salts, especially KCl, on the streaming rate in plasmalemma-permeabilized cells (SHIMMEN and TAZAWA 1983) and also in tonoplast-free cells, which contain much endoplasm (TOMINAGA *et al.* 1987). Both cell models are more intact in terms of cytoplasmic integrity and more sensitive to Ca^{2+} than tonoplast-free cells, which contain poor endoplasm.

2. Materials and Methods

Chara australis R. BROWN (= *C. corallina* KLEIN ex WILD., em.R.D.W.) was cultured indoors as described before (TOMINAGA and TAZAWA 1981 a). Before each experiment young internodal cells were isolated and stored in APW containing 0.1 mM each of KCl, NaCl, and $CaCl_2$. The osmotic pressure of the vacuolar sap was modified by vacuolar perfusion with media of different osmolalities. The method of perfusion was carried out according to TOMINAGA *et al.* (1987). The basic ACS was similar to the natural cell sap of *N. flexilis*, and contained 90 mM KCl, 40 mM NaCl, 15 mM $CaCl_2$, and 10 mM $MgCl_2$ (TAZAWA *et al.* 1974). It is called 1 Nf medium. Media containing n times as much salts as 1 Nf medium are called n Nf medium. In cells perfused with Nf media the tonoplast remained intact, since the media contained enough Ca^{2+}. We call cells perfused with Nf media cells having ACS. These are the same as cells having ACS in other papers (TAZAWA 1968, 1975, TOMINAGA and TAZAWA 1981 b).

The plasmalemma was permeabilized according to SHIMMEN and TAZAWA (1983) with a minor modification (TOMINAGA *et al.* 1987). Internodal cells were placed on a slide glass cooled on ice and superfused with a slightly hypertonic cold Mg medium (5 mM EGTA, 6 mM $MgCl_2$, 30 mM PIPES, 64 mM K^+, 150–200 mM sorbitol, pH 7.1) instead of being dipped into the cold Mg medium. Control of the cytoplasmic chemical composition is possible by superfusing cells with media of known compositions (SHIMMEN and TAZAWA 1983). Mg · ATP medium (Mg medium + 1 mM ATP) was used to induce cytoplasmic streaming.

The tonoplast-free cells containing much endoplasm were prepared by perfusing the vacuole slowly with a slightly hypotonic medium containing EGTA (TOMINAGA *et al.* 1987). Tonoplast-free cells prepared by a usual method (TAZAWA *et al.* 1976) is less Ca^{2+}-sensitive with respect to the inhibition of streaming (TOMINAGA and TAZAWA 1981 a, TOMINAGA *et al.* 1983). On the other hand, permeabilized cells are sensitive to Ca^{2+} (TOMINAGA *et al.* 1983). The less Ca^{2+}-sensitive tonoplast-free cells have less endoplasm which is washed out of the cell as a result of the usual perfusion method, while the tonoplast-free cells which contain more intact endoplasm are as sensitive to Ca^{2+} as permeabilized cells (TOMINAGA *et al.* 1987). In order to see whether the endoplasm plays some role in KCl-sensitivity as in Ca^{2+}-sensitivity, Ca^{2+}-sensitive tonoplast-free cells were prepared. Figure 1 schematically shows the difference in endoplasmic structure between the Ca^{2+}-sensitive and less Ca^{2+}-sensitive tonoplast-free cell models. Hereafter we call them endoplasm-rich and endoplasm-poor tonoplast-free cells, respectively. To check the Ca^{2+}-sensitivity of either permeabilized or tonoplast-free cells, two concentrations of Ca^{2+}, 10 µM and 1 mM, were used (TOMINAGA *et al.* 1987).

To see the effect of K^+ and Cl^- on the streaming, KCl was added to Mg · ATP medium (0–200 mM sorbitol) and tested in either permeabilized or tonoplast-free cells. Besides KCl, the following salts were used: NaCl, choline chloride ($(CH_3)_3NClCH_2CH_2OH$), K-methanesulfonate (CH_3SO_3K, KMS), and K_2SO_4. They were added to the Mg · ATP medium and applied to permeabilized cells.

To estimate the concentration of KCl in the cytoplasm ($[KCl]_c$) from the osmotic pressure of the vacuole, we assumed that van't Hoff's law is valid for the cytoplasm with the formula:

$$\Pi_c (V_c - b) = \text{constant} \quad (1)$$

where Π_c and V_c represent the osmotic pressure and the volume of the cytoplasm, respectively, and b is the non-solvent volume of the cytoplasm. In the endoplasm of *Nitella*, b is practically zero (YONEDA and KAMIYA 1969). At equilibrium, Π_c should be equal to the osmotic pressure of the vacuole (Π_v). Then

$$\Pi_c V_c = \Pi_v V_c = \text{constant} \quad (2)$$

Thus Π_c can be controlled by changing Π_v. Normal *Chara* cells have

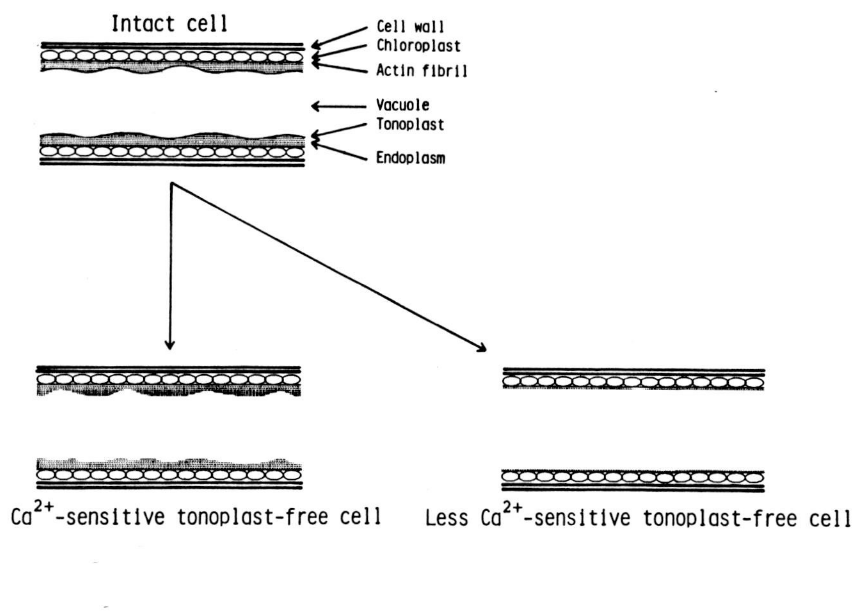

Fig. 1. Schematic representation of two types of tonoplast-free cells. Ca^{2+}-sensitive tonoplast-free cell is endoplasm-rich, while less Ca^{2+}-sensitive one is endoplasm-poor

a Π_v of 250 mOsm. Cytoplasmic concentrations of K^+ and Cl^- are estimated to be 108 mM and 29 mM, respectively (SAKANO and TAZAWA 1984). Then we can calculate $[K^+]_c$ and $[Cl^-]_c$ for various Π_v from the data (Table 2). Since the cytoplasmic streaming of characean cells is not affected by ionic compositions but by the osmolalities of the vacuole (TOMINAGA and TAZAWA 1981 b), we used an ACS used for *Nitella flexilis* (Nf medium) also for *Chara*, although the ionic compositions of vacuoles are slightly different between the two species. One Nf corresponds to 300 mOsm which is 1.2-fold hypertonic for *Chara*.

The osmolalities of media were checked with a vapor pressure osmometer (Wescor, Model 5100CXR).

TFP was kindly supplied by Yoshitomi Pharmaceutical Industry. All perfusion experiments were carried out at room temperature (20–25 °C).

3. Results

3.1. Effect of Osmotic Pressure of Vacuolar Sap on Streaming Rate

Table 1 shows the rate of streaming in *Chara* cells in relation to the osmolality of the cell sap. We could not measure the rate of streaming at 1.0-fold osmolality. In the cells perfused with isotonic medium, the streaming was very unstable and often stopped irreversibly (data not shown). It may be due to the fact that the endoplasm of the present material (*Chara*) is less viscous compared with *N. flexilis* previously used (TOMINAGA and TAZAWA 1981 b). As a result the tonoplast may be broken during perfusion. Then Ca^{2+} in the perfusion medium may invade into the endoplasm and disturb the streaming. Heightening the sap's osmolality up to 1.2-fold resulted in dehydration of the endoplasm and stabilized the streaming due to increase in endoplasm viscosity. Consequently only the rates of streaming at 1.2-fold and more hypertonic osmolalities are listed in Table 1. Normal *Chara* cells have a Π_v of

Table 1. *Rate of cytoplasmic streaming (in µm/sec), before (v_0) and after (v) perfusion with media of various osmotic pressures in Chara cells having ACS*

	Osmotic pressure (%)[1]				
	100	120	200	250	300
v (µm/sec)	62 (v_0)	34	16	3.5	1.3
± SE	3	4	2	0.8	0.6
n[2]	12	5	8	6	6
v/v_0 (%)		55	25	5.6	2.1

[1] The osmotic pressure of vacuolar perfusion media is expressed as % of the normal one. 1 Nf medium has the osmotic pressure of 300 mOsm which corresponds to 1.2-fold the osmotic pressure of the cell sap (250 mOsm) in *C. australis*

[2] Number of cells

Table 2. *Estimated concentrations of K^+ and Cl^- in the cytoplasm and rates of cytoplasmic streaming in Chara cells having ACS of various osmolalities*

Osmolality %	$[K^+]$ mM	$[Cl^-]$ mM	v/v_0 %
100[1]	108[2]	29[2]	100[3]
120	129	34	55
150	162	43	
200	216	57	25
250	270	72	5.6
300	324	86	2.1

[1] The osmolality of 100% corresponds to 250 mOsm

[2] The data from SAKANO and TAZAWA (1984)

[3] The rate of cytoplasmic streaming at 100% osmolality was 62 µm/sec (Table 1)

250 mOsm, and $[K^+]_c$ and $[Cl^-]_c$ are estimated to be 108 mM and 29 mM, respectively (SAKANO and TAZAWA 1984). An increase in Π_v to 2-fold the natural one causes a decrease in the streaming rate to 25% of the normal rate in *Chara*. This is comparable to the value found for *Nitella* (28%, TOMINAGA and TAZAWA 1981 b). Thus the relationship between the sap's osmolality and the streaming rate is very similar between the two species.

A 3-fold increase in the sap's osmolality from 250 mOsm to 750 mOsm decreases the rate of streaming down to 2.1% of the normal rate (Tables 1 and 2). If the Eq. (2) is valid, $[K^+]_c$ and $[Cl^-]_c$ exposed to the 750 mOsm cell sap should increase from 108 and 29 mM to 324 and 86 mM, respectively (Table 2). However, we previously reported that in tonoplast-free *Chara* cells, K^+ at 322 mM (3.0-fold) with Cl^- at 312 mM (11-fold) decreased the rate only to 38% (TOMINAGA and TAZAWA 1981 b). Thus the streaming of tonoplast-free cells seems to be less sensitive to K^+ and Cl^- than the cells having ACS.

3.2. Effect of KCl on the Streaming Rate in Permeabilized Cells

The K^+ and Cl^- concentrations that inhibit streaming in permeabilized cells were different from those in ACS-having cells and tonoplast-free cells. When a permeabilized cell bathed in Mg medium was superfused at time zero with Mg · ATP medium, streaming was reactivated (Fig. 2). Upon the addition of 110 mM KCl to the Mg · ATP medium (totally 300 mOsm) the reactivated (but not fully activated yet) streaming stopped. The total $[K^+]_c$ and $[CL^-]_c$ were 174 mM and 122 mM,

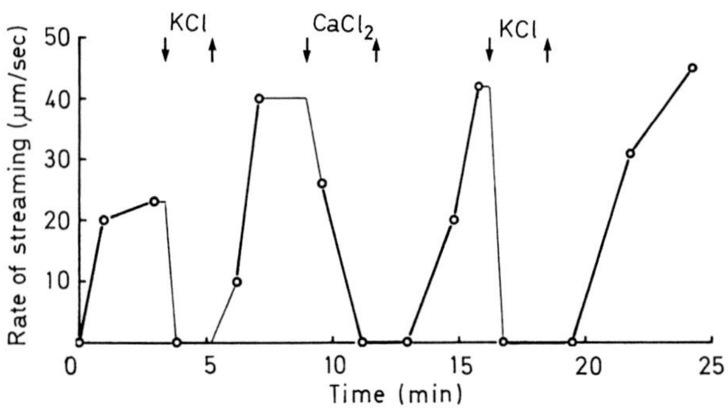

Fig. 2. The reversible inhibition of cytoplasmic streaming by KCl and $CaCl_2$ in a permeabilized *Chara* cell. At time zero a permeabilized cell bathed in Mg medium was superfused with Mg · ATP medium to reactivate the streaming. The concentrations of KCl and $CaCl_2$ added to Mg · ATP medium were 110 and 1 mM, respectively

Table 3. *Concentrations of K^+ and Cl^- of the perfusion media and the rate of cytoplasmic streaming in permeabilized Chara cells*

Added KCl to Mg · ATP medium (mM)	Total conc. $[K^+]$ (mM)	Total conc. $[Cl^-]$ (mM)	v/v_0 ± SE (n^2) (%)	
0	64 (.59)[1]	12 (.42)	100	
50	114 (1.1)	62 (2.2)	100	0 (5)
55	119 (1.1)	67 (2.3)	73	12 (4)
100	164 (1.5)	112 (3.9)	0	0 (7)
110	174 (1.6)	122 (4.3)	0	0 (6)

[1] Numbers in parentheses indicate the ratios of $[K^+]_c$ and $[Cl^-]_c$ in permeabilized cells to $[K^+]_c$ and $[Cl^-]_c$ in normal cells which are 108 mM and 29 mM, respectively (SAKANO and TAZAWA 1984)

[2] Number of cells

respectively (Table 3). After the removal of the added KCl (the KCl was replaced with sorbitol in the Mg · ATP medium) the streaming recovered. Subsequent superfusion with EGTA-free Mg · ATP medium including 1 mM $CaCl_2$ completely and reversibly inhibited the streaming. The second KCl treatment affected the streaming similarly as the first one. Thus KCl and Ca^{2+} treatments did not leave any aftereffect on the sensitivity to each other. The addition of 100 mM KCl also caused a complete cessation of the streaming in a few minutes (Fig. 6). A partial inhibition of the streaming was observed by addition of 55 mM KCl (73 ± 12%, 4 cells), but no inhibition was found when 50 mM KCl was added. Thus the critical concentration of K^+ and Cl^- for the inhibition of the streaming were 114–119 mM and 62–67 mM, respectively (Table 3). The K^+ and Cl^- concentrations that inhibit streaming in permeabilized cells are different from those in cells having ACS and endoplasm-poor tonoplast-free cells (Fig. 3).

3.3. Effect of KCl on the Streaming Rate in Endoplasm-rich Tonoplast-free Cells

The Ca^{2+}-sensitive tonoplast-free cells were as sensitive to KCl as permeabilized cells. In Fig. 4, the streaming of a *Chara* internodal cell was inhibited by 44% after perfusion with a 1.2-fold hypertonic solution. A slow perfusion of Mg · ATP medium removed the tonoplast and the streaming returned over the normal rate (45 μm/sec). The next perfusion with the Mg · ATP medium to which 110 mM KCl was added completely inhibited the streaming. The inhibition was reversed by removing the 110 mM KCl. The second treatment of KCl again caused complete streaming cessation. Perfusion gradually swept away most endoplasmic particles from the cell. When we replaced the KCl with sorbitol, we detected few flowing particles and consequently could not check the reversibility of the streaming in the second KCl treatment. Comparing Fig. 4 with Fig. 2, we can conclude that tonoplast-free cells are as

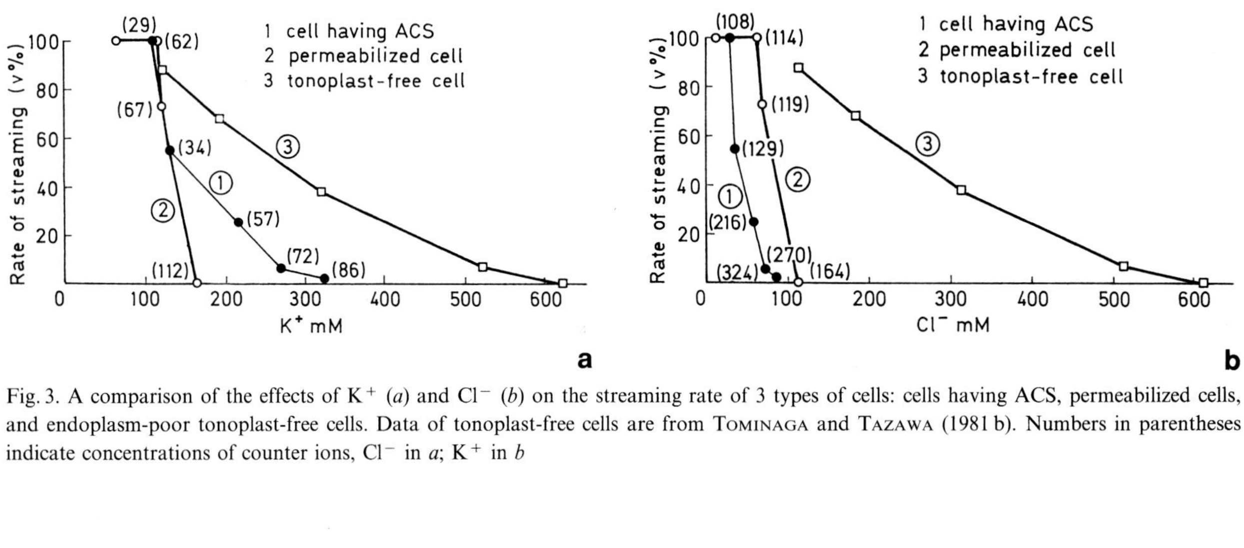

Fig. 3. A comparison of the effects of K^+ (*a*) and Cl^- (*b*) on the streaming rate of 3 types of cells: cells having ACS, permeabilized cells, and endoplasm-poor tonoplast-free cells. Data of tonoplast-free cells are from TOMINAGA and TAZAWA (1981 b). Numbers in parentheses indicate concentrations of counter ions, Cl^- in *a*; K^+ in *b*

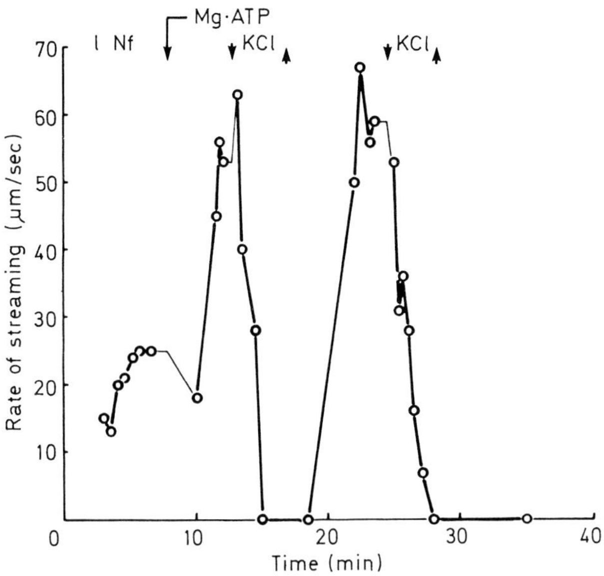

Fig. 4. The effect of a high concentration of KCl on cytoplasmic streaming in a endoplasm-rich tonoplast-free cell. At time zero the vacuole was perfused with a hypertonic solution, 1 Nf. The streaming was inhibited from the original rate, 45 μm/sec by 44% as a consequence of dehydration of the cytoplasm caused by hypertonicity, and recovered when the tonoplast was removed by slowly perfusing the vauole with hypotonic Mg · ATP medium. The recovered streaming was reversibly inhibited by perfusing with Mg · ATP medium including 110 mM KCl. After the second application of 110 mM KCl, no recovery was observed

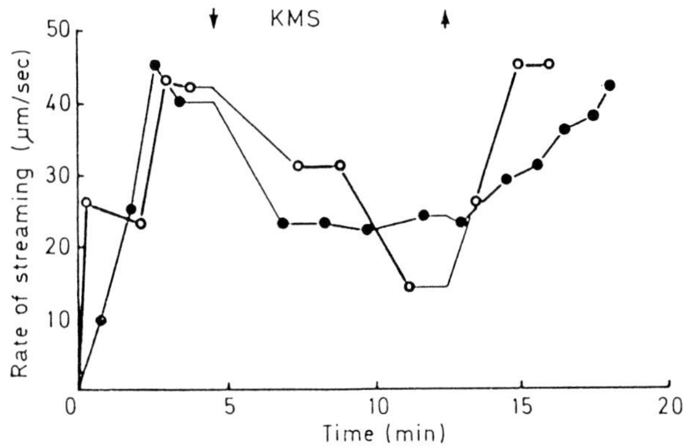

Fig. 5. The effect of potassium methanesulfonate (KMS) on streaming in two permeabilized cells. At time zero 2 permeabilized cells bathed in Mg medium were superfused with Mg · ATP medium. The recovered streaming was inhibited by superfusing the cells with Mg · ATP medium including 125 mM KMS. The inhibited streaming was recovered by removing the KMS

sensitive to KCl as permeabilized cells, but only if the endoplasm is kept intact. Thus the loss of KCl-sensitivity in tonoplast-free cells (Fig. 3, curve 3) is not caused by the disintegration of the tonoplast but by the dispersion of the endoplasm.

3.4. Effects of Other K-Salts and Chlorides on Streaming in Permeabilized Cells

We designed experiments in order to determine whether the ionic strength or the ion species is important for the inhibition of streaming. The K^+ and Cl^- concentrations needed to decrease the streaming rate to 25% of the normal rate in *Chara* cells having ACS were estimated to be 216 and 57 mM, respectively (Table 2), while the $[K^+]$ and $[Cl^-]$ at which the streaming was completely inhibited in permeabilized *Chara* cells were 164 and 112 mM, respectively (Table 3). Thus, Cl^- seems to be more intimately correlated to the streaming inhibition caused by an increase in the vacuolar osmotic pressure above 1.2 fold. To definitely prove this, experiments were designed to see the effects of various K^+-salts and chlorides on streaming in permeabilized cells. Following salts were used: 110 mM NaCl, 110 mM choline chloride, 125 m M KMS, and 86 mM K_2SO_4. The osmolalities of these salt solutions are nearly equal to that of 110 mM KCl (ca. 0.2 Osm). The streaming was completely and reversibly stopped, when either NaCl, choline-Cl, or K_2SO_4 was added to the superfusing Mg · ATP medium. But when 125 mM KMS was added, the streaming of a cell was not affected at call (data not shown) and the streaming rate on the average of 3 cells only decreased to 62%. Figure 5 shows two examples of permeabilized cells. The streaming was induced by superfusing the cells with Mg · ATP medium, and then was inhibited respectively to 33 and 53% upon addition of KMS to the Mg · ATP medium. The results on KMS are consistent with the previous ones obtained in tonoplast-free *Chara* cells (TOMINAGA and TAZAWA 1981 b).

It should be mentioned that application of NaCl to permeabilized cells always broke the tonoplast. If we stained the vacuole with neutral red before permeabilization, we could observe dispersion of the red color from the vacuole to the cytoplasm upon application of NaCl before the streaming stopped. The breakage of the tonoplast caused mixing of vacuolar components with the cytoplasm and impaired the motile system as was shown by the loss of the Ca^{2+}-sensitivity (TOMINAGA and TAZAWA 1981 a, TOMINAGA *et al.* 1983). Thus a high concentration of Na^+ in the cytoplasm produced ill effects on the tonoplast and prevents observations on the effect of NaCl per se on cytoplasmic streaming.

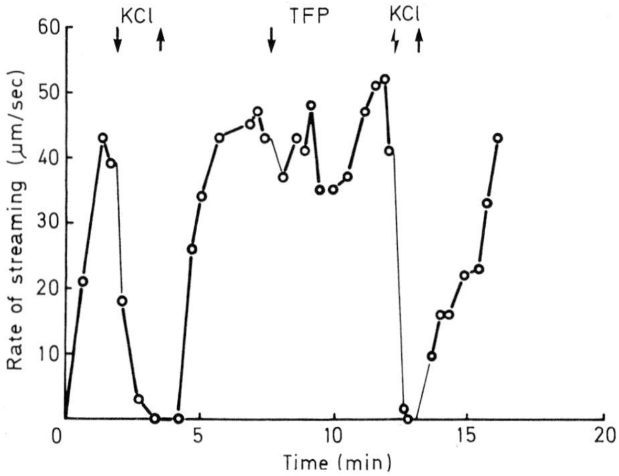

Fig. 6. The effect of TFP on the recovery of streaming after KCl treatment. At time zero a permeabilized cell bathed in Mg medium was superfused with Mg · ATP medium to reactivate the streaming. Addition of 100 mM KCl to Mg · ATP medium inhibited the streaming. The streaming was recovered by removing the KCl. Addition of 25 μM TFP to Mg · ATP medium had little effect on the streaming. The drug was removed by superfusing with 100 mM KCl added to Mg · ATP medium. The pretreatment of the drug did not interfere with the reversible inhibition of streaming induced by the KCl treatment

3.5. Effect of Trifluoperazine on the Recovery of Streaming After KCl Treatment

It is known that the recovery of streaming from the Ca^{2+} inhibition is completely impaired by calmodulin-binding drugs (TOMINAGA *et al.* 1985). In order to see whether the KCl-induced cessation is similar to the Ca^{2+}-induced one, TFP was applied to permeabilized cells before adding 100 mM KCl to the superfusing Mg · ATP medium. The streaming once stopped by KCl recovered to 73 ± 1% (4 cells) of the initial rate after the removal of KCl (Fig. 6). Thus the mechanism of the KCl effect is different from that of the Ca^{2+} effect.

4. Discussion

4.1. High Concentrations of KCl in the Endoplasm May Increase the Sliding Resistance

In plasmalemma-permeabilized cells, the lowest concentrations of K^+ and Cl^- needed to stop the cytoplasmic streaming were 164 and 112 mM, respectively. In the endoplasm-poor tonoplast-free cells, the streaming is inhibited only by 30% with 192 mM K^+ and 182 mM Cl^- (TOMINAGA and TAZAWA 1981 b). But the endoplasm-rich tonoplast-free cells were shown to be as sensitive to KCl as the permeabilized cells. The streaming was completely and reversibly stopped with 174 mM K^+ and 122 mM Cl^- (Fig. 4). Thus the lower sensitivity to KCl in endoplasm-poor tonoplast-free cells (Figs. 3 and 4) is not ascribed to a loss of the tonoplast but to that of the integrity of the endoplasmic structure which is well retained in permeabilized cells. A problem, which remains to be answered, is that the streaming of cells having ACS does not stop even at concentrations of 216 mM K^+ and 57 mM Cl^-. The discrepancies of the KCl effect among cells having ACS, permeabilized cells, and endoplasm-poor tonoplast-free cells are shown in Fig. 3.

As mentioned in Materials and Methods perfusion of *Chara* vacuole with isotonic ACS disturbed cytoplasmic streaming. Consequently we regarded the rates of streaming before perfusion as the rates for the isotonic osmolality. Then we can not exclude the effect of vacuolar perfusion per se on the streaming because we could not measure the streaming rate after perfusion with isotonic medium. However, the data obtained by the perfusion method of the previous (TOMINAGA and TAZAWA 1981 b) and present experiments are similar to those obtained by the transcellular osmosis (KURODA 1956).

In Fig. 3 *a*, curve 1 shows that the relation between the estimated $[K^+]_c$ and the streaming rate in cells having ACS is very close to curve 2 showing the same relation in permeabilized cells for conditions where the $[K^+]$ is lower than 129 mM (1.2 fold).

Above 129 mM K^+, not only K^+ but also Cl^- are effective in inhibiting the streaming. As for K^+, the streaming does not stop even at 324 mM in cells having ACS while it stops at 164 mM in permeabilized cells. Such a difference can not be explained, if K^+ is the sole factor in inhibiting streaming. On the other hand, the streaming was stopped or strongly inhibited at Cl^- concentration of about 90 mM in both cells having ACS (86 mM) and permeabilized cells (112 mM) (Fig. 3). WILLIAMSON (1975) reported that Cl^- at 110 mM is strongly inhibitory to the streaming. Although the streaming was sensitive not only to choline-Cl but also to K_2SO_4, we need not to consider SO_4^{2-} as a regulatory factor of the cytoplasmic streaming in intact characean cells, because the concentration of SO_4^{2-} in *Chara* cells is assumed to be low (HOAGLAND and DAVIS 1922). Thus, the streaming inhibition caused by an increase in vacuolar osmotic pressue is attributed to increases in both $[K^+]_c$ and $[Cl^-]_c$.

When intact cells are perfused with hypertonic ACS, an increase in the sliding resistance is observed along with a decrease in the streaming rate (TOMINAGA and TAZAWA 1981 b). We assumed that the increase in sap osmolality might cause not only an increase in the ionic concentrations of the cytoplasm but also an increase in the viscosity at the sol-gel interface through dehydration of the cytoplasm (TOMINAGA and TAZAWA 1981 b). The present results suggest that dehydration of the cytoplasm will increase the sliding resistance by increasing the $[K^+]_c$ and $[Cl^-]_c$.

4.2. *Presence of a KCl-Sensitizing Endoplasmic Factor*

High concentrations of KCl do not directly affect the actomyosin system, since the streaming in endoplasm-poor tonoplast-free cells is not stopped even by 500 mM and more concentrated KCl (Fig. 3, TOMINAGA and TAZAWA 1981 b). Thus it is unlikely that high concentrations of either K^+ or Cl^- directly affect the actomyosin system (TOMINAGA and TAZAWA 1981 b). How do K^+ and Cl^- exert their inhibitory effects on the streaming rate in permeabilized cells? (Table 3, Fig. 3) A possible mechanism is that high $[K^+]_c$ and $[Cl^-]_c$ induce a release of Ca^{2+} from the cell organelles sequestering Ca^{2+}.

However, this mechanism is unlikely, because the K^+ and Cl^- inhibition occurred even when KCl was added to Mg·ATP medium containing a large amount of EGTA. TFP hardly affected the recovery of streaming which had been stopped by KCl (Fig. 6). Cytoplasmic streaming in permeabilized cells is stopped by 1 μM or more Ca^{2+}. Removal of Ca^{2+} completely recovers the streaming (TOMINAGA *et al.* 1983). Application of 25 μM TFP before Ca^{2+} treatment completely inhibits this recovery. It is suggested that the drug makes the Ca^{2+}-induced cessation irreversible through inhibition of calmodulin (TOMINAGA *et al.* 1985). In order to make sure that the KCl-induced cessation is different from the Ca^{2+}-induced cessation, TFP was applied to permeabilized cells before adding 100 mM KCl to the superfusing Mg·ATP medium. The streaming once stopped by KCl recovered to 73 ± 1% (4 cells) after removing KCl. The report that high concentrations of KCl inhibit calcium binding to calmodulin (OGAWA and TANOKURA 1984) is not applicable to the inhibition of streaming by KCl.

Our present study clearly showed that some endoplasmic factors sensitizing the streaming to KCl are lost in endoplasm-poor tonoplast-free cells which are prepared according to either TAZAWA *et al.* (1976) or TOMINAGA and TAZAWA (1981 b). The streaming in endoplasm-poor tonoplast-free cells is insensitive to both KCl and Ca^{2+}, while the streaming in both permeabilized cells and endoplasm-rich tonoplast-free ones (Figs. 2 and 4) is very sensitive to both KCl and Ca^{2+}. These facts do not necessarily mean that the same factor is responsible both for KCl and for Ca^{2+} sensitization, but suggest that KCl-sensitizing factor localizes in the endoplasm as the Ca^{2+}-sensitizing one. Although at present the mechanism of KCl inhibition of the cytoplasmic streaming remains unknown, we can say at least that the KCl-sensitizing factor in the endoplasm should be a high molecular mass substance which can not pass through the cell wall, because the permeabilized cells having the cell wall remain sensitive to KCl.

Acknowledgement

We wish to thank Dr. R. WAYNE of Cornell University for his kind check on English of the manuscript. We also thank Dr. T. SHIMMEN for his valuable comments on the manuscript. We are grateful to Dr. Y. OKAZAKI of Osaka Medical College for helping us in the measurement of osmotic pressure of solutions with Wescor osmometer. This research was partly supported by a Grant in aid for Scientific Research from the Ministry of Education, Science and Culture of Japan. Y.T. is grateful to the Japan Society for Promotion of Science for the Grants for Research Visits.

References

HOAGLAND DR, DAVIS AR (1922) The composition of the cell sap of the plant in relation to the absorption of ions. J Gen Physiol 5: 629–646

KAMIYA N, KURODA K (1956) Velocity distribution of the protoplasmic streaming in *Nitella* cells. Bot Mag (Tokyo) 69: 544–554

KURODA K (1956) Recherches sur la relation entre la pression osmotique de la cellule et le courant protoplasmique. Cytologia 21: 200–207

OGAWA Y, TANOKURA M (1984) Calcium binding to calmodulin: effects of ionic strength, Mg^{2+}, pH and temperature. J Biochem 95: 19–28

SAKANO K, TAZAWA M (1984) Intracellular distribution of free amino acids between the vacuolar and extravacuolar compartments in internodal cells of *Chara australis*. Plant Cell Physiol 25: 1477–1486

SHIMMEN T, TAZAWA M (1983) Control of cytoplasmic streaming by ATP, Mg^{2+} and cytochalasin B in permeabilized *Characeae* cell. Protoplasma 115: 18–24

TAZAWA M (1968) Motive force of the cytoplasmic streaming in *Nitella*. Protoplasma 65: 207–222

— (1975) Cytoplasmic streaming and ionic regulation in *Nitella flexilis* having artificial cell saps of low ionic strength. Bot Mag Tokyo 88: 219–229

— KISHIMOTO U, KIKUYAMA M (1974) Potassium, sodium and chloride in the protoplasm of *Characeae*. Plant Cell Physiol 15: 103–110

— KIKUYAMA M, SHIMMEN T (1976) Electric characteristics and cytoplasmic streaming of *Characeae* cells lacking tonoplast. Cell Struct Funct 1: 165–176

TOMINAGA Y, MUTO S, SHIMMEN T, TAZAWA M (1985) Calmodulin and Ca^{2+}-controlled cytoplasmic streaming in characean cells. Cell Struct Funct 10: 315–325

— SHIMMEN T, TAZAWA M (1983) Control of cytoplasmic streaming by extracellular Ca^{2+} in permeabilized *Nitella* cells. Protoplasma 116: 75–77

— TAZAWA M (1981 a) Reversible inhibition of cytoplasmic streaming by intracellular Ca^{2+} in tonoplast-free cells of *Chara australis*. Protoplasma 109: 103–111

— — (1981 b) Motive force and rate of cytoplasmic streaming in *Characeae* cells in relation to osmotic pressure and ionic strength of the vacuole and the cytoplasm. Protoplasma 109: 113–125

— WAYNE R, TUNG HY, TAZAWA M (1987) Phosphorylation-dephosphorylation is involved in Ca^{2+}-controlled cytoplasmic streaming of characean cells. Protoplasma 136: 161–169

WILLIAMSON RE (1975) Cytoplasmic streaming in *Chara*: a cell model activated by ATP and inhibited by cytochalasin B. J Cell Sci 17: 655–668

YONEDA M, KAMIYA N (1969) Osmotic volume change of cytoplasmic drops isolated *in vitro* from internodal cells of *Characeae*. Plant Cell Physiol 10: 821–826

Protoplasma (1988) [Suppl 1]: 35–41

Cytoplasmic Streaming and Distribution of Actomyosin Fibrils in the Aphanoplasmodium of *Stemonitis herbatica*

M. Ishigami* and K. Motomochi

Department of Biology, Faculty of Education, Shiga University, Otsu, Shiga

Received Jannuary 5, 1988
Accepted May 10, 1988

Dedicated to Professor Dr. Noburo Kamiya on the occasion of his 75th birthday

Summary

The manner of cytoplasmic streaming and distribution of actomyosin fibrils were investigated in the plasmodia of *Stemonitis herbatica*. Twenty specimens of *Stemonitis* aphanoplasmodia, one of the three general types of plasmodium, were examined to observe streaming direction and velocity at 15-sec intervals for about 80 min. Many specimens (thirteen out of twenty) displayed both unidirectional and bidirectional flows intermittently intermingled. In some cases (five specimens), only the intermittent bidirectional flow was observed, and in the rest (two specimens), only the intermittent unidirectional flow was seen. All of these three streaming types showed a definite pause during each cycle, whose length ranged between 0.37 and 0.61 of the cycle length. The averaged values of maximum streaming velocities diverged from 21 to 111 μm/sec for twenty specimens.

In the coralloid plasmodium, which is produced from aphanoplasmodium at the end of vegetative stage, the direction of cytoplasmic streaming changes alternately without any distinct pause. The averages of maximum velocities ranged from 44 to 83 μm/sec for five specimens.

Well-developed actomyosin fibrils were distributed in tiny anterior portions of the aphanoplasmodium, and also in nodular portions which were formed by gathering of many slender veins in the posterior region. The main veins were diffusely stained with rhodamine-phalloidin.

Keywords: Aphanoplasmodium; *Stemonitis herbatica*; Cytoplasmic streaming; Actomyosin fibril; Rhodamine-phalloidin.

Abbreviations: EDTA ethylene diamine tetraacetic acid; DTT dithiothreitol; PBS phosphate buffered saline.

1. Introduction

The phaneroplasmodium, one of three general types of plasmodium (Alexopoulos 1960, Gray and Alexopoulos 1968), exhibits dynamic and continuous shuttle streaming. This type of cytoplasmic streaming is remarkable both with respect to the high velocity of flow and the great amount of cytoplasm carried, as compared with other types of streaming in plant and fungus cells. This is why the phaneroplasmodium, especially that of *Physarum polycephalum*, has been used to study the cytoplasmic streaming physiologically, biochemically, and morphologically (*cf.*, Kamiya 1959, 1981). It has been generally accepted that the streaming is caused by the local difference of intraplasmodial pressure generated by the fibrillar system of actomyosin (Wohlfarth-Bottermann 1964, Kamiya 1973, Ishigami 1986, Ishigami and Hatano 1986, Ishigami *et al.* 1981, 1987). In contrast to a great store of knowledge about cytoplasmic streaming in the phaneroplasmodium, we have little information about this type of movement in the presumably more primitive types of plasmodia, that is, aphanoplasmodium and protoplasmodium. The aphanoplasmodium, typically formed at the vegetative stage by members of the Order Stemonitales, characteristically consists of a reticulum of thin, flattened, non-granular veins lacking a gelled ectoplasm, and has no distinct fan-like sheet of cytoplasm at the front which is usually observed in the phaneroplasmodium of *Physarum* (Alexopoulos 1959, McManus 1961, Haskins 1974). A reversible flow of cytoplasm was briefly described for the plasmodium of *Stemonitis*, and the bundles of microfilaments (actomyosin fibrils) in the subplasmalemmal region of the cell were briefly described by electron microscopy (Haskins and Hinchee 1974). To obtain a more gen-

* Correspondence and Reprints: Department of Biology, Faculty of Education, Shiga University, Otsu, Shiga 520, Japan.

eral understanding of the streaming mechanism in the Myxomycete plasmodia it is important to elucidate the mechanism of streaming in aphanoplasmodia.

In the present paper, we describe the unique manner of cytoplasmic streaming in the aphanoplasmodia of *Stemonitis herbatica*, and the distribution of cytoplasmic actomyosin fibrils revealed with rhodamine-phalloidin.

2. Materials and Methods

2.1. Culture

We used plasmodia of *Stemonitis herbatica* produced from sporangia, which were collected in Otsu, Japan in July 1979. The spores germinated to produce haploid cells in a few days after being placed, together with *Escherichia coli*, on agar plates buffered with 0.01 M phosphate (pH 6.5). The haploid cells fed on *E. coli* and produced transparent plasmodia which were delicately reticulated. The plasmodia grew into aphanoplasmodia consisting of a plasmodial network, and finally entered the coralloid stage which generated many clusters of sporangia (ALEXOPOULOS 1959, GRAY and ALEXOPOULOS 1968). Subsequent cultures of the vegetative aphanoplasmodia were maintained by transferring a small piece of plasmodium onto a new agar plate at 20 °C with oatmeal.

2.2. Observation of Cytoplasmic Streaming

We used for observation twenty specimens of the vegetative plasmodia and five of the coralloid ones. Cytoplasmic streaming in veins of about 30–50 μm in diameter was monitored by a inverted phase-contrast microscope (Olympus IMT) equipped with a video camera (National WV-1550). Microscopic images were recorded for about 80 min with a video recorder (Victor HR-2200). Direction and velocity of streaming were measured on images reproduced on a video monitor (Victor V-14R).

2.3. Fluorescence Microscopy

For fluorescence microscopy, pieces of aphanoplasmodium with agar foundation were fixed for 5 min in an ice-cold fixative solution containing 2% glutaraldehyde, 2% paraformaldehyde, 50 mM KCl, 25 mM EDTA, 10 mM DTT, 20 mM K-phosphate buffer (pH 7.0), and 1% Triton X-100. This fixative solution was previously shown to be effective in stabilizing instantly the cytoplasmic actomyosin fibrils in the plasmodia of *Physarum* (ISHIGAMI 1986). The specimens were transferred into a fixative solution free of Triton X-100. After 30 min, they were thoroughly washed in PBS, stained for 30 min with rhodamine-phalloidin (Molecular Probes, Inc., Junction City, U.S.A.) diluted by 20-fold with PBS. Observations were made with a fluorescence microscope (Nikon Optiphoto EFDA), and photomicrographs were taken with a Nikon UFX-II camera on Kodak Tri-X films, which were developed in Konidolsuper (Konsihiroku Inc., Tokyo) corresponding to ASA 1,600.

3. Results

3.1. Vegetative Aphanoplasmodium

The vegetative aphanoplasmodium is composed largely of ramified veins as shown in Fig. 1 *a*. The thicker veins

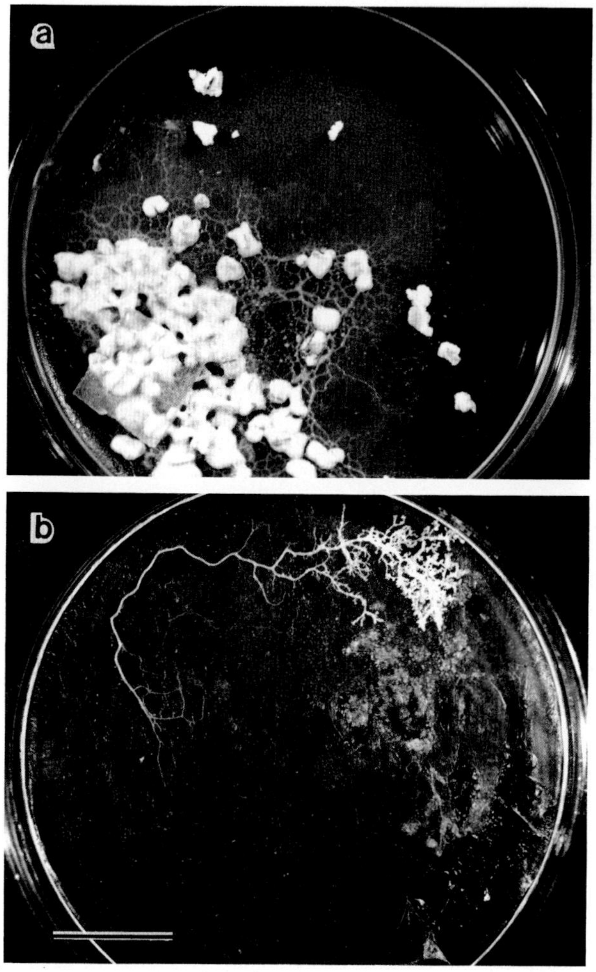

Fig. 1. *a* An aphanoplasmodium of *Stemonitis herbatica*, feeding on oatmeal on agar plate. It consists largely of ramified veins, and of tiny fan-like portions at the advancing front. *b* A coralloid plasmodium, produced usually at the end of the assimilation phase, apparently shows an anterior-and-posterior polarity. Bar represents 20 mm

are usually ornamented with slender satellite ones (*cf.*, Fig. 6). With twenty plasmodia, directions and velocities of streaming in veins of about 30–50 μm in diameter were checked at 15-sec intervals for about 80 min. The typical pattern of streaming is shown in Fig. 2 *a*, where intermittent unidirectional and bidirectional flow is seen at the same locus. This type of streaming was observed in thirteen out of twenty specimens. In some cases (five specimens), only intermittent bidirectional flow was observed, and in the rest (two specimens), intermittent unidirectional flow was exclusively recognized (Fig. 2 *b*, *c*). The duration of a single cycle was nearly constant in each specimen, and ranged from 3 to 8 min for twenty plasmodia (mean value with a standard deviation: 5.4 ± 0.6 min). Without excep-

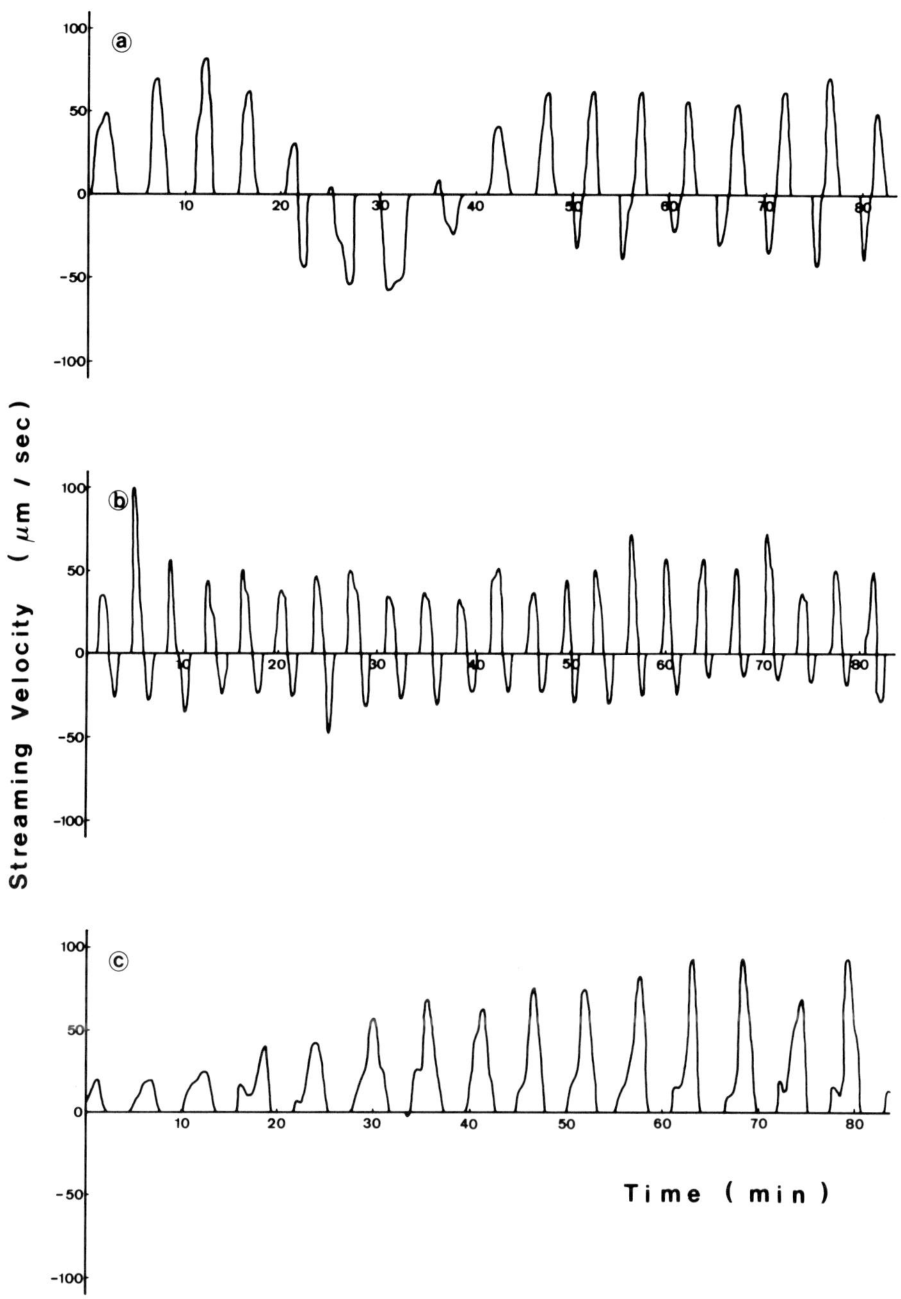

Fig. 2. Streaming patterns in the aphanoplasmodia. Direction and velocity of the streaming in twenty specimens were examined at 15-sec intervals for about 80 min. *a* The typical pattern of streaming, where unidirectional and bidirectional flow occur intermittently. This type of streaming was observed in thirteen specimens. *b* The pattern of intermittent bidirectional flow, observed in five specimens. *c* Intermittent unidirectional flow. This type was seen in two specimens. It is worth noting that a pause always occurs in each cycle. The streaming velocity in *Stemonitis* aphanoplasmodia is far lower than *Physarum* phaneroplasmodia

tion, all streaming exhibits a lengthy pause during each cycle, ranging from 0.37 to 0.61 of the duration of a cycle (Mean ± SD: 0.44 ± 0.05). It was quite often observed that the streaming in the main vein was initiated by the influx of cytoplasm from the satellite veins. Streaming velocities were found to vary greatly with respect to the sample, time, and phase of the streaming cycle. The averaged values of the maximum velocities in a cycle diverge from 21 to 111 μm/sec among twenty specimens (Mean ± SD: 44 ± 22 μm/sec). The maximum instantaneous velocity of all the 301 cycles, examined for twenty specimens, was 177 μm/sec.

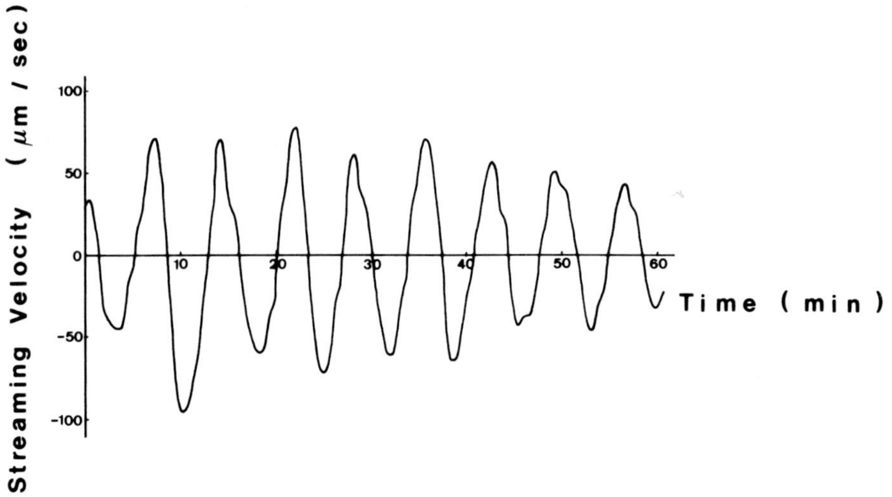

Fig. 3. Streaming pattern in the coralloid plasmodium, where the direction of flow alternates in a rhythmic manner without any pause. This type of streaming is similar to that of *Physarum*, although the streaming velocity is lower

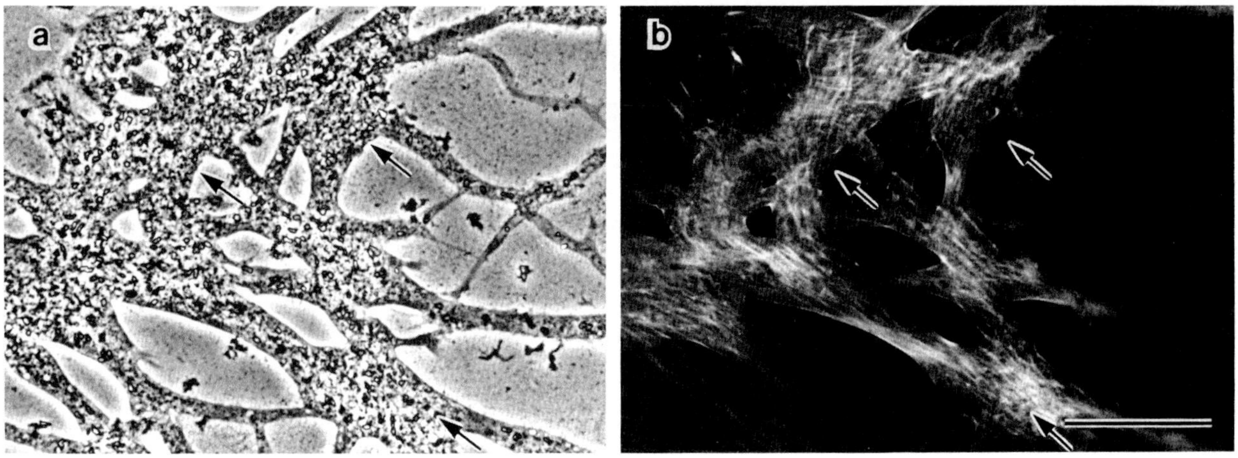

Fig. 4. Actomyosin fibrils in a small sheet-like portion at the migrating front (arrows), revealed with rhodamine-phalloidin staining. A three-dimensional network is composed of many fibrils. *a* Phase-contrast micrograph. *b* Fluorescence micrograph. Bar represents 100 μm

3.2. Coralloid Plasmodium

The vegetative aphanoplasmodium turns morphologically into a coralloid plasmodium at the end of assimilation phase. This plasmodium apparently has a front-and-rear polarity (Fig. 1 *b*), and migrates actively on an agar surface. The veins of five coralloid plasmodia, about 50 μm wide, were examined at 15-sec intervals for about 60 min with regard to the streaming direction and velocity. Figure 3 shows the typical manner of streaming in coralloid plasmodia, where the direction of flow alternates in a rhythmic pattern without any distinct pause. All five coralloid plasmodia showed this type of streaming, which is identical with the shuttle streaming observed in the phaneroplasmodium of *Physarum*. Mean periods of a cycle were distributed between 4.5 and 7.1 min (Mean ± SD: 5.9 ± 0.9 min), and the averages of maximum velocities ranged from 44 to 83 μm/sec (Mean ± SD: 62 ± 13 μm/sec) among five plasmodia. The maximum instantaneous velocity of all the 48 cycles, examined for five specimens, was 110 μm/sec.

3.3. Distribution of Actomyosin Fibrils in Aphanoplasmodium

The growing aphanoplasmodium consists of ramified veins for the most part, and of a tiny fan-like portion at the advancing front. Many cytoplasmic lumps (designated as "nodular portion"), which many slender veins converge at, exist among the ramified veins. They show an appearance of coagulated cytoplasmic mass,

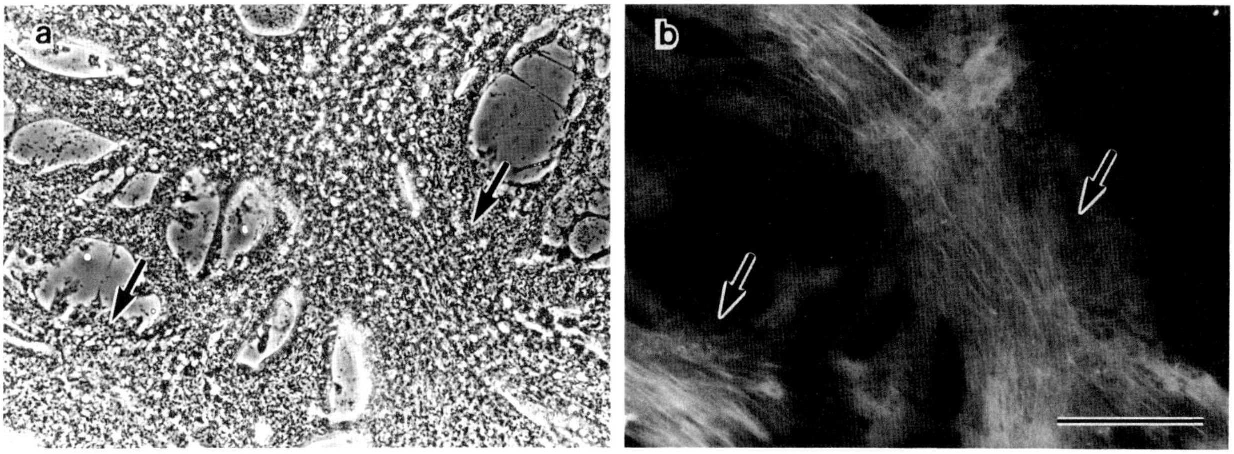

Fig. 5. Actomyosin fibrils in nodular portions (arrows) which seem to be formed by the coalescence of many slender veins in the posterior region. Many fibrils can be seen running parallel to each other. *a* Phase-contrast micrograph. *b* Fluorescence micrograph. Bar represents 50 µm

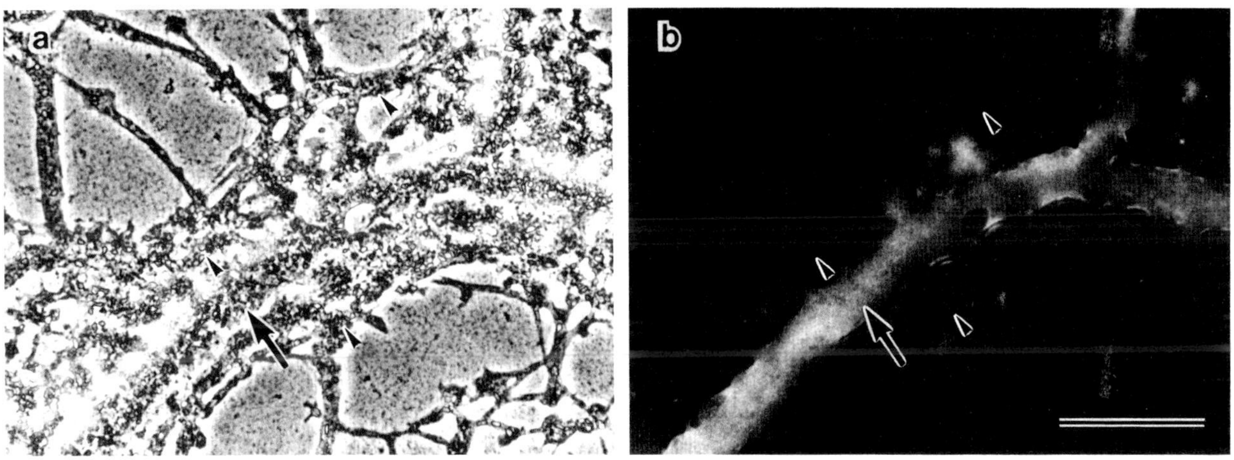

Fig. 6. A main vein (arrows) of aphanoplasmodium, stained diffusely with rhodamine-phalloidin. The secondary veins (arrowheads), which are obvious in the phase-contrast micrograph, are stained very weakly. *a* Phase-contrast micrograph. *b* Fluorescence micrograph. Bar represents 100 µm

100–200 µm in size. For fluorescence microscopy, 50 pieces of aphanoplasmodium containing an anterior portion, prepared from 10 culture dishes, were fixed and stained with rhodamine-phalloidin, a specific reagent binding to polymerized actin (WULF *et al.* 1979, FÜCHTBAUER *et al.* 1983, AMATA *et al.* 1983). Well-developed actomyosin fibrils were clearly observed in about 60% of examined anterior portions, as shown in Fig. 4. The fibrils were observed also in the nodular portions (Fig. 5). Usually, the fibrils were about 50–100 µm long and 1–2 µm wide. The main veins were diffusely stained with no distinct fibrillar appearance, and the secondary veins were stained weakly (Fig. 6).

4. Discussion

We have described the manner of cytoplasmic streaming in the *Stemonitis* aphanoplasmodium. Some aspects of the streaming are different from those observed in the phaneroplasmodium of *Physarum*. They are: (1) only unidirectional flow continues for fairly long time, more than 80 min in some cases; (2) there always is a pause within each cycle, lasting for about a half of a cycle length; (3) the streaming velocity at the maximum phase is considerably slower when compared to that in *Physarum* phaneroplasmodium, which is usually several hundred micrometers per second. These features give an impression that cytoplasmic streaming in the aphanoplasmodium is not as highly organized as in the phaneroplasmodium.

We have classified the streaming patterns of the aphanoplasmodium into three groups (*cf.*, Fig. 2 *a*, *b*, and *c*). There is, however, a possibility that these patterns alternate with one another during periods lasting longer than 80 min. Unidirectional streaming has been previously reported in a plasmodium, the size of a black speck barely visible to the naked eye, of *Cribraria violacea* (MCMANUS 1966). However, further investigations should be carried out using other species to determine whether unidirectional flow is specific to the presumably more primitive proto- and aphanoplasmodia (GRAY and ALEXOPOULOS 1968).

In the anterior region of *Physarum* phaneroplasmodium, the cytoplasmic actomyosin fibrils form when the cytoplasm is streaming backward and degenerate in the reverse direction. In contrast, the fibrils in the posterior regions show no rhythmic changes coupled with the alternation of streaming direction (KAMIYA 1973, ISHIGAMI 1986, ISHIGAMI and HATANO 1986, ISHIGAMI *et al.* 1981). It was also shown, using the plasmodial fragment system, that the fibrils develop with the progress of contraction, and that they degenerate on the relaxation process, disintegrating almost completely at the maximum relaxation (ISHIGAMI *et al.* 1987). Based on these facts, it is generally accepted that the periodical contraction of the anterior region, caused through the generation-degeneration cycle of the actomyosin fibrils, is directly responsible for the shuttle streaming: that is, the endoplasm flows forward when the internal pressure of anterior region becomes lower than that of the posterior, which remains nearly constant; and flows backward when the pressure in the anterior becomes higher (YOSHIMOTO and KAMIYA 1978, ISHIGAMI *et al.* 1987). In this work, we observed many actomyosin fibrils in about 60% of the front portions of *Stemonitis* aphanoplasmodia, and no fibrils in the rest. These results may offer an idea that the cyclic generation-degeneration of the fibrils takes place in the frontal portion as coupled with the contraction-relaxation. On the other hand, polymerized actin was evenly distributed in the posterior region as revealed with rhodamine-phalloidin staining. Thus, it may be inferred that the cytoplasmic streaming in the aphanoplasmodium caused by the same mechanism as in the phaneroplasmodium; namely, the periodical contraction of the anterior region causes the shuttle streaming also in the aphanoplasmodium. To discuss this problem more precisely, dynamics of the actomyosin fibrils in the anterior have to be elucidated as related to the force generation.

Should the streaming in the aphanoplasmodium occur by the manner mentioned above, it would be very interesting what condition of internal pressure produces the pause of streaming. Under phase-contrast optics, pausing veins were flattened untidily, and the particles included in the cytoplasm remained almost completely motionless. These observations may support one possibility that the internal pressures in the regions around a pausing vein are totally in equilibrium at a low level of contraction.

We also observed many actomyosin fibrils in the nodular portion of the aphanoplasmodium. The functions of these fibrils, and of the nodular portion *per se* have not been understood as yet.

Acknowledgements

We wish to thank Dr. H. HAGIWARA of National Science Museum, Tokyo for identifying the material used.

References

ALEXOPOULOS CJ (1959) The laboratory cultivation of *Stemonitis*. Am J Bot 46: 140–142

— (1960) Gross morphology of the plasmodium and its possible significance in the relationships among the Myxomycetes. Mycologia 52: 1–20

AMATO PA, UNANUE ER, TAYLOR DL (1983) Distribution of actin in spreading macrophages: a comparative study on living and fixed cells. J Cell Biol 96: 750–761

FÜCHTBAUER A, JOCKUSCH BM, MARUTA H, KILIMANN MW, ISENBERG G (1983) Disruption of microfilament organization after injection of F-actin capping proteins into living tissue culture cells. Nature 304: 361–364

GRAY WD, ALEXOPOULOS CJ (1968) Biology of the Myxomycetes. Ronald Press, New York.

HASKINS EF (1974) *Stemonitis flavogenita* (Myxomycetes)—plasmodial phase (Aphanoplasmodium). Film E 2000 des Inst. Wiss. Film, Göttingen

— HINCHEE AA (1974) Light- and ultra-microscopical observations on the surface structure of the protoplasmodium, aphanoplasmodium, and phaneroplasmodium (Myxomycetes). Can J Bot 52: 1835–1839

ISHIGAMI M (1986) Dynamic aspects of the contractile system in *Physarum* plasmodium. I. Changes in spatial organization of the cytoplasmic fibrils according to the contraction-relaxation cycle. Cell Motil Cytoskeleton 6: 439–447

— HATANO S (1986) Dynamic aspects of the contractile system in *Physarum* plasmodium. II. Contractility of Triton cell models in accordance with the contraction and relaxation phases of the plasmodia. Cell Motil Cytoskeleton 6: 448–457

— KURODA K, HATANO S (1987) Dynamic aspects of the contractile system in *Physarum* plasmodium. III. Cyclic contraction-relaxation of the plasmodial fragment in accordance with the generation-degeneration of cytoplasmic actomyosin fibrils. J Cell Biol 105: 381–386

— NAGAI R, KURODA K (1981) A polarized light and electron microscopic study of birefringent fibrils in *Physarum* plasmodia. Protoplasma 109: 91–102

KAMIYA N (1959) Protoplasmic streaming. In: HEILBRUNN LV, WEBER F *et al* (eds) Protoplasmatologia, vol 8. Springer, Wien, part 3 a

— (1973) Contractile characteristics of the myxomycete plasmodium. Proc 4th Int Biophys Congr (Moscow) Symp III Biophysics of Motility, pp 447–465

— (1981) Physical and chemical basis of cytoplasmic streaming. Ann Rev Plant Physiol 32: 205–236

MCMANUA SMA (1961) Culture of *Stemonitis fusca* on glass. Am J Bot 48: 582–588

— (1966) Cultivation on agar of the plasmodia of *Licea biforis, Licea veriabilis*, and *Cribraria violacea*. Mycologia 58: 479–483

WOHLFARTH-BOTTERMANN KE (1964) Differentiations of the ground cytoplasm and their significance for the generation of the motive force of ameboid movement. In: ALLEN RD, KAMIYA N (eds) Primitive motile systems in cell biology. Academic Press, New York London, pp 79–109

WULF E, DEBOBEN A, BAUTZ FA, FAULSTICH H, WIELAND Th (1979) Fluorescent phallotoxin, a tool for the visualization of cellular actin. Proc Natl Acad Sci USA 76: 4498–4502

YOSHIMOTO Y, KAMIYA N (1978) Studies on contraction rhythm of the plasmodial strand. IV. Site of active oscillation in an advancing plasmodium. Protoplasma 95: 123–133

Protoplasma (1988) [Suppl 1]: 42–47

Motile Behaviour of Endoplasmic Drops Isolated from *Nitella* Cells

F. Miura** and K. Kuroda*

Department of Biology, Faculty of Science, Osaka University, Osaka

Received February 29, 1988
Accepted April 26, 1988

Dedicated to Professor Dr. Noburo Kamiya on the occasion of his 75th birthday

Summary

Endoplasmic drops isolated from the internodal cells of *Nitella flexilis* by the "negative pressure method" developed by Kamiya and Kuroda (Proc Japan Acad 33: 149–152, 1957) showed no mass streaming at first. However, after 1 day incubation in an artificial medium, they began to show mass streaming with sliding at the interface between the outermost layer of the endoplasmic mass and the inner surface of the membrane. New cell wall deposition at the outer surface of the membrane was detected by staining with Calcofluore White ST. The endoplasmic drops could survive outside the cell for 7 days in the artificial medium.

Keywords: Cell wall formation; Culture of endoplasm; Endoplasmic drop; Mass streaming; *Nitella flexilis*.

Abbreviations: DTT dithiothreitol; IAA indole-3-acetic acid; AMS agar-medium sheet; BS basal solution; ED endoplasmic drop; IM incubation medium.

1. Introduction

Since the report by Kamiya and Kuroda (1956) that the motive force of the cytoplasmic streaming in *Nitella* is generated at the interface between the ectoplasm and the outermost layer of the streaming endoplasm, a number of studies have revealed that the surface of the chloroplast files carries bundles of F-actin and that the myosin molecule is present in the endoplasm (for review, see Kamiya 1981, 1986). They all indicated the functional significance of the subcortical F-actin fibrils anchored on the ectoplasm. In other words, the endoplasm alone is not capable of driving the streaming. To elucidate what actually does occur, it is important to study the behaviour of the pure endoplasm.

Part of a *Nitella* cell can be amputated by the "negative-pressure method" (Kamiya and Kuroda 1957 a), to obtain large drops of the endoplasm alone which can survive for 10–50 hours in an artificial solution of cationic constitution similar to that of natural cell sap. The endoplasm which streams vigorously in the intact cell, does not exhibit mass streaming in the isolated drop. However, both chloroplasts and nuclei included in the drop rotate very actively around their axes (Kamiya and Kuroda 1957 b).

The endoplasm has been shown to contain moving fibrils which are scattered in the endoplasm at first after the isolation. After 1 day's incubation of the drops with an appropriate medium, some endoplasmic drops begin to show local cytoplasmic streaming (Kuroda 1964). The actin bundles are thought to gradually become arranged in an orderly fashion on the inner surface of the membrane.

The present paper reports on how the mass streaming developed in isolated endoplasmic drops after incubation by an improved technique and describes the mechanisms of the intracellular motion in *Nitella* cells.

2. Materials and Methods

Nitella flexilis was kept in the laboratory in Knop's solution and the basal solution (see below), diluted to 1/1,000 of the standard concentration, under ca. 2,000 lux light from white fluorescent lamps for alternate periods of 10 hours light and 14 hours darkness at room temperature. Under these conditions, the osmotic pressure of the plant was equivalent to 0.35 M sucrose. Consequently, the basal solution was used in 1.3-fold that reported by Kamiya and Kuroda (1957 a). In the previous case, the osmotic pressure of the plant was 0.26 M and the basal solution consisted of 0.08 M KNO_3, 0.05 M NaCl and 0.004 M $Ca(NO_3)_2$.

* Correspondence and Reprints: Department of Biology, Faculty of Science, Osaka University, Toyonaka, Osaka 560, Japan.

** Present address: Kobeya Baking Co., Ltd., Osaka, Japan.

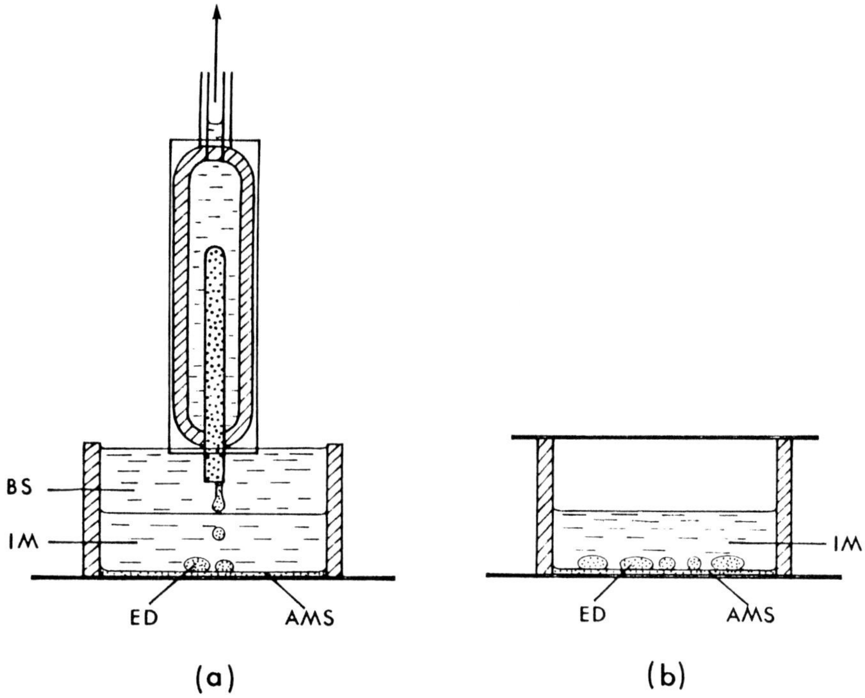

Fig. 1. Apparatus for the effusion of the endoplasm from the *Nitella* cell using a "negative pressure chamber" (*a*) and that for the incubation cuvette (*b*). The basal solution on the top layer in the cuvette (*a*) is carefully removed with a pipette and the endoplasmic drops are incubated in the cuvette covered with a glass cover-slip (*b*). *BS* Basal solution, *IM* incubation medium, *AMS* agar-medium sheet, *ED* endoplasmic drop

Young internodal cells, mostly 2–3 cm long, were freed from neighbouring cells and washed with running water to mechanically eliminate as much as possible the foreign organisms attached to the cell surface. The cell thus prepared was set in a narrow groove in a chamber, as shown in Fig. 1 *a*, filled with the basal solution. The lower part of the cell was put in a cuvette filled with the same solution layered on an incubation medium containing Knop's solution diluted to 1/2 the standard solution, Heller's minor mineral salts solution, 1 mg/l each of IAA, nicotinic acid, folic acid, thiamine HCl, vitamin B_{12}, vitamin C and biotin, and also including 3 mM DTT, 0.2% casamino acid, 2% bovine serum albumin, 10% coconut water, 50 mM glucose and 300 mM sorbitol. The media and cuvettes were autoclaved at 112 °C except for the heat-labile substances.

One of the important things was to place a 1% agar film containing the above incubation medium at the bottom of the cuvette. The agar-medium sheet has a distinct advantage over glass, because it keeps the endoplasm in a more favourable condition by serving as a cushion for the endoplasmic drops against physical and chemical shocks.

While applying weak negative hydrostatic pressure of about 2 cm of mercury to chamber, we cut the cell in the solution. The endoplasm in the amputated cell continued to stream and flowed out into the solution forming a continuous strand of endoplasm or intermittent endoplasmic drops (KAMIYA and KURODA 1957 a). As the effusion of the endoplasm occurred most smoothly in the basal solution, we cut the cell in this solution overlaid on the incubation medium. The effusing endoplasm enters the surrounding solution as drops having clear boundary membranes. The drops fell through the boundary between the basal solution and the incubation medium onto the bottom of the cuvette covered with the agar-medium sheet. Next the basal solution on the top layer in the cuvette was carefully removed with a pipette leaving the incubation medium and the agar-medium sheet with the sessile endoplasmic drops (Fig. 1 *b*). These drops were incubated at 18 ± 1 °C under continuous weak light (ca. 50 lux).

The investigation of cytoplasmic motility in a drop on the bottom of the cuvette was carried out with an inverted microscope (Nikon MD, or Olympus) equipped with 10 × (NA 0.30), 20 × (NA 0.40) and 40 × (NA 0.80) objective lenses. The movements were recorded with a Nikon 35-mm camera or 16-mm cine camera (Bolex H16M or Kodak Special-II) and a selected portion was analyzed with a Kodak Analyst.

To examine the new cell wall deposit at the surface, the incubated endoplasmic drops were stained for 5 min with 0.2% Calcofluore White ST (American Cyanamid Co., NJ, U.S.A.) (NAGATA and TAKEBE 1970) in the basal solution. After washing out the excess dye with a large amount of the basal solution, the drops were mounted on a quartz slide and observed with a Zeiss microscope (Photo II) equipped with an Osrum HBO 200 mercury lamp, UG 1 and BG 38 excitation filters and barrier filters 41 and -65. All photographs of the samples stained with Calcofluore White ST were taken, developed and printed under the same conditions.

The criterion of life was the autonomous rotation of chloroplasts included in the drops.

3. Results

As reported by KAMIYA and KURODA (1957 b), we could ascertain no mass streaming in the isolated endoplasmic drops in which the chloroplasts rotated vigorously around their own axes at the rate of about 1–2 revolutions per second. This shows that the endoplasm alone is inert with respect to mass streaming. However, when an improved incubation technique was used, some drops showed after 1 day's incubation an unidirectional rotation which continued for more than 3 hours. Figure 2 shows a series of pictures reproduced from a strip of cinematographic film. Chloroplasts in the drop circulated around the center of the drop with

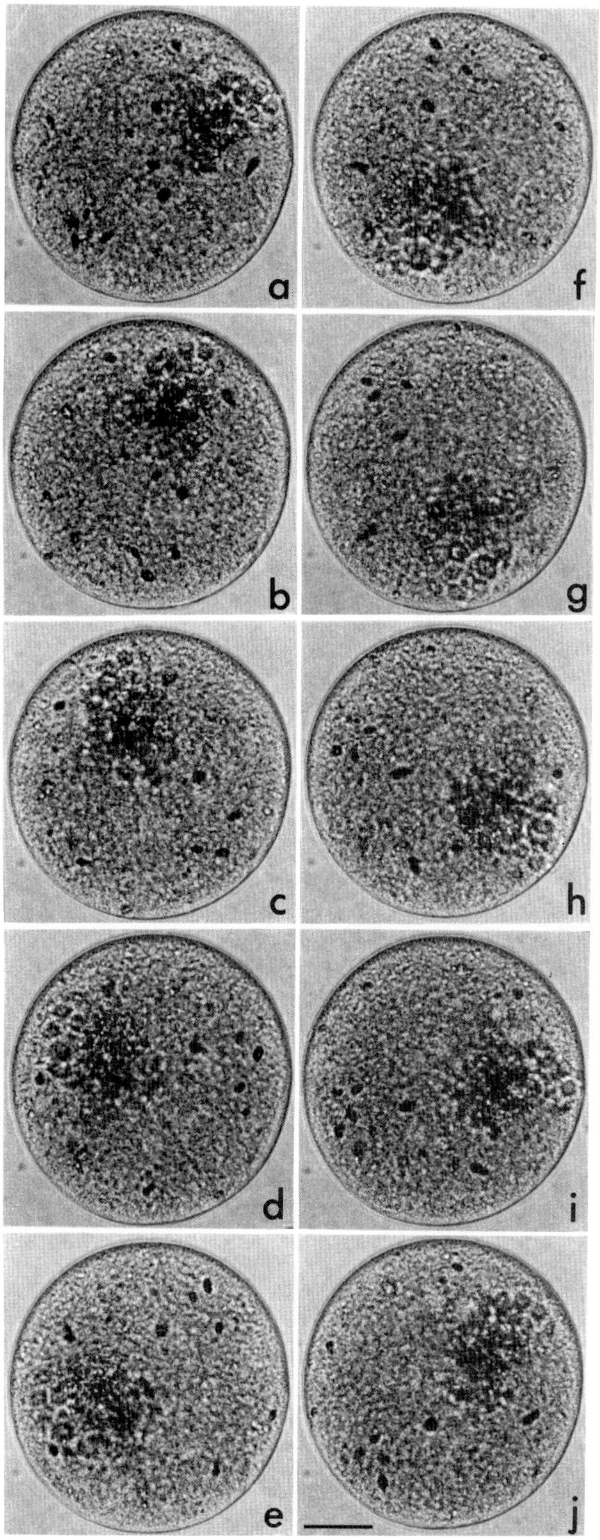

Fig. 2. Successive stages of endoplasmic rotation in a drop incubated for 1 day *in vitro*. Each interval is 20 sec. Reproduced from 16 mm cine film. Bar: 50 μm

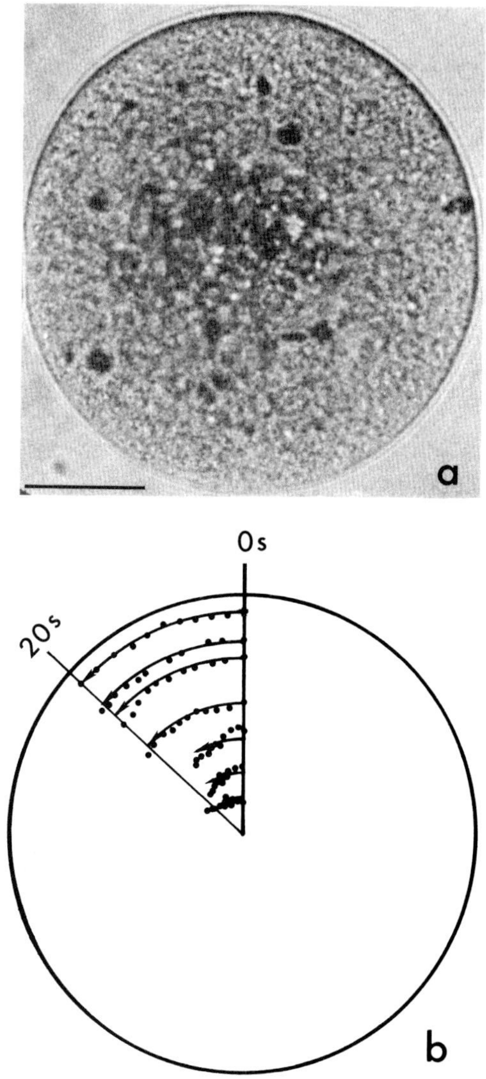

Fig. 3. Velocity profile (*b*) of the rotation of endoplasm in a drop (*a*) incubated for 1 day *in vitro*. Bar: 50 μm

a constant angle velocity showing rotation around their own axes (Fig. 3). No shearing in the endoplasm occurred; the endoplasm rotates as a mass inside the cortex. The motile behaviours of 6 different drops were consistent with Figs. 2 and 3. Only the velocity varied a littel from drop to drop. These observations show that the newly formed membrane began to act as ectoplasm after 1 day's incubation. Some drops which showed the mass streaming displayed a protuberance from the surface to the substrate. Consequently this mass rotation was not a rolling of the drop.

Some drop showed mass streaming in one direction, then became immobile and next began to display mass streaming in the opposite direction. This cycle of mass rotation in reverse direction continued for several times with the period of about 150 seconds.

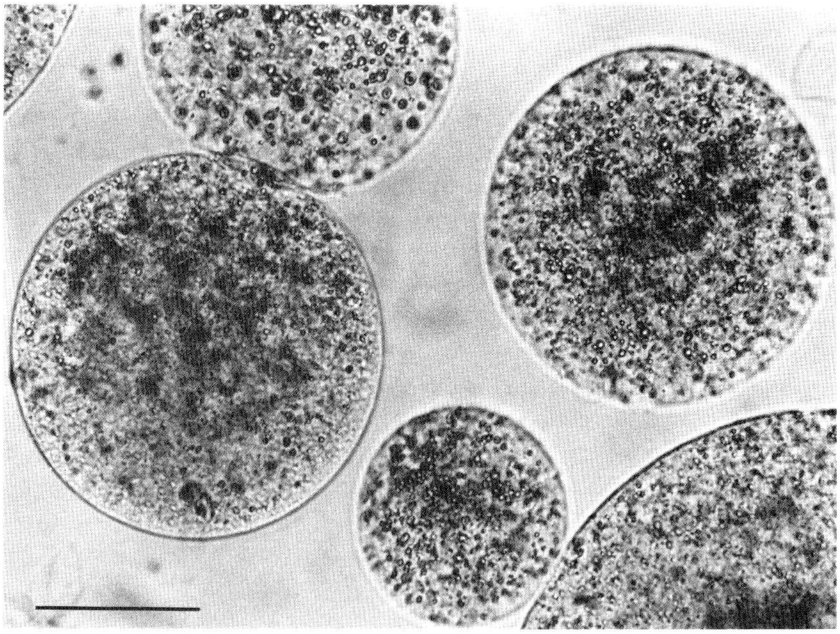

Fig. 4. Endoplasmic drops incubated for 6 days in the incubation medium *in vitro*. The chloroplasts contained in the drops rotated around their own axes. Bar: 100 µm

The isolated drops could show the chloroplast rotations even on the 7th day of incubation. This means that the endoplasmic drop could survive out of the cell in an artificial medium for 7 days. As shown in Fig. 4, in this cuvette all drops were in the good condition on the 6th day.

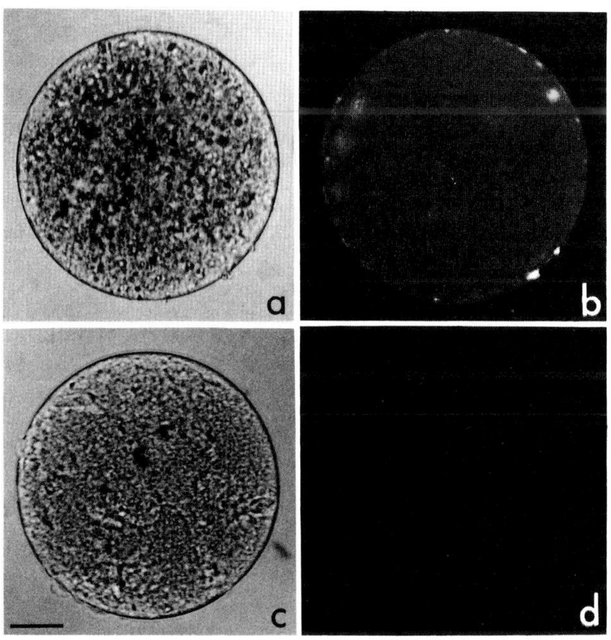

Fig. 5. Endoplasmic drops isolated from *Nitella* internodal cells. Ordinary light (*a*) and fluorescence (*b*) micrographs of a 2-day incubated endoplasmic drop, stained with Calcofluore White ST, regenerating a cell wall. (*c*) and (*d*) are ordinary light and fluorescence micrographs of a newly isolated endoplasmic drop stained with Calcofluore White ST. No cell wall formation is detected. Bar: 50 µm

To understand the properties of the newly formed membrane, we tried to stain it with Calcofluore White ST which stains cellulose fibrils in the cell wall. Immediately after the endoplasmic drop was effused from the internodal cell, there was no fluorescence on the surface of the drop (Fig. 5 *c, d*). But new cell wall formation was detectable through faint staining with Calcofluor White ST on the second day of incubation (Fig. 5 *a, b*). This led us to assume that the newly formed "surface precipitation membrane" demarcating the endoplasm from the surrounding solution would become a membrane resembling a "cell membrane".

4. Discussion

The endoplasmic drop obtained by the "negative pressure method" consists almost entirely of endoplasm alone, because the cortical gel layer in which most of the chloroplasts carrying the actin bundles are imbedded remains completely within the cell showing no indication of injury. Consequently, the surface membrane of the endoplasmic drop does never contain the original cell membrane. Although there are several reports (INOUE *et al.* 1973 b, SAKANO and TAZAWA 1986, LÜHRING 1986) on the origin of the newly formed endoplasmic membrane, it is not yet completely understood. However, as supposed by KURODA (1964) the nature of the membrane may be more like a tonoplast than the cell membrane at first.

In the newly isolated endoplasmic drops, we could not ascertain mass streaming except for the rotatory movement of chloroplasts contained in the drop. However, moving fibrils (JAROSCH 1956) consisting of F-actin (HIGASHI-FUJIME 1980) have been found in isolated pure endoplasm not containing cortical gel plasm (KURODA 1964). If these motile fibrils which are at first scattered around in the drop could become oriented in an ordered fashion, arranging their polarity inside the newly formed membrane of the drop during the incubation, mass streaming might occur in the endoplasmic drop. KURODA (1964) reported the development of local streaming in the isolated endoplasmic drop. In the present study, we observed the endoplasmic rotation which reverses its direction periodically and the mass streaming of the endoplasm in drops incubated for 1 day. This reversing of the rotation may be explained by the organization process of the inner surface of the membrane which interacts with the rotating endoplasm. This movement may be a transitional one from the local cytoplasmic streaming to the organized rotational mass streaming.

The reversion of the cytoplasmic rotation was reported by YOTSUYANAGI (1953) in the small cytoplasmic droplets formed by the sudden plasmolysis of *Elodea* leaf cells. The movement of granules became very active through disordered movement and began to rotate at 2–3 revolutions in one direction. After that, the rate decreased and the movement almost stopped, then started again in the opposite direction. After repetition of this process several times, stable rotation continued at one direction. Furthermore, rotational movements in small fragments of protoplasm squeezed out of the internodal cell of *Characeae* have been reported by many (for review, see KAMIYA 1959). However, these fragments contain both endoplasm and ectoplasm. To understand the mechanism of the cytoplasmic streaming, it is important to study the behaviour of the endoplasm alone. The present work is the first to show the total endoplasmic rotation as mass streaming in the endoplasmic drop. The bulk of the endoplasm is thought to be dragged by the shearing force generated at the boundary between the newly developed ectoplasm and the outermost layer of the endoplasm.

The tension at the surface of the endoplasmic drop of *Nitella* has been reported to be extremely small, that is, in the range of 0.002–0.004 dynes/cm, which corresponds to the tension of the membrane of white yolk spheres (KAMIYA and KURODA 1958). This tension increases, however, up to 0.1 dynes/cm when the endoplasmic drop is kept in a low ionic strength solution for 2–3 hours (UEDA *et al.* 1973). On the other hand, the tension at the surface of the cell membrane of the protoplast freshly isolated enzymatically from the cells of *Nitella* is known to be 0.2 dynes/cm (KURODA 1980). These findings suggest that, within several hours, the surface membrane formed at the effused endoplasmic drop becomes the membrane having a tension force of the same order as that of the cell membrane. Furthermore, the membrane potential of the endoplasmic drops decreases with time and the membrane of the drops becomes excitable like the intact *Nitella* cell membrane (INOUE *et al.* 1973 a).

Nitella protoplasts deposite the cell wall at the outer surface of the cell membrane after 2 days of incubation (KURODA 1980). As shown in Fig. 5, the outer surface of the endoplasmic drops, which do not have the original cell membrane, was stained with Calcofluore White ST, indicating the beginning of the formation of a new cell wall.

The findings described thus far clearly show that the surface membrane of the endoplasmic drops grew gradually into a cell membrane which can deposit the cell wall outside and organize the motile fibrils inside in order to generate a motive force. Thus, moving F-actin fibrils might be oriented on the inner surface of the membrane and may interact with myosin in the endoplasm (CHEN and KAMIYA 1975, KURODA and KAMIYA 1975). The ultrastructures of the underlayer of the membrane remain to be clarified.

What also needs to be examined is the behaviour of the endoplasm with respect to its viability. As the internodal cells of *Characeae* are coenocytes, they contain several thousand nuclei which are in the endoplasm and flow together with endoplasm in the intact cells. Hence, all endoplasmic drops contain some dozens of nuclei. The fact that the endoplasmic drops could regenerate a new cell wall led us to assume that the endoplasmic drop isolated from the cell might be capable of growing into an intact cell, if we could make it survive indefinitely. We have been able to prolong the survival time up to 7 days, but survival periods are limited by contamination from bacteria and fungi. Improvement of the aseptic procedure for the incubation is urgently required. One of the other problems is an amelioration of the medium. Continuous changes of the medium might be an efficient technique for culturing the endoplasmic drops.

To the best of our knowledge, this is the first report of a culture of the endoplasm alone in an artificial medium. We are continuing our work with the aim of trying to regenerate from the endoplasmic drop a whole cell which exhibits the rotational cytoplasmic streaming around the central vacuole and can divide into two daughter cells.

Acknowledgements

We wish to express our gratitude to Professor T. Hayashi of Ibaragi University for his valuable advice on the culture of the endoplasm.

References

Chen JCW, Kamiya N (1975) Localization of myosin in the internodal cell of *Nitella* as suggested by differential treatment with N-ethylmaleimide. Cell Struct Funct 1: 1–9

Higashi-Fujime S (1980) Active movement *in vitro* of bundles of microfilaments isolated from *Nitella* cell. J Cell Biol 87: 569–578

Inoue I, Ishida N, Kobatake Y (1973 a) Studies of excitable membrane formed on the surface of protoplasmic drops isolated from *Nitella*. IV. Excitability of the drop membrane in various compositions of the external salt solution. Biochim Biophys Acta 330: 27–38

— Ueda T, Kobatake Y (1973 b) Structure of excitable membranes formed on the surface of protoplasmic drops isolated from *Nitella*. I. Conformation of surface membrane determined from the refractive index and from enzyme actions. Biochim Biophys Acta 298: 653–663

Jarosch R (1956) Plasmaströmung und Chloroplastenrotation bei Characeen. Phyton (Argentina) 6: 87–107

Kamiya N (1959) Protoplasmic streaming. In: Heilbrunn LV, Weber F *et al* (eds) Protoplasmatologia, vol 8. Springer, Wien, part 3 a

— (1981) Physical and chemical basis of cytoplasmic streaming. Ann Rev Plant Physiol 32: 205–236

— (1986) Cytoplasmic streaming in giant algal cells: A historical survey of experimental approaches. Bot Mag (Tokyo) 99: 441–467

— Kuroda K (1956) Velocity distribution of the protoplasmic streaming in *Nitella* cells. Bot Mag (Tokyo) 69: 544–554

— — (1957 a) Cell operation in *Nitella*. I. Cell amputation and effusion of the endoplasm. Proc Jap Acad 33: 149–152

— — (1957 b) Cell operation in *Nitella*. II. Behaviour of isolated endoplasm. Proc Jap Acad 33: 201–205

— — (1958) Cell operation in *Nitella*. IV. Tension at the surface of the effused endoplasmic drops. Proc Jap Acad 34: 435–438

Kuroda K (1964) Behavior of naked cytoplasmic drops isolated from plant cells. In: Allen RD, Kamiya N (ed) Primitive motile systems in cell biology. Academic Press, New York, pp 31–41

— (1980) Giant protoplast from *Nitella* cells. Cell Biol Int Rep 4: 195–199

— Kamiya N (1975) Active movement of *Nitella* chloroplast *in vitro*. Proc Jap Acad 51: 774–777

Lühring H (1986) Recording of single K^+ channels in the membrane of cytoplasmic drop of *Chara australis*. Protoplasma 133: 19–28

Nagata K, Takebe I (1970) Cell wall regeneration and cell division in isolated tobacco mesophyll protoplasts. Planta 92: 301–308

Sakano K, Tazawa M (1986) Tonoplast origin of the envelope membrane of cytoplasmic droplets prepared from *Chara* internodal cells. Protoplasma 131: 247–249

Ueda T, Inoue I, Kobatake Y (1973) Studies of excitable membrane formed on the surface of protoplasmic drops isolated from *Nitella*. II. Tension at the surface of protoplasmic drops. Biochim Biophys Acta 318: 326–334

Yotsuyanagi Y (1953) Recherches sur les phénomènes moteurs dans les fragments de protoplasme isolés. I. Movement rotatoire et le processus de son apparition. Cytologia 18: 146–156

Cell Movement—Contraction and Migration

Protoplasma (1988) [Suppl 1]: 51–56

Patterns in Intracellular ATP Distribution and Rhythmic Contraction in Relation to Amoeboid Locomotion in the Plasmodium of *Physarum polycephalum*

T. Ueda*, T. Nakagaki, and Y. Kobatake

Faculty of Pharmaceutical Sciences, Hokkaido University, Sapporo

Received December 2, 1987
Accepted January 22, 1988

Dedicated to Professor Dr. Noburo Kamiya on the occasion of his 75th birthday

Summary

Locomotion velocity in plasmodia of *Physarum polycephalum* decreased in the presence of sugars, salts or amino acids. Accompanying this locomotion inhibition, patterns both in intracellular ATP distribution and in rhythmic contraction were modified in two distinct ways. In the first case, both intracellular ATP concentration and the amplitude of rhythmic contraction decreased uniformly throughout a plasmodium (amplitude modulation). In the second, the ATP distribution became humpy, and the oscillation patterns segregated into many domains (pattern modulation). When two plasmodia coalesced, polar ATP pattern disappeared and the protoplasm in the advancing fronts moved away. Thus, our previous proposal that self-organization of chemical patterns forms a basis for amoeboid cell behavior is also applicable to amoeboid locomotion of the *Physarum* plasmodium.

Keywords: Amoeboid locomotion; Oscillation; ATP; *Physarum*; Pattern formation.

1. Introduction

The plasmodium of *Physarum polycephalum* is a giant amoeboid cell in which the naked protoplasm extends on a substratum like a sheet, each component part oscillating coordinately (Kamiya 1959, Kessler 1982). Owing to its large size, spatial and temporal distributions of such intracellular components as ATP and cAMP were determined (Ueda *et al.* 1986, 1987). Spatio-temporal organization of thickness oscillation and its changes in response to local external stimulation were also analyzed by using computer image processing (Matsumoto *et al.* 1986, Mori *et al.* 1986). Here we study amoeboid locomotion in the *Physarum* plasmodium by inhibiting the locomotion with various chemicals and by modifying cell organization with coalescence. We compare various levels of the locomotion activity with various patterns both in ATP distribution and in oscillation dynamics. The locomotion is inhibited by two mechanisms; an amplitude modulation which leads to a decrease in the activity of each contractile element, and a pattern modulation which lowers the degree of coordination in a population of oscillatory elements.

* Correspondence and Reprints: Faculty of Pharmaceutical Sciences, Hokkaido University, Sapporo 060, Japan.

2. Material and Methods

2.1. Organism

The plasmodium of *Physarum polycephalum* was fed with oatflakes on 1% agar gel. The organisms were allowed to move on moist filter paper (Whatmann, No. 1) overnight in troughs ($22 \times 33\,cm^2$), and the frontal parts from several troughs were further placed on large moist filter paper. They fused each other a few hours later, and a large plasmodium extended, the periphery of advancing front reaching more than 60 cm.

2.2. Measurements of Stationary Locomotion Velocity

Frontal part of large plasmodium as above was cut into more than 30 pieces ($2 \times 1\,cm^2$ each) together with filter paper, and each piece was placed on an agar gel containing chemicals to be tested in a narrow ditch ($2 \times 20\,cm^2$). The organism migrated at a constant speed over several hours, the locomotion velocity being 3–4 µm/sec on water-agar. Experiments were done duplicately at a given concentration for a series of dilution, and repeated at least twice. Results were averages.

2.3. Coalescence

Two plasmodia as above were allowed to move on water-agar face to face from two ends of a measuring trough. Eventually, they met and coalesced spontaneously (Fig. 5).

2.4. Measurements of Intracellular ATP Distribution

The spatial distribution of intracellular ATP concentration was determined chemiluminescently (BEUTLER and BALUDA 1964). A single plasmodium was cut into 30–50 pieces, and the amount of ATP extracted with hot buffer (10 mM Tris-HCl, pH 7.1) in each piece was assayed chemiluminescently with use of firefly lantern extracts (UEDA et al. 1987). The protein was assayed by the Lowry method. The ATP concentration is the amount of ATP divided by that of protein. The ATP concentration of 30 μg/mg protein is calculated to be 0.62 mM by assuming that 1 mg protein corresponds to 95 mg water. This value is similar to those reported previously: 0.4 mM by TCA extraction method (HATANO and TAKEUCHI 1960), 0.2–0.5 mM by ^{31}P-NMR measurements (KOHAMA et al. 1984) and 0.2–0.7 mM by chemiluminescence method (YOSHIMOTO et al. 1981).

2.5. Spatio-temporal Patterns of Rhythmic Contraction

Spatio-temporal organizations of the rhythmic contraction in plasmodia was visualized by applying an image processing with use of a microcomputer (MATSUMOTO et al. 1986, UEDA et al. 1986). Brightness level at each point in videoimages, which reflected the thickness of plasmodium, oscillated with a period of a few minute, and the spatial distribution of 40 × 60 oscillations was visualized by small squares with graded shades of 17 levels, as shown in Fig. 2. Experiments were repeated at least three times, and results presented were typical. Temperature was 22 ± 1 °C.

2.6. Chemicals

Agar (Difco, bacto agar), various amino acids (Nakarai Chemicals Ltd.), firefly-lantern extracts (Sigma, FLE-50), and other chemicals were used as delivered.

3. Results

3.1. Inhibition of Locomotion with Chemicals

Many amino acids at mM concentrations inhibited the locomotion of plasmodia (Fig. 1 *A*). The inhibition depends on the concentration and fits the Hill-type plots:

$$v = v_0/(1 + (KC)^n). \quad (1)$$

Here v and v_0 are locomotion velocities in the presence (at the concentration C) and absence of chemicals, respectively. K and n are parameters, expressing sensitivity and cooperativity, respectively, and their values for various amino acids are summarized in Table 1. Low cooperativity with n = 1/2 is found among those amino acids having either negative charges at the side chain (asp and gln) or bulky polar group at the beta carbon (thr and tyr), while high cooperativitiy with n = 2 among those having either positive charges (arg) or methyl group at the beta carbon (ile, val and pro). Also, D-serine has ten-fold lower affinity and lower n value than L-serine. In short, sensitivity and cooperativity as expressed in Eq. (1) are strongly affected both by charged state at the side chain and by molecular configuration at the alpha and beta carbons. Some amino acids (the group II in Table 1) inhibited the locomotion only partially (60 to 90%).

On the other hand, sugars and salts inhibited the locomotion only at high concentrations (Fig. 1 *B*). The locomotion velocity v decreases linearly with increasing concentration:

$$v = v_0 (1 - C/K_c). \quad (2)$$

The parameter K_c indicates the critical concentration for complete suppression. To the first approximation, the values of K_c are related to the osmolarity of the solution. Therefore, effects of sugars and salts are through thermodynamic colligative properties of the solution.

3.2. Effects of Chemicals on the Oscillation Patterns

Typical oscillation patterns in plasmodia in the presence and absence of chemicals are shown in Fig. 2.

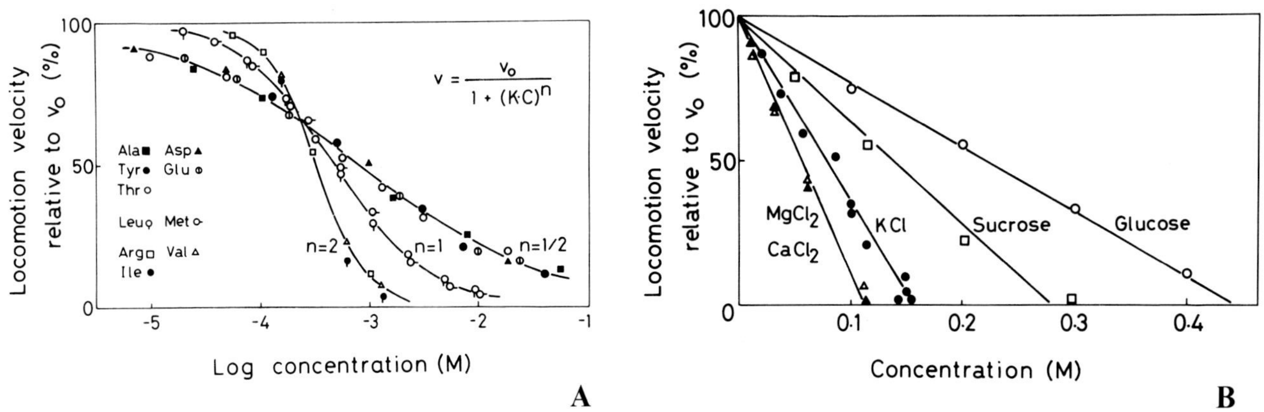

Fig. 1. Locomotion velocity as a function of concentration of chemicals. *A* In the presence of amino acids. The concentration is in log scale, *B* in the presence of salts and sugars. Dose-reponse relationship follows the Hill-type plot for amino acids, whereas salts and sugars at high concentrations decrease the locomotion linearly

Table 1. *Sensitivity (K) and cooperativity (n) found among various amino acids*

	Group I		Group II	
	Amino acids	Concentration [M] at 50% inhibition (= 1/K)	Amino acids	Concentration [M] at 50% inhibition
n = 1	Tryptophan	1.5×10^{-4}	Cystein	1.5×10^{-4}
	Lysine·HCl	1.5×10^{-4}	Serine	1.5×10^{-4}
	Methionine	6×10^{-4}	Histidine	6×10^{-4}
	Leucine	6×10^{-4}	Glycine	1.0×10^{-3}
	Asparagine	8×10^{-4}		
	Glutamine	1.0×10^{-3}		
n = 2	Isoleucine	5×10^{-4}		
	Arginine	5×10^{-4}		
	Valine	5×10^{-4}		
	Proline	5×10^{-3}		
n = 1/2	Alanine	6×10^{-4}	Phenylalanine	1×10^{-2}
	Glutamic acid	6×10^{-4}		
	Aspartic acid	7×10^{-4}		
	Threonine	6×10^{-4}		
	Tyrosine	1×10^{-3}		
	D-Serine	2×10^{-3}		

Parameters n and K in the group I are defined by the equation $v/v_0 = 1/(1 + (KC)^n)$, where v and v_0 are locomotion velocities in the presence and abscence of chemicals, respectively, and C the concentration of chemicals. The group II amino acids are listed at the corresponding n values determined at low concentrations

When the organism migrates on water-agar (*A*), phases of the oscillation at the front and at the rear are almost opposite, although the oscillation entrains into a single frequency. Effects of chemicals appear in two distinct ways; amplitude and pattern modulations. The former case is shown in Fig. 2 *B*. In the presence of 1 mM tryptophan, oscillations almost fade away, and yet polar pattern is still visible. The amplitude of oscillation decreased as the concentration of a chemical increased, and the oscillation became noisy as the amplitude diminished (data not shown). Similar responses as shown in Fig. 2 *B* were observed for all the group-I amino acids in Table 1 and for all the sugars and salts tested (Fig. 1 *B*).

Pattern modulation with an amino acid is shown in Fig. 2 *C*. Phenylalanine (1 mM) breaks lateral synchrony and coordination between the rear and front in the control (Fig. 2 *A*). Here a single plasmodium separates into many regions in terms of the phase distribution, while the amplitude of oscillation remains as large as the control level. The amino acids of the group II (Table 1) caused similar changes in oscillation patterns. Size of separated domains (Fig. 2 *C*) differed from one chemical to another, and also depended on the concentration.

3.3. Intracellular ATP Distribution in Relation to Locomotion

Corresponding to amplitude and pattern modulations in thickness oscillation, effects of chemicals on intracellular ATP distribution are also classified into two types; a decrease in the ATP concentration and a modulation of ATP distribution, as shown in Fig. 3. In the absence of chemicals, ATP distributes twice as high at the front as at the rear (*A*). For some chemicals, this polar distribution flattens uniformly (*B*, *C*, and *D*), and all the group-I amino acids, sugars and salts induced similar effects.

On the other hand, some amino acids induce humpy ATP distributions (*e.g.*, *E* and *F*). Regions in which the ATP concentration is as high as that at the front in the control (20—30 μg/mg protein) appear here and there throughout the plasmodium. As a result, the polarity of ATP distribution disappears. Very often, several new advancing fronts emerged at regions rich in ATP. The amino acids of the group II induced a similar humpy ATP distribution.

In the case of amplitude modulation, locomotion velocity fell on a single curve as a function of intracellular ATP concentration at the front (Fig. 4). Above the

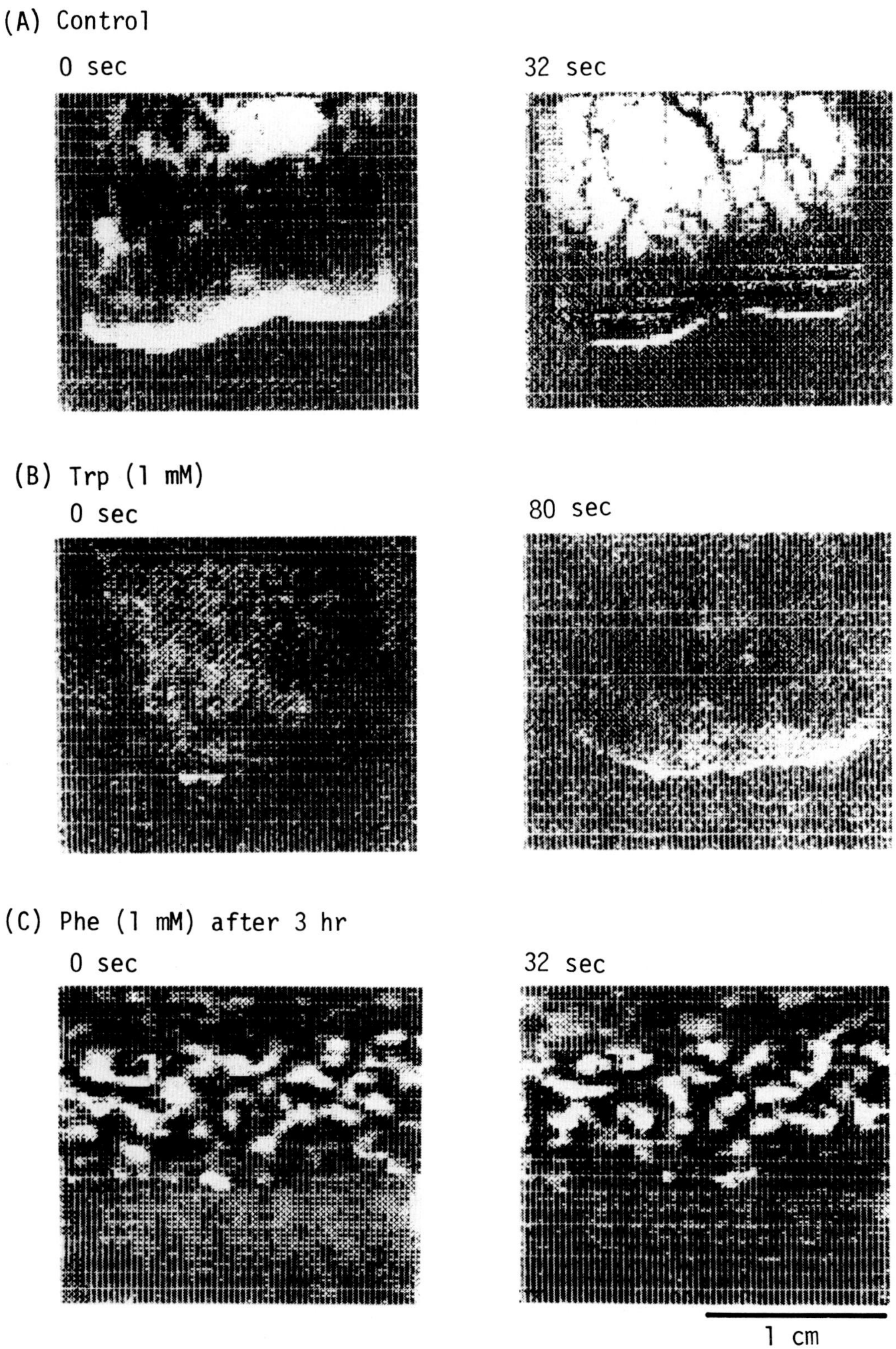

Fig. 2. Changes in oscillation patterns in the presence of chemicals. (*A*) water agar, (*B*) in the presence of 1 mM tryptophan, (*C*) in the presence of 1 mM phenylalanine. Effects of chemicals on the oscillation pattern are classified into two groups, a decrease in the oscillation amplitude (*B*) and a modulation of the pattern (*C*)

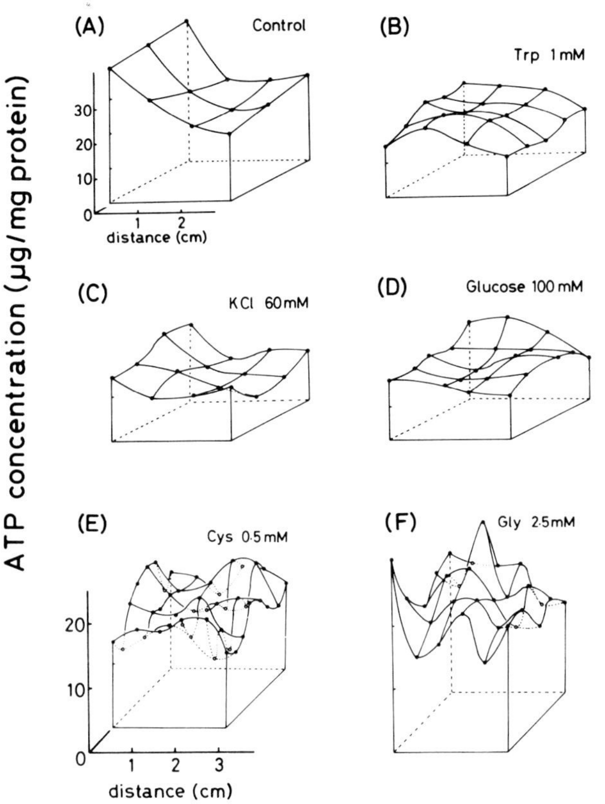

Fig. 3. Changes in intracellular ATP distribution in the presence of various chemicals. Scales to *A—D* and to *E—F* are shown in *A* and *E*, respectively. The ATP distribution either flattens (*B—D*) or becomes undulated (*E*, *F*), corresponding to amplitude and pattern modulations respectively in Fig. 2

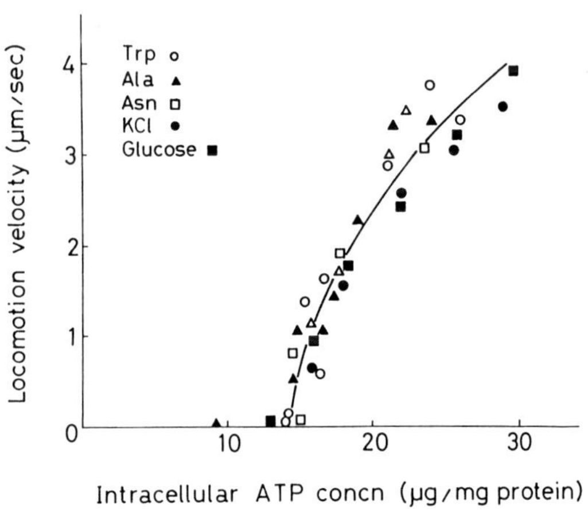

Fig. 4. Locomotion velocity as a function of intracellular ATP concentration. The relationship shows strong cooperativity similar to critical phenomena

threshold ATP_{th}, the locomotion velocity v increases cooperatively as follows:

$$v \propto (ATP - ATP_{th})^a \qquad (3)$$

where ATP denotes the intracellular ATP concentration. The parameter, a, took a value of 0.5–0.8, and ATP_{th} was about half the ATP concentration at the control.

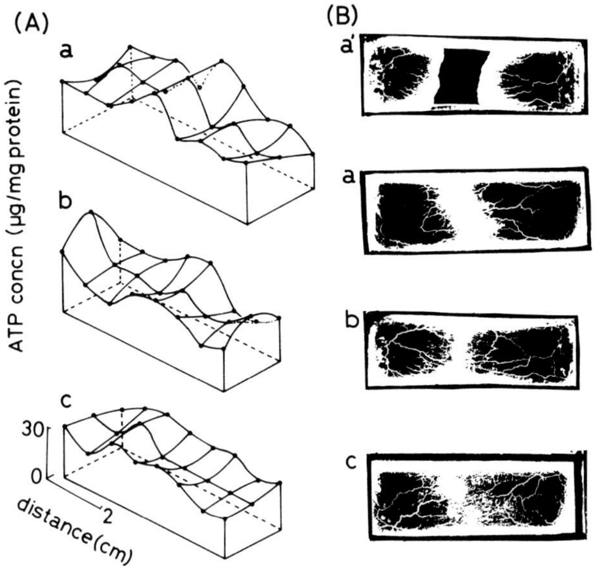

Fig. 5. Changes in intracellular ATP distribution and cell morphology accompanying coalescence. *B-a'* before collision. *A-a, b, c* and *B-a, b, c* ATP distribution and photographs, respectively, at 20-min intervals just after two plasmodia met. The protoplasm flows toward the high ATP region. See text in detail

3.4. Changes in ATP Distribution Accompanying Coalescence

When a migrating plasmodium collides at a solid wall, it changes the direction of locomotion. Accompanying these, intracellular ATP distribution also changes (UEDA *et al.* 1987). What happens when two migrating plasmodia collide and fuse each other? Changes in cell morphology and ATP distribution accompanying coalescence are shown in Fig. 5. Before coalescence, protoplasm at the advancing front was thick, and the ATP concentration there was high (*a*). Just after coalescence, however, protoplasm at the frontal part was carried away from there and began accumulating around the periphery, as if the organism is trying to clime the wall of a trough (*B-b*). At this stage, ATP was low at the middle and high at the ends (*A-b*). Later, the protoplasm distributed rather uniformly with no apparent advancing fronts, and the ATP distribution was no longer polar, but undulated (*c*).

4. Discussion

As shown above, locomotion and the intracellular flow of protoplasm in the *Physarum* plasmodium is closely related to, or governed by patterns of the intracellular

ATP distribution. Actually, the protoplasm flows toward the microinjected high ATP region. This, however, does not mean that ATP concentration regulates directly the contraction or motility, because it is shown that the ATP concentration within the "caffeine drops" is sufficient to maintain the active movement (Sato *et al.* 1981). How are various ATP patterns self-organized in the *Physarum* plasmodium? Notion of a "dissipative structure" in a system far away from equilibrium (Nicolis and Prigogine 1977) seems applicable to the formation of chemical patterns in the *Physarum* plasmodium (Ueda *et al.* 1987). Theoretically, it is shown that chemical reactions with autocatalytic nature oscillate and also form spatial patterns when coupled spatially with diffusion. The system can bifurcate from one pattern to another, depending on the kinetic parameters and boundary conditions. A well-known example is the activator-inhibitor model in morphogenesis (Meinhardt 1982). These characteristics also explain why amoeboid cell's behavior such as *Amoeba proteus, Dictyostelium discoideum*, macrophages, leukocytes etc. is oscillatory (Satoh *et al.* 1985).

Based on the above model, our results are interpreted as follows: environmental factors such as the presence or absence of chemicals alter kinetic parameters, which leads to bifurcation among dissipative structures. The bifurcation seems to appear as either pattern or amplitude modulations in the *Physarum* plasmodium. In the former case, high ATP regions appear here and there (Fig. 3 *E* and *F*). This does not allow the cooperative flow of protoplasm toward the frontal part, but permits many local flows toward local high ATP regions, and hence the locomotion as a whole is inhibited. In the latter, intracellular ATP concentration decreases, and the amplitude of rhythmic contraction decreases due to loss of energy. Change in ATP distribution accompanying coalescence (Fig. 5) may be due to a change in boundary conditions. Recently, formation of spatio-temporal patterns is demonstrated in cell-free systems such as the glycolytic system (Muller *et al.* 1986) or mitochondria (Holmuhamedov 1986). It might be possible to mimick cellular responses in these cell-free systems.

References

Beutler E, Baluda M (1964) Simplified determination of blood adenosine triphosphate using the firefly system. Blood 23: 688–698

Hatano S, Takeuchi I (1960) ATP content in myxomycete plasmodium and its levels in relation to some external conditions. Protoplasma 52: 169–183

Holmuhamedov EL (1986) Oscillating dissipative structures in mitochondrial suspensions. Eur J Biochem 158: 543–546

Kamiya N (1959) Protoplasmic streaming. In: Heilbrunn LV, Weber F *et al* (eds) Protoplasmatologia, vol 8. Springer, Wien, part 3 a

Kessler D (1982) Plasmodial structure and motility. In: Aldrich HC, Daniel JW (eds) Cell biology of Physarum and Didymium, vol 1. Academic Press, New York, pp 145–208

Kohama K, Tanokura M, Yamada K (1984) ^{31}P-nuclear magnetic resonance studies of intact plasmodia of *Physarum polycephalum*. FEBS Lett 176: 161–165

Matsumoto K, Ueda T, Kobatake Y (1986) Propagation of phase wave in relation to tactic responses by the plasmodium of *Physarum polycephalum*. J Theor Biol 122: 339–345

Meinhardt H (1982) Models of biological pattern formation. Academic Press, London

Mori Y, Matsumoto K, Ueda T, Kobatake Y (1986) Spatio-temporal organization of intracellular ATP content and oscillation patterns in response to blue light by *Physarum polycephalum*. Protoplasma 135: 31–37

Muller SC, Plesser T, Hess B (1986) Two-dimensional spectroscopy and pseudo-color representation of chemical reaction patterns. Naturwissenschaften 73: 165–179

Nicolis G, Prigogine I (1977) Self-organization in nonequilibrium systems. J Wiley, New York

Sato H, Hatano S, Sato Y (1981) Contractility and protoplasmic streaming preserved in artificially induced plasmodial fragments, the "caffeine drops". Protoplasma 109: 187–208

Satoh H, Ueda T, Kobatake Y (1985) Oscillations in cell shape and size during locomotion and in contractile activities of *Physarum polycephalum, Dictyostelium discoideum, Amoeba proteus* and macrophages. Exp Cell Res 156: 79–90

Ueda T, Matsumoto K, Akitaya T, Kobatake Y (1986) Spatial and temporal organization of intracellular adenine nucleotides and cyclic nucleotides in relation to rhythmic motility in *Physarum* plasmodium. Exp Cell Res 162: 486–494

— Mori Y, Kobatake Y (1987) Patterns in the distribution of intracellular ATP concentration in relation to coordination of amoeboid cell behavior in *Physarum polycephalum*. Exp Cell Res 169: 191–201

Yoshimoto Y, Sakai T, Kamiya N (1981) ATP oscillation in *Physarum* plasmodium. Protoplasma 109: 159–168

Protoplasma (1988) [Suppl 1]: 57–63

Morphology and Visco-Elastic Properties of *Physarum* Strands During the Steady-State of Their Contractile Behavior

W. Naib-Majani[1], V. A. Teplov[2], and Z. Baranowski[3,*]

[1] Institute of Cytology, University of Bonn, [2] Institute of Biophysics, USSR Academy of Sciences, Pushchino, and [3] Nencki Institute of Experimental Biology, Polish Academy of Sciences, Warszawa

Received December 21, 1987
Accepted March 1, 1988

Dedicated to Professor Dr. Noburo Kamiya on the occasion of his 75th birthday

Summary

The inhibition of both the cytochrome and the alternative pathways of electron transport in mitochondria of *Physarum polycephalum* with the aid of cyanide (KCN) and salicylhydroxamic acid (SHAM) results in the cessation of the contraction-relaxation cycle of plasmodial strands. The cessation of the oscillations is accompanied by a vacuolization of the protoplasm, pronounced fibrillogenesis of plasmodial actomyosin and an increase of the coefficients of viscosity and elasticity. Similar changes in the morphology of the contractile apparatus and the visco-elastic properties of the plasmodial strands are observed in the presence of 2% ethanol. However, in contrast to the inhibitors of respiration 2% ethanol leads to an enhancement of the oscillatory activity. These observations enable us to assume that the cessation of the contraction-relaxation cycle evoked by the inhibition of respiration is not the result of the blockage of the actomyosin system but is a consequence of the modification of a self-oscillating system.

Keywords: Cytoplasmic actomyosin; Oscillations; Cyanide; Ethanol; Viscosity; Young's modulus; *Physarum polycephalum.*

1. Introduction

Plasmodia of *Physarum polycephalum* show various types of periodic phenomena among which shuttle streaming and the contraction-relaxation cycle are the most spectacular ones. Since the isolation of an actomyosin-like protein from an extract of *Physarum polycephalum* plasmodia (Loewy 1952) and the visualization of actomyosin fibrils *in situ* (Wohlfarth-Bottermann 1962) it is generally accepted that the oscillatory contractile activity is realized by a Mg-ATP dependent interaction between plasmodial actin and myosin (Matsumura *et al.* 1980).

Application of tensiometric methods to study *Physarum* contractility (Kamiya 1970) has allowed to analyse the contraction-relaxation cycle. Although the parameters of the oscillations of the force generated by isolated plasmodial strands, for example frequency and amplitude, may be changed due to the experimental procedure, it has appeared that the oscillatory regime is very stable. The temporal arrest of the periodic contractility was noticed only during mitosis (Gotoh and Kuroda 1982, Kessler and Achenbach 1985).

The biochemical nature and location of the oscillator are still a matter of speculation. Wohlfarth-Bottermann (1979) has discussed four possible locations, namely cytoplasmic actomyosin, the energy supply system, the inner Ca^{2+} stores and the plasmalemma. Recently investigations have concentrated upon the elucidation of the regulatory role of energy metabolism (Satoh *et al.* 1982, 1984, Korohoda *et al.* 1983, Shraideh *et al.* 1983, Baranowski 1985a, Baranowski *et al.* 1983). The finding that the blockage of respiration in *Physarum* leads to a reproducible and reversible cessation of the oscillations of the force (Baranowski 1985b) has focused our attention on the role of respiration.

Physarum mitochondria exhibit two pathways for electron transport. The cytochrome pathway with a terminal oxidase that is inhibited by cyanide and the alternative one which branches at the level of ubiquinone and is inhibited by salicylhydroxamic acid—SHAM

* Correspondence and Reprints: Nencki Institute of Experimental Biology, 3 Pasteur str., 02-093 Warszawa, Poland.

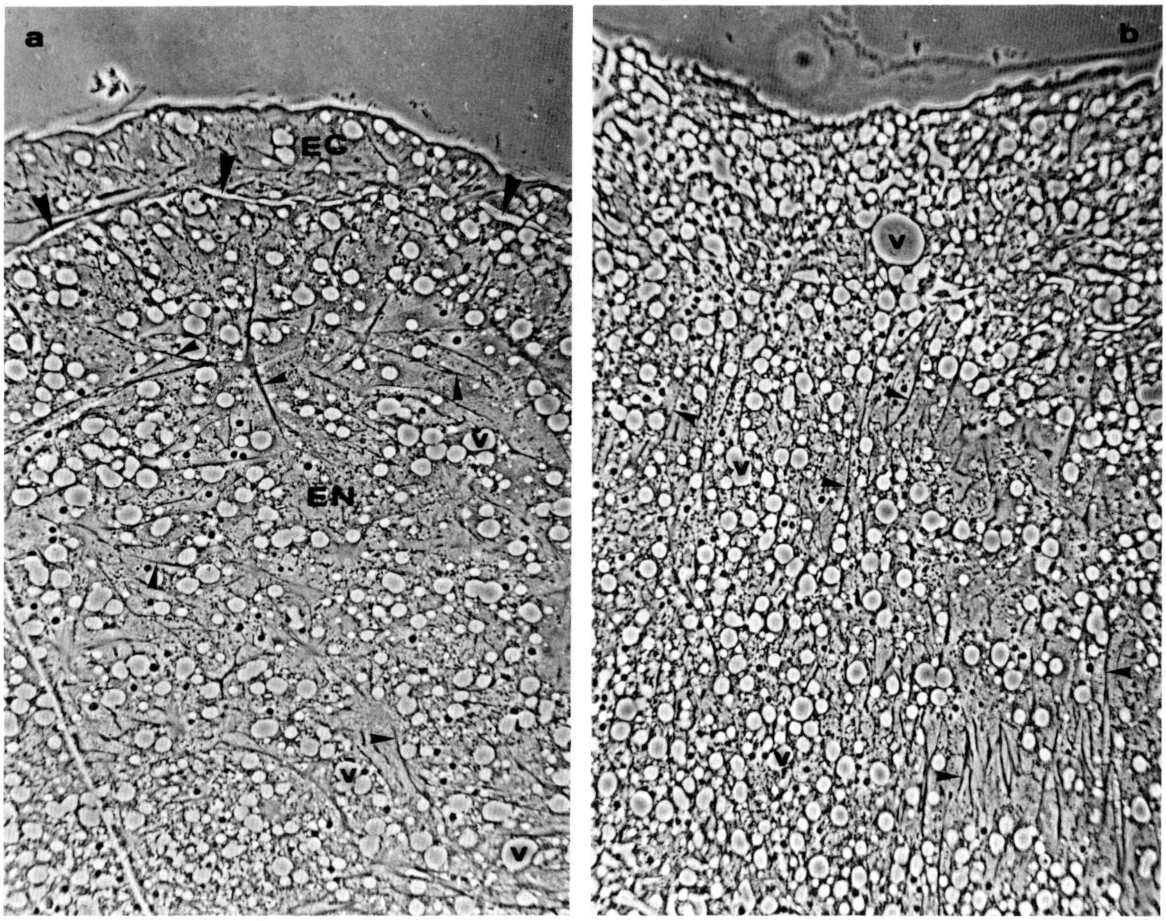

Fig. 1. Phase-contrast micrographs of semithin cross (*a*) and longitudinal (*b*) sections of a *Physarum* strand fixed in the steady-state of its contractile activity. The steady-state *i.e.* reversible cessation of the contraction-relaxation cycle was evoked by KCN and SHAM. Note the vacuolization of the protoplasm and the polygonal pattern of actomyosin fibrils. The circular arrangement of the fibrils is shown in the longitudinal, tangential section (*b*) of the strand. ◀ Invagination system; ◂ actomyosin fibrils; *EC* ectoplasm; *EN* endoplasm; *v* vacuoles. Magn.: *a* and *b* × 300

(BARANOWSKI 1985a, BEYLINA *et al.* 1988). Therefore, the blockage of the respiration of *Physarum* plasmodia is evoked by the simultaneous application of both inhibitors, *i.e.*, KCN and SHAM.

In the presented work we studied the morphology and the visco-elastic properties of the plasmodial strands in order to judge whether inhibition of respiration:

(a) alters the actomyosin system in such a manner that the periodic signal coming from the hypothetical oscillator is not expressed in form of the contraction-relaxation cycle, or

(b) modifies the system responsible for self-oscillations.

In the latter case the reversible cessation of the oscillations makes a convenient starting-point for further investigations of the nature of the oscillator in *Physarum*.

2. Material and Methods

Strands of *Physarum polycephalum* (strain ATCC 44912) isolated from the phaneroplasmodia that were freely migrating on a surface of a piece of moist filter paper were used for the study.

The respirator inhibitors SHAM and KCN as well as ethanol were dissolved in salt solution containing 1 mM $CaCl_2$, 2 mM NaCl, 1 mM KCl and 3 mM $MgCl_2$. The test solutions containing i) 7 mM SHAM and 5 mM KCN, or ii) 2% ethanol were adjusted to pH 7.0 with the aid of KOH.

Light and electron microscopy were carried out as follows: a) fixation by replacing the test solution containing the inhibitors of respiration with 1% OsO_4 + 0.5% $HgCl_2$, pH 6.4 for 1 hour (PARDUCZ 1952), b) rinsing off the fixative with Tyrode solution, c) dehydratation of the speciments in a graded series of ethanol, d) poststaining with 1% phosphotungstic acid and 0.5% uranyl acetate in 70% ethanol, and e) embedding in styrene methacrylate (KUSHIDA 1961).

Tensiometric investigations were carried out under isometric conditions (WOHLFARTH-BOTTERMANN 1975). The method for the es-

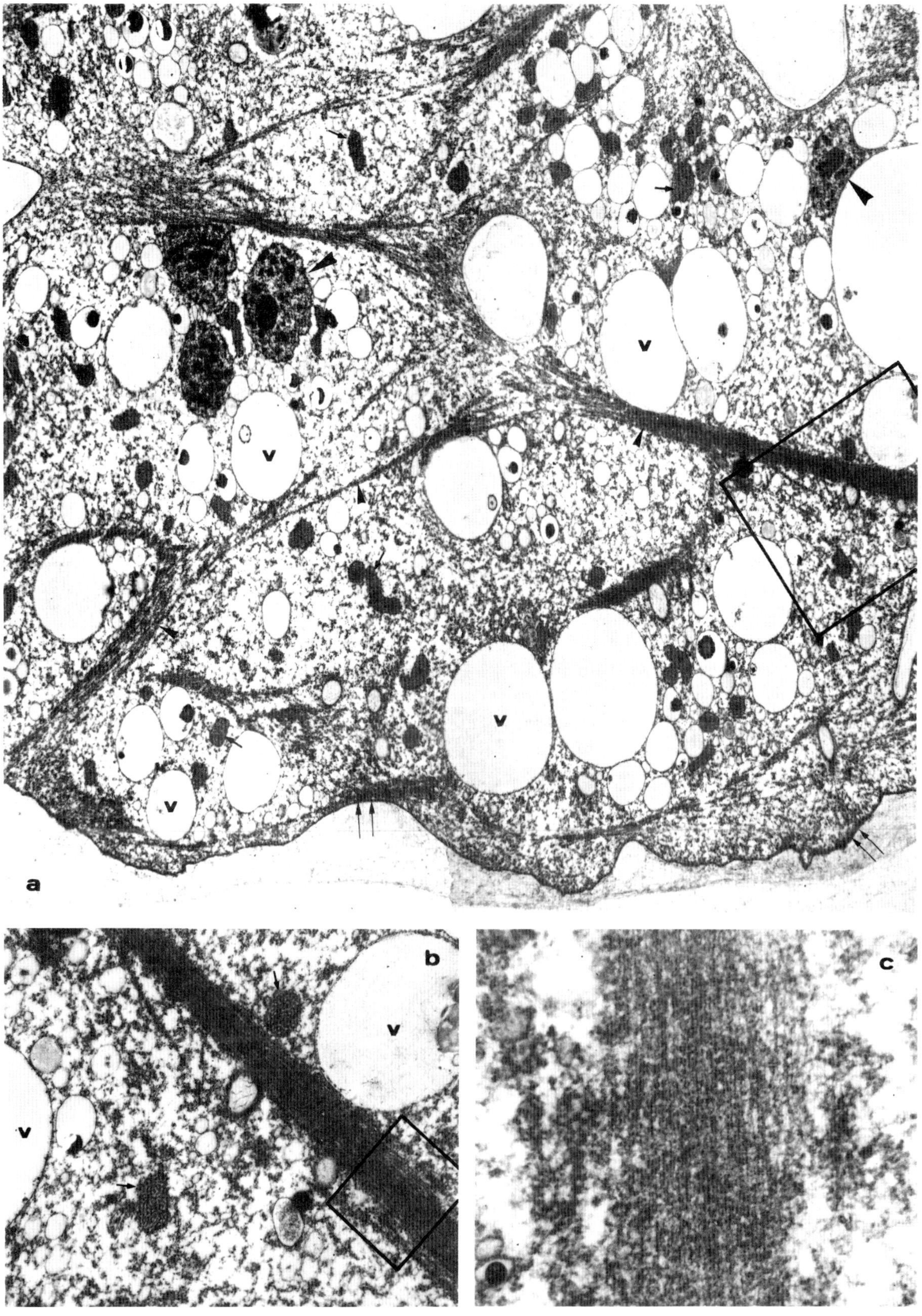

Fig. 2. Ultrathin section through a cortical region of a plasmodial strand treated with KCN and SHAM and fixed in the steady-state of its contractile activity. Note the polygonal pattern of actomyosin fibrils and their close contact with the plasmalemma (*a*). The parallel arrangement of actomyosin filaments is shown in *b* and *c* which are higher magnifications of the framed region in *a*. ◀ Nuclei; ← mitochondria; ◂ actomyosin fibrills; *v* vacuoles; ⇇ plasmalemma. Magn.: *a* × 4,800, *b* × 11,000, *c* × 42,000

timation of the visco-elastic properties of the plasmodial strands is described in the Results.

3. Results

3.1. Morphological Investigations

Figure 1 shows a phase-contrast micrographs of semithin cross (*a*) and longitudinal (*b*) sections of a plasmodial strand treated with KCN and SHAM. The fixation was performed when the tensiometric control displays a steady-state of contractile activity (cp, Fig. 3 *a*). There are two distinct morphological symptoms following the inhibition of respiration:

(i) vacuolization of the protoplasm, and

(ii) fibrillogenesis of plasmodial actomyosin.

The plasmodial strands preserve the system of plasmalemma invaginations that usually forms a borderline between the ecto- and endoplasm (WOHLFARTH-BOTTERMANN 1974). The irregular shape of the plasmalemma invaginations enables us to distinguish this system from the newly formed vacuoles. The latter have a more or less regular circular shape and they are seen all over the sections.

In the present morphological observations we have focused our attention on the changes of the arrangement of fibrillar structures which are visible at the light microscopy level. Immunocytochemical investigations have revealed that these structures represent the state of aggregation of plasmodial actomyosin (NAIB-MAJANI *et al.* 1982, 1983, 1984; OSBORN *et al.* 1983). The fibrillar structures, usually seen only in the cortical area of the strand (EC) and in close contact with the system of the plasmalemma invaginations, are randomly distributed all over the sections after treatment with respiration inhibitors. The appearance and the random distribution of the fibrillar structures point out that actin or actomyosin fibrils have formed as three dimensional network as a response to the inhibition of respiration. The polygonal pattern of this network is shown in the electron micrographs of the cortical region of the strand (see Fig. 2). In Fig. 2 *a* it is clearly seen that the network is in close contact with the plasmalemma. The fibrils which form the network show parallel arrangement of actin filaments (see Fig. 2 *c*).

A similar fibrillogenesis is observed as a response to the treatment of the plasmodia with ethanol (PIES and WOHLFARTH-BOTTERMANN 1985). However, as it will be shown below, the contractile behavior is entirely different when ethanol or the inhibitors of respiration have been applied.

3.2. Tensiometric Investigations

Cyanide and SHAM, as it was previously reported (BARANOWSKI 1985 b), evoke a reversible cessation of the oscillations (see Fig. 3 *a*), while 2% ethanol causes an enhancement of the oscillatory activity (see Fig. 3 *b*). At this concentration of ethanol we observed a prolongation of the period of the oscillations, and an increase of the force level and the amplitude.

Different pattern of contractility noticed in samples with similar morphology might be the result of a blockage of the oscillator or to changes of the contractile properties of the actomyosin system which under these circumstances is not able to follow the oscillator. The impediment in function of the contractile apparatus

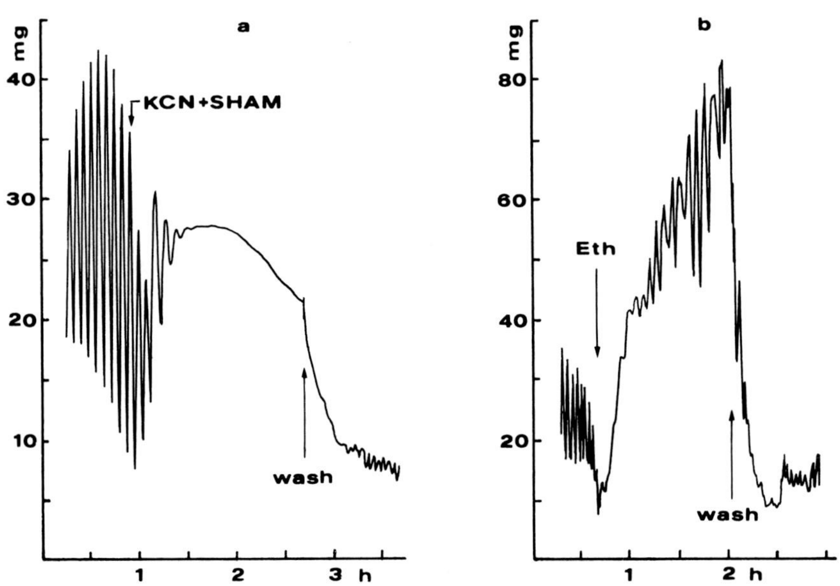

Fig. 3. The longitudinal contractile behavior of plasmodial strands under isometric conditions as a response to (*a*) KCN and SHAM, and (*b*) 2% ethanol treatment. Note the reversible cessation of the contraction-relaxation cycle *i.e.* the steady-state in contractile behavior of the strand in the presence of the inhibitors of respiration (*a*) and an enhancement of oscillations in the presence of ethanol (*b*)

could be reflected by changes of the visco-elastic properties of the samples.

3.3. Visco-Elastic Properties

The visco-elastic parameters are estimated on the basis of transient stretch-release induced changes of the longitudinal force generated by the strand under isometric conditions. A quick stretch of the strand induces an instantaneous rise in the force followed by its exponential decrease until a new, steady level of the force is achieved. A return to the initial length (quick release) causes an exact opposite change in the force. Examples of stretch-release measurements are shown in Fig. 4 *b, c, d*. The presented tensiograms show the changes in the force generated by the same strand in response to equal stretch stimuli (circa 1% of the initial length) under following experimental conditions:

(a) in salt solution (control),
(b) in 2% ethanol, and
(c) during steady-state conditions evoked by KCN and SHAM.

It was demonstrated that the viso-elastic properties of the strand during time intervals much shorter than the period of self-oscillations are well approximated by the model shown in Fig. 4 *a* (TEPLOV and ROMANOVSKY 1987). The following formulas which estimate Young's moduli (E_1 and E_2) and the coefficient of viscosity (η) can be obtained on the basis of this model:

$$E_1 = \frac{F_1 \cdot l}{\triangle l \cdot s} \quad (1)$$

$$E_2 = \frac{F_2 \cdot l}{\triangle l \cdot s} \quad (2)$$

$$\eta = \tau \cdot E_2 \quad (3)$$

where: l is the initial length of the strand, $\triangle l$ is the value of the strand deformation, s is a cross section of the strand, F_1 is the asymptote of the exponential decrease of the force, F_2 is obtained according to the relationship $F_2 = F(t)_{t=0} - F_1$, τ is the time constant for the exponential decrease of the force. The time constant τ and F_2 were calculated from the slope and intercept of the best fit straight line approximating the function ln $[F(t) - F_1]$. The coefficient of correlation "r" of a linear regression was chosen as a measure of the accuracy of the fitting of the model shown in Fig. 4 *a* to the experimental data. In experimental conditions (*a*—*c*) "r" was found to be in the range of 0.992 and 0.999. It denotes no qualitative differences in the viscoelasticity of the plasmodial strands treated with ethanol or KCN with SHAM comparing to the control conditions. Therefore, the model is applicable in these cases.

Table 1 shows that except for Young's modulus E_2, in the presence of KCN and SHAM, all the other param-

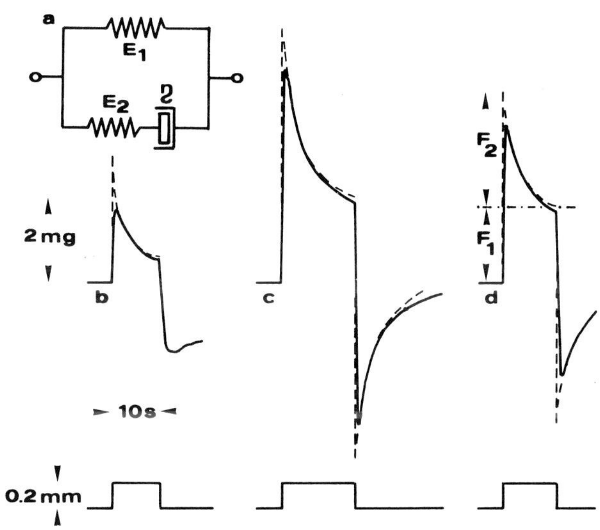

Fig. 4. *a* Model of visco-elastic properties of Physarum plasmodial strands. E_1 and E_2 are the elements of elasticity, η is the element of viscosity. *b*—*d* Force responses of the same plasmodial strand induced by equal stretch-release stimuli (0.2 mm *i.e.* 1% of the initial length of the strand), *b* in salt solution (control), *c* after treatment with 2% ethanol, and *d* in the steady-state evoked by KCN and SHAM. Dotted line denotes the approximation of the force responses according to the model shown in *a*. Further explanation see Results

Table 1. *Changes in the visco-elastic properties of the same plasmodial strand of Physarum polycephalum exposed to the successive replacement of various solutions.* The numbers represent the mean values ± S.D. (n = 5)

Visco-elastic parameters	Salt sol. (control)	2% Ethanol	KCN + SHAM	Dimension
τ	2.6 ± 0.7	3.9 ± 0.3	3.2 ± 0.4	s
E_1	0.3 ± 0.2	1.2 ± 0.2	1.0 ± 0.1	10^5 dyne · cm^{-2}
E_2	1.1 ± 0.3	2.2 ± 0.6	1.5 ± 0.2	10^5 dyne · cm^{-2}
η	2.8 ± 0.7	8.7 ± 3.0	4.9 ± 0.5	10^5 dyne · s · cm^{-2}

eters of viscoelasticity rise substantially in comparison to the control. The degree of this increase is higher in the presence of ethanol than that observed in the steady-state of the contractile activity.

4. Discussion

In the present paper we demonstrate that the cessation of the contraction-relaxation cycle in response to the inhibition of respiration is accompanied by fibrillogenesis, changes in the arrangement of the fibrils and an increase in the viscosity and elasticity of the strands. Alterations of the instantaneous Young's modulus during the course of the spontaneous contraction-relaxation cycle were studied by Kamiya and Yoshimoto (1972). Extending these investigations Teplov and Romanovsky (1987) showed that the viscoelastic properties of the plasmodial strand are satisfactory approximated by the dashpot-spring model shown in Fig. 4 *a*. Using quick (30 ms) stretch-release stimuli with a duration of 1 s they demonstrated that both Young's modulus E_1 and the viscosity η oscillate in phase with the force generated by the plasmodial strand while Young's modulus E_2, remains constant during the course of the contraction-relaxation cycle. Therefore it may be assumed that E_1 reflects the elasticity of the plasmodial actomyosin system while E_2 represents the elasticity of the other constituents of the cell which do not participate in the generation of the force. In the simplest case variations of E_1 may reveal the changes in the amount of formed actomyosin complexes.

However, the information which may be drawn from morphological observations and the investigations of viscoelastic properties of the strands are not sufficient to judge whether the cessation of the contraction-relaxation cycle is related to the disturbances in the function of the oscillator or just to the blockage of the actomyosin system. Following the aim of this research (see Introduction a, b) we tried to solve this problem by comparing the effects of the inhibitors of respiration and 2% ethanol on contractile activity. Pies and Wohlfarth-Bottermann (1985) examined the morphology and contractility of plasmodial strands treated with 1.5–5% ethanol and different concentrations of Ca^{2+}. Ethanol in the presence of millimolar concentrations of Ca^{2+} induces fibrillogenesis manifested by the appearance of a large amount of fibrils in the endoplasmic channel. On the other hand, as it is presented in this paper, 2% ethanol also causes an increase of the parameters of viscoelasticity. Persistence of the contraction-relaxation cycle in strands treated with 2% ethanol and the lack of this cycle in strands treated with KCN with SHAM suggest that the inhibition of respiration modifies the system responsible for the self-oscillations in *Physarum*. In other words we did not find any evidence that in the presence of KCN and SHAM the actomyosin system is not able to oscillate in response to a periodic signal coming from the hypothetical oscillator. The results presented herein suggest that respiration is involved in the self-regulating dynamic unit which is responsible for the oscillatory activity in *Physarum polycephalum*.

Acknowledgements

This work was partially supported by Polish Academy of Sciences CPBP-04.01 and Alexander von Humboldt Stiftung. We would like to thank Dr. Jerzy Sikora for helpful discussion.

References

Baranowski Z (1985 a) Alternative pathway of respiration in *Physarum polycephalum* plasmodia. Cell Biol Int Reps 9: 85–90

— (1985 b) Consequence of impeding in mitochondrial functions in *Physarum polycephalum*. III. Reversible cessation of contraction-relaxation cycle and temperature sensitivity of the alternative respiration. Eur J Cell Biol 39: 283–289

— Shraideh Z, Korohoda W, Wohlfarth-Bottermann KE (1983) Consequence of impeding in mitochondrial function in *Physarum polycephalum*. II. Influence on "de novo" generation of contractile activity and response to glucose and blue light. Eur J Cell Biol 31: 187–196

Beylina SI, Hrebenda B, Cieślawska M, Baranowski Z (1988) Cyanide-resistant respiration of *Physarum polycephalum* in course of starvation. Int Cell Biol Reps 12: 29–34

Gotoh K, Kuroda K (1982) Motive force of cytoplasmic streaming during plasmodial mitosis of *Physarum polycephalum*. Cell Motility 2: 173–181

Kamiya N (1970) Contractile properties of the plasmodial strand. Proc Japan Acad 46: 1026–1031

— Yoshimoto Y (1972) Dynamic characteristic of the cytoplasm. A study on the plasmodial strand of a Myxomycete. In: Hamaguchi K (ed) Aspects of cellular and molecular physiology. Tokyo Univ Press, Tokyo, pp 167–189

Kessler D, Achenbach F (1985) Change of contractile behavior in plasmodia of *Physarum polycephalum* during mitosis. Protoplasma 124: 71–79

Korohoda W, Shraideh Z, Baranowski Z, Wohlfarth-Bottermann KE (1983) Energy metabolic regulation of oscillatory contraction activity in *Physarum polycephalum*. Cell Tissue Res 231: 675–691

Kushida H (1961) A styrene methacrylate resin embedding method for ultrathin sectioning. J Electron Microsc 10: 16–19

Loewy AG (1952) An actomyosin-like substance from the plasmodium of a Myxomycete. J Cell Comp Physiol 40: 127–156

Matsumura F, Yoshimoto Y, Kamiya N (1980) Tension generation by actomyosin thread from a non-muscle system. Nature 285: 169–171

Naib-Majani W, Stockem W, Wohlfarth-Bottermann KE (1982) Immunocytochemistry of the acellular slime mold *Phy-*

sarum polycephalum. II. Spatial organization of cytoplasmic actin. Eur J Cell Biol 28: 103–114

— OSBORN M, WEBER K, WOHLFARTH-BOTTERMANN KE, HINSSEN H, STOCKEM W (1983) Immunocytochemistry of the acellular slime mould Physarum polycephalum. I. Preparation, morphology, and reliability of results concerning cytoplasmic actomyosin patterns in sandwiched plasmodia. J Cell Sci 60: 13–28

— ACHENBACH F, WEBER K, WOHLFARTH-BOTTERMANN KE, STOCKEM W (1984) Immunocytochemistry of the acellular slime mold *Physarum polycephalum*. IV. Differentiation and dynamics of the polygonal actomyosin system. Differentiation 26: 11–22

OSBORN M, WEBER K, NAIB-MAJANI W, HINSSEN H, STOCKEM W, WOHLFARTH-BOTTERMANN KE (1983) Immunocytochemistry of the acellular slime mold *Physarum polycephalum*. III. Distribution of myosin and the actin-modulating protein (fragmin) in sandwiched plasmodia. Eur J Cell Biol 29: 179–186

OSTER GF, ODELL GM (1984) Mechanics of cytogels I: oscillations in *Physarum*. Cell Motil 4: 469–503

PARDUCZ B (1952) Eine neue Schnellfärbemethode im Dienste der Protistenforschung und des Unterrichts. Ann Hist-Nat Mus Nat Hung 2: 5–15

PIES NJ, WOHLFARTH-BOTTERMANN KE (1985) Enrichment of fibrillar cytoplasmic actomyosin in protoplasmic strands of *Physarum polycephalum* for the production of cell-free models. Cell Tissue Res 239: 365–374

SATOH H, UEDA T, KOBATAKE Y (1982) Primary oscillator of contractional rhythm in the plasmodium of *Physarum polycephalum*: role of mitochondria. Cell Struct Funct 7: 275–283

— — — (1984) Rhythmic contraction in the plasmodium of the Myxomycete *Physarum polycephalum* in relation to the mitochondrial function. Cell Struct Funct 9: 37–44

SHRAIDEH Z, BARANOWSKI Z, KOROHODA W, WOHLFARTH-BOTTERMANN KE (1983) Consequences of impeding in mitochondrial function in *Physarum polycephalum*. I. Reversible effects of anoxia, KCN, and influence of the Ca^{2+}ionophore A-23187. Eur J Cell Biol 31: 175–186

TEPLOV VA, ROMANOVSKY YM (1987) Mechanochemical distributed autooscillations in cell motility. Autowave phenomena in contractile activity of *Physarum polycephalum* plasmodium. Biol Res Centre USSR Acad Sci Press, Pushchino, pp 1–48 (in Russian)

WOHLFARTH-BOTTERMANN KE (1962) Weitreichende, fibrilläre, Protoplasmadifferenzierungen und ihre Bedeutung für die Protoplasmaströmung. I. Elektronenmikroskopischer Nachweis und Feinstruktur. Protoplasma 54: 514–539

— (1974) Plasmalemma invaginations as characteristic constituents of plasmodia of *Physarum polycephalum*. J Cell Sci 16: 23–27

— (1975) Tensiometric demonstrations of endogenous, oscillating contractions in *Physarum polycephalum*. Z Pflanzenphysiol 76: 14–27

— (1979) Oscillatory contraction activity in *Physarum*. J Exp Biol 81: 15–32

Protoplasma (1988) [Suppl 1]: 64–71

Pseudopodial Tension in *Amphisorus hemprichii*, a Giant Red Sea Foraminifer

J. L Travis[1, *], S. S. Bowser[2], J. G. Calvin[3], and J. J. Lee[4]

[1] Department of Biological Sciences, State University of New York, Albany, New York, [2] Wadsworth Center for Laboratories and Research, New York State Department of Health, Albany, New York, [3] Department of Biology, Vassar College, Poughkeepsie, New York, and [4] Department of Biology, City College of New York, New York City, New York

Received February 1, 1988
Accepted March 1, 1988

Dedicated to Professor Dr. Noburo Kamiya on the occasion of his 75th birthday

Summary

We used the giant discoid Red Sea foraminifer ("foram"), *Amphisorus hemprichii* as a model system to study the development and maintenance of traction forces in foram pseudopodial networks (*i.e.*, reticulopodia). *Amphisorus* extends pseudopodia radially from the margins of its shell. When plated on a flexible rubber substrate, wrinkles develop where pseudopodia initially contact the substrate and pull against it. Numerous microtubules extend throughout the networks, whereas F-actin-containing plaques are only seen along the ventral surface of the pseudopodia, in regions in close contact with the substrate. Pseudopodial tension was relieved by severing the pseudopodia with a glass needle or by treating the networks with colchicine or cytochalasin B. Together with earlier observations these findings suggest that microtubules and microfilaments interact to generate tensile forces within the pseudopodia; these forces may be used for cell attachment or locomotion.

Keywords: Amphisorus hemprichii; Cell adhesion; Foraminifera; Microtubules; Microfilaments; Reticulopodia; Tension.

1. Introduction

Foram reticulopodia are excellent model systems with which to study various aspects of microtubule-associated cell motility. For example, video-enhanced contrast microscopy has been used to observe directly microtubule movements and microtubule-associated transport in the reticulopodia of benthic foramis (Allen *et al.* 1981, Travis *et al.* 1983, Bowser and Rieder *et al.* 1985, Travis and Bowser 1988). Although the active role of microtubules in many facets of foram motility is well documented, the mechanism by which these organisms locomote remains unclear.

In order to pull their relatively large cell bodies along a substrate, it is thought that forams use their pseudopodia to attach to and exert a rearwards traction force against the substratum (*e.g.*, Arnold 1953). Such a force would place the pseudopodia under tension, a contention supported by the observation that laser ablation of pseudopodial cytoplasm induces an apparently elastic recoil (Travis and Bowser 1986 b). However, such a recoil could also represent an active motile response to the laser-surgical injury.

In the present studies we used a deformable rubber substrate (Harris *et al.* 1980) to localize, in a noninvasive manner, the sites where traction forces are exerted on the substrate by the reticulopodia of *Amphisorus hemprichii*. We also examined these reticulopodia ultrastructurally, and employed anti-cytoskeletal drugs and micromanipulation methods to test whether the microtubule and actin-containing microfilament systems are involved in the production and maintenance of pseudopodial traction forces. Portions of this work appeared in abstract form (Travis and Bowser 1987).

2. Materials and Methods

2.1. Organisms

The soritiid foraminifer, *Amphisorus hemprichii* was collected by scuba divers at depths of 25–50 m in the Red Sea at Elat, Israel. The forams were maintained at 21–23 °C in an artificial Red Sea water (RSW) of 40‰ salinity (1,250 mosm/kg H_2O; Aquarium Systems,

* Correspondence and Reprints: Department of Biological Sciences, State University of New York at Albany, 1400 Washington Avenue, Albany, NY 12222, U.S.A.

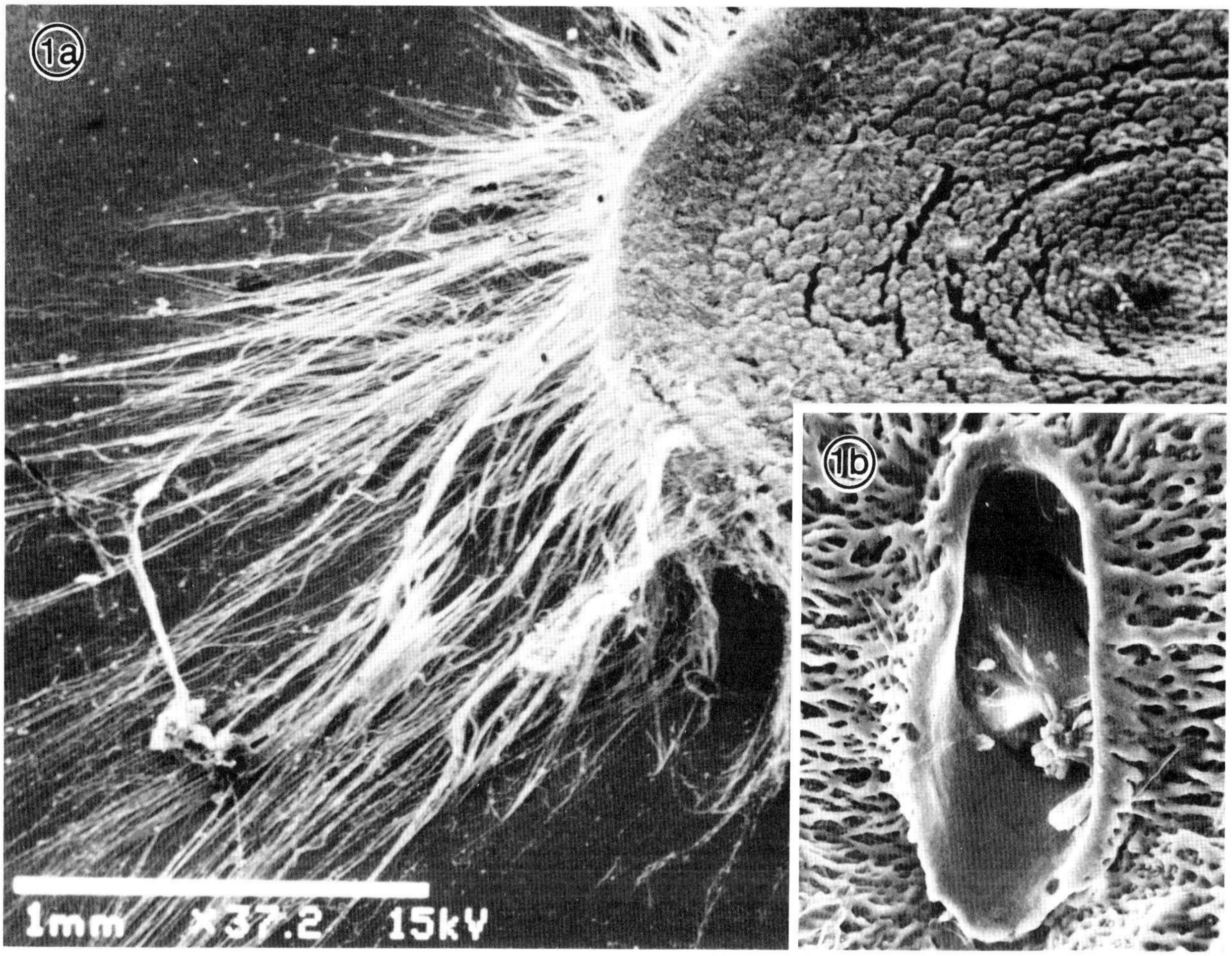

Fig. 1. The giant Red Sea foraminifer, *Amphisorus hemprichii*. (*a*) Scanning electron micrograph demonstrating the extensive pseudopodial network. Magnification: 57×. (*b*) Higher magnification view of a marginal aperture. Magnification: 1,100×

Mentor, OH). The forams were kept on a 14 hr light/10 hr dark photoperiod; light intensity was maintained between 30–70 μEin M^{-2} S^{-1}. *Dunaliella salina* was supplied as prey. Individual *Amphisorus* were removed from culture, cleaned of adherent debris with fine camel hair brushes, and placed on cover slips with or without a flexible rubber coating (see below). All experiments were performed in a Ca^{2+}-depleted RSW containing approximately 470 mM NaCl, 60 mM $MgCl_2$, 3.1 mM Na_2SO_4, 9.66 mM KCl, 2.4 mM $NaHCO_3$, 10 mM Tris at pH 8.0.

2.2. Electron Microscopy

Organisms were placed on alcian blue 8GX-coated, acid-etched glass coverslips immersed in Ca^{2+}-depleted RSW in microbiological grade plastic petri dishes. After extending pseudopodial networks, the forams were fixed for 1 hour in 2% glutaraldehyde and 0.05% tannic acid buffered in a solution of 120 mM Pipes, 50 mM Hepes, 20 mM EGTA and 2 mM $MgCl_2$ (PHEM). Tests were decalcified by overnight incubation in 100 mM EGTA, briefly post-fixed in 0.5% OsO_4, and then dehydrated in ethanol. Some specimens were critical-point-dried from liquid CO_2, sputter coated with 10 nm of Au-Pd, and examined in an Amray 1200 C scanning electron microscope.

For transmission electron microscopy, dehydrated organisms were flat embedded in Epon 812. Portions of the embedded reticulopodia were cut from the wafers and oriented on Epon stubs. 60–80 nm sections were cut, stained with uranyl acetate and lead citrate, and examined at 80 kV in a JEOL 100 B. Serial 0.25 μm sections were also examined with the Wadsworth Center's AEI high-voltage electron microscope operated at 800–1,000 kV.

2.3. Fluorescence Microscopy

For tetramethylrhodamine (TMR-)phalloidin staining, *Amphisorus* were initially fixed for 1 minute in 2% glutaraldehyde in PHEM, followed by 1 hr with 0.2% glutaraldehyde in the same buffer. Free aldehydes were reduced with 0.5% $NaBH_4$ in phosphate-buffered saline and then stained with 3×10^{-7} M TMR-phalloidin (Molecular Probes, Junction City, OR) as described in TRAVIS and BOWSER (1986 a). These specimens were then mounted in glycerol containing 4% n-propyl gallate and examined with Zeiss IM-35 epifluorescence optics. Micrographs were taken on Kodak Tri-X film which was push processed (ASA 1200) with Diafine.

2.4. Silicone Rubber Substrata and Drug Studies

Thin sheets of heat cross-linked silicone rubber (30,000 cP polydimethylsiloxane, Dow Corning) were prepared on 22 mm^2 glass coverslips (Harris *et al.* 1980). The coated coverslips were tacked to the bottom of 60 mm dia. petri dishes by small dollops of plasticene. After filling the dishes with Ca^{2+}-depleted RSW, the forams were plated on the rubber sheets and observed using darkfield or phase contrast inverted optics. Micromanipulations were performed by hand. For the drug experiments, 1 mM colchicine or 10–20 µM cytochalasin B (final concentrations) were added to the petri during observation. The experiments were recorded on videotape. Videomicrographs were obtained either by photographing the images from the display monitor or with a Polaroid Freeze Frame Video Recorder (Polaroid Corporation, Cambridge, MA).

3. Results

3.1. Morphology of *Amphisorus* Reticulopodia

Within 1–3 hours after plating, *Amphisorus* extended a radial array of reticulopodia (Fig. 1 *a*) which originated from apertures distributed along the test margin (Fig. 1 *b*). Although frequently attaining lengths over 1 cm, the motility and ultrastructure of *Amphisorus* pseudopodial networks were typical of all forams studied to date (Anderson and Bé 1979). Therefore we will only illustrate features pertinent to this discussion and refer the reader elsewhere for recently published detailed descriptions of other reticulopodia (*e.g.*, Travis and Allen 1981, Travis *et al.* 1983, Koonce *et al.* 1986, Travis and Bowser 1988).

TMR-phalloidin localized filamentous actin to sites where pseuodpodial trunks initially contacted the substrate (Fig. 2 *a, b*) (see also Travis and Bowser 1986 a, b). In these regions, the first few serial longitudinal sections through the ventral aspect of the pseudopodia contained dense filamentous plaques corresponding to the TRM-phalloidin stained foci (Fig. 3 *a*). Elsewhere, the reticulopodia contained numerous bundles of cross-linked microtubules (Fig. 3 *b*) (but no plaques. In stereo high-voltage EM views of 0.25 µm sections, microtubules were frequently seen associated with the filament plaques (not shown) (see Bowser *et al.* 1988).

3.2. Pseudopodial Tension

The networks extended by *Amphisorus* on thin silicone rubber sheets appeared identical to those extended on glass or plastic surfaces. Traction forces generated by the reticulopodia could be visualized, often with the naked eye, by the production of wrinkles in the sheets (Fig. 4). The predominant wrinkle pattern appeared as tandem folds oriented perpendicular to the pseudopodia's long axes (*i.e.*, the compression wrinkles described by Harris and coworkers 1980). Compression wrinkles formed beneath and behind the sites of substrate attachment, and these were stretched into characteristic chevron shapes as the pseudopodia continued to pull against against the adhesion sites (Fig. 4). Additional traction further deformed the sheet, and severe deformation resulted in the production of long radial tension wrinkles (not shown) in areas of the sheet outlying the attachment sites. Tension wrinkles form in response to the tension generated in the stretched sheet, their folds are oriented parallel to the axes of tension created, and hence are also parallel to the long axes of the pseudopodia (Harris *et al.* 1980).

A given pseudopodium wrinkled the rubber sheet over a period of several minutes, and the wrinkles typically persisted for 2 to 25 min. The particular pattern of wrinkles at an adhesion site changed with time, with new wrinkles forming as the pseudopodia continued to pull on the rubber substrate. Occasionally, a wrinkle relaxed momentarily, before being pulled back to its original shape.

Whereas only a subset of the pseudopodia produced wrinkles, it was possible to correlate the location of the wrinkles a specimen's motile history. In locomoting specimens, intense compression wrinkles were formed where the leading pseudopodia contacted the substratum. Occasionally wrinkles were also seen in the trailing pseudopodia, as might be expected if they were pulled by the advancing cell body. Stationary forams also generated wrinkles; in such cases the wrinkles were associated with the numerous radially distributed pseudopodia.

To test whether traction forces developed within the reticulopodium depended on pseudopodial continuity between the cell body and the point of cell-substrate

Fig. 2. Corresponding phase contrast (*a*) and epifluorescence (*b*) micrographs of TMR-phalloidin stained *Amphisorus* reticulopodia, demonstrating discrete actin filament plaques occurring at sites where converged trunk pseudopodia contact the substratum. Magnification: ca. 2,300 ×

Fig. 3. Transmission electron micrographs of *Amphisorus* reticulopodia. (*a*) Thin section through the ventral surface of a pseudopodium, at a region of cell-substrate attachment near the cell body. Actin adhesion plaques appear as an electron-dense filamentous matrix. Pseudopodia sectioned more dorsally, or distal to the cell body (*b*), contain abundant microtubules, but no filament plaques. Periodic crossbridges linking adjacent microtubules, and microtubules to the organelles and plasma membrane, are clearly visible, even in this survey micrograph. Magnification: approximately 6,000 ×

Fig. 4. Compression wrinkles under a trunk pseudopodium, localizing the site where traction force is applied to the substrate. Magnification: 460 ×

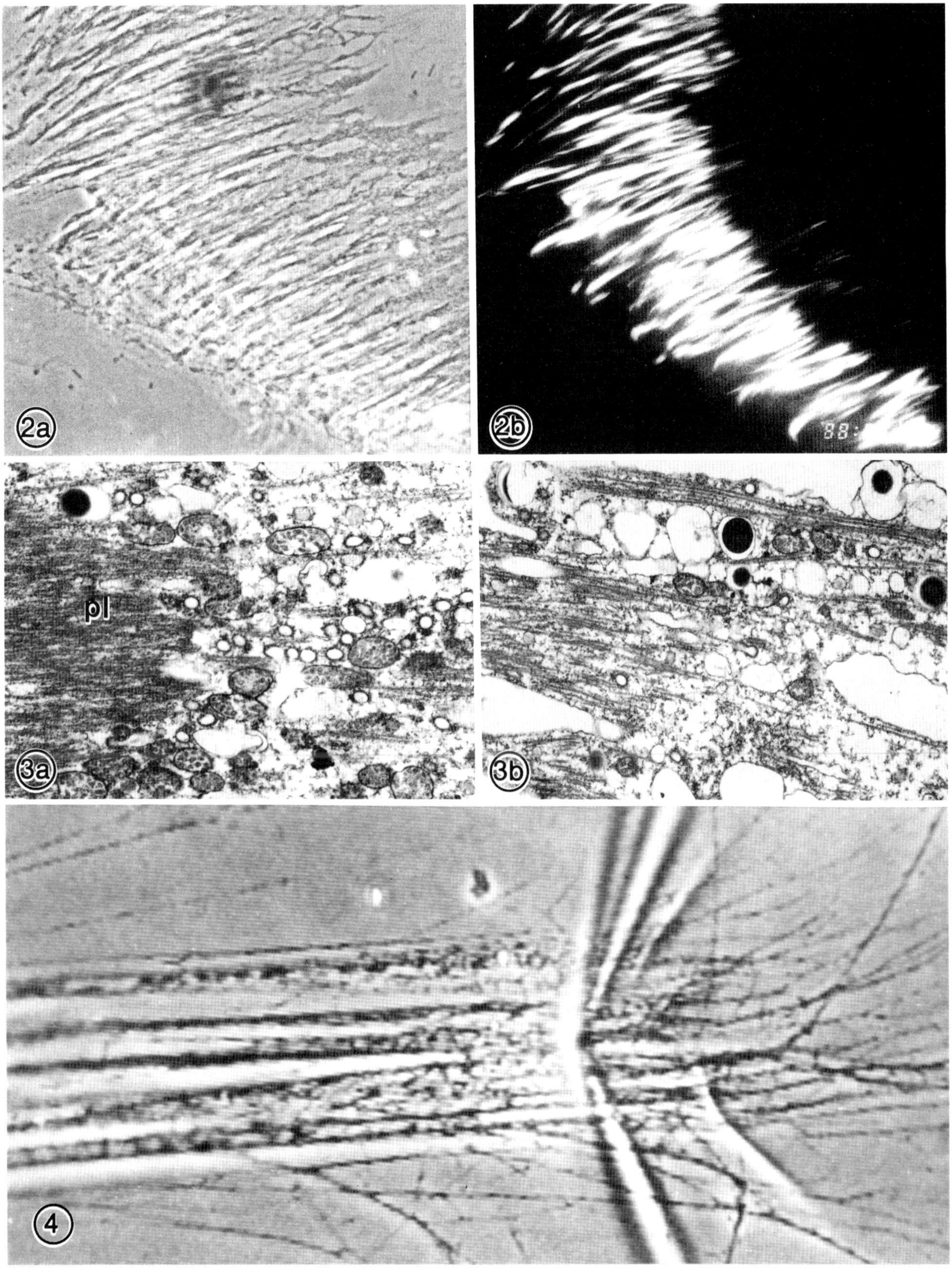

Figs. 2–4

adhesion, we cut tensed pseudopodia. As can be seen in Fig. 5, microsurgical severing of a pseudopodium results in the rapid loss of both compression and tension wrinkles from the silicone rubber substrate (Fig. 5 *a*–*d*). Pulling on the severed pseudopodium with a microneedle caused wrinkles to reappear at approximately the same locations, demonstrating that pseudopod adhesion is unaffected by the amputation (Fig. 5 *d*).

3.3. Cytoskeleton Inhibitors Disrupt Pseudopodial Tension

It is well established that in allogromiid forams, anti-microtubule agents (*e.g.*, colchicine, nocodazole) disassemble reticulopodial microtubules; concomitantly, cytoplasm collects at branch points in the network and within 10 minutes all motility stops (see TRAVIS and BOWSER 1985 a). Exposing *Amphisorus* to 1 mM colchicine resulted in the same sequence of morphological changes, culminating in the arrest of motility. Following colchicine treatment the pseudopodia appeared flaccid and were readily flailed by the slightest fluid motion, further indicating that reticulopodial microtubules had disassembled.

In the experiment illustrated in Fig. 6, 1 mM colchicine causes the compression wrinkles to relax within 3 min 20 sec. This release of tension occurred while motility was lost and gross morphological changes became apparent. After washing out the colchicine, normal pseudopodial shape and function returned and, in some cases, wrinkles reappeared at their original sites. Together these observations suggest that pseudopodial tension, but not cell-substrate adhesion, depends on an intact microtubule cytoskeleton.

Cytochalasins are fungal metabolites which disrupt or disassemble actin filament arrays in diverse cell types (MCCLEAN-FLETCHER and POLLARD 1980, BROWN and SPUDICH 1981, SCHLIWA 1982). In allogromiid forams, cytochalasins B and D induce the loss of actin-containing adhesion plaques and the subsequent coalescence of pseudopodial cytoplasm into large, irregularly-shaped bodies that display rapid but confused motility (TRAVIS and BOWSER 1986 a). Identical changes in reticulopodial morphology and motility were seen when *Amphisorus* was treated with 1–2×10^{-5} M cytochalasin B. As shown in Fig. 7, wrinkles in the silicone rubber substrate gradually relaxed as the pseudopodia began to coalesce into cytoplasmic droplets.

4. Discussion

The present study indicates that tensile forces, of sufficient magnitude to be detected by the rubber sheet method, are typically applied to the substrate at points where trunk pseudopodia initially contact the substrate. The function of this pseudopodial tension is only partially clear. When crawling across a solid surface, benthic forams, like most cells, appear to firmly attach portions of their pseudopodia to the substratum and subsequently apply a rearward traction force (ARNOLD 1953, HARRIS *et al.* 1980, 1981). In order for forams with radially-dispersed reticulopodia to move, this rearward traction force must be generated vectorially. Such vectorial patterns of force production have been observed. At other times, however, tension was generated simultaneously in radially opposed pseudopodia—a condition which is unlikely to effect locomotion, but may instead help to anchor the foram to the substrate. In nature forams are frequently exposed to bioturbation and hydrodynamic forces. The radial generation of pseudopodial tension and concomitantly enhanced "grip" would have significant adaptive value by resisting such forces. Time lapse studies will be necessary to correlate more fully the pseudopodial tension with locomotion.

Traction forces may also be involved in phagocytic processes in forams. For example, the distal pseudopodia in some forams are able to rip apart soft food materials, a rending activity probably mediated by locally-generated pseudopodial tension (BOWSER 1985). Related activities may enable forams to disperse clumps of algal prey, or crawl through sediment.

Pseudopodial tension may also be involved in reticulopodial morphogenesis and shape maintenance. For example, ALLEN (1964) noted that tension between *Allogromia*'s cell body and pseudopod attachment sites

Fig. 5. Pseudopodial tension is lost upon severing. In (*a*) a trunk reticulopodium has pulled against the rubber sheet at two adhesion sites. The tension wrinkles largely disappear (*b*, *c*) during amputation of the reticulopod with a glass microneedle. Pulling the proximal portion of the excised pseudopod (*d*) produces new wrinkles at the original sites of pseudopod-substrate adhesion. Magnification: approximately 100 ×

Fig. 6. Colchicine disrupts *Amphisorus* reticulopodia and releases pseudopodial tension. (*a*) 2 min 23.5 sec after the administration of 1 mM colchicine the wrinkle pattern is virtually indistinguishable from that observed before adding the drug (not shown). By 2 min 48 sec after administration of the drug (*b*), a marked diminuation of tension is evidenced by the reduced size and apparent displacement of the wrinkle. The wrinkle is completely gone by 3 min 16 sec after the addition of the drug (*c*). All pseudopodial motility and intracellular movement has stopped by 4 min 21 sec (*d*). Magnification: ca. 125 ×

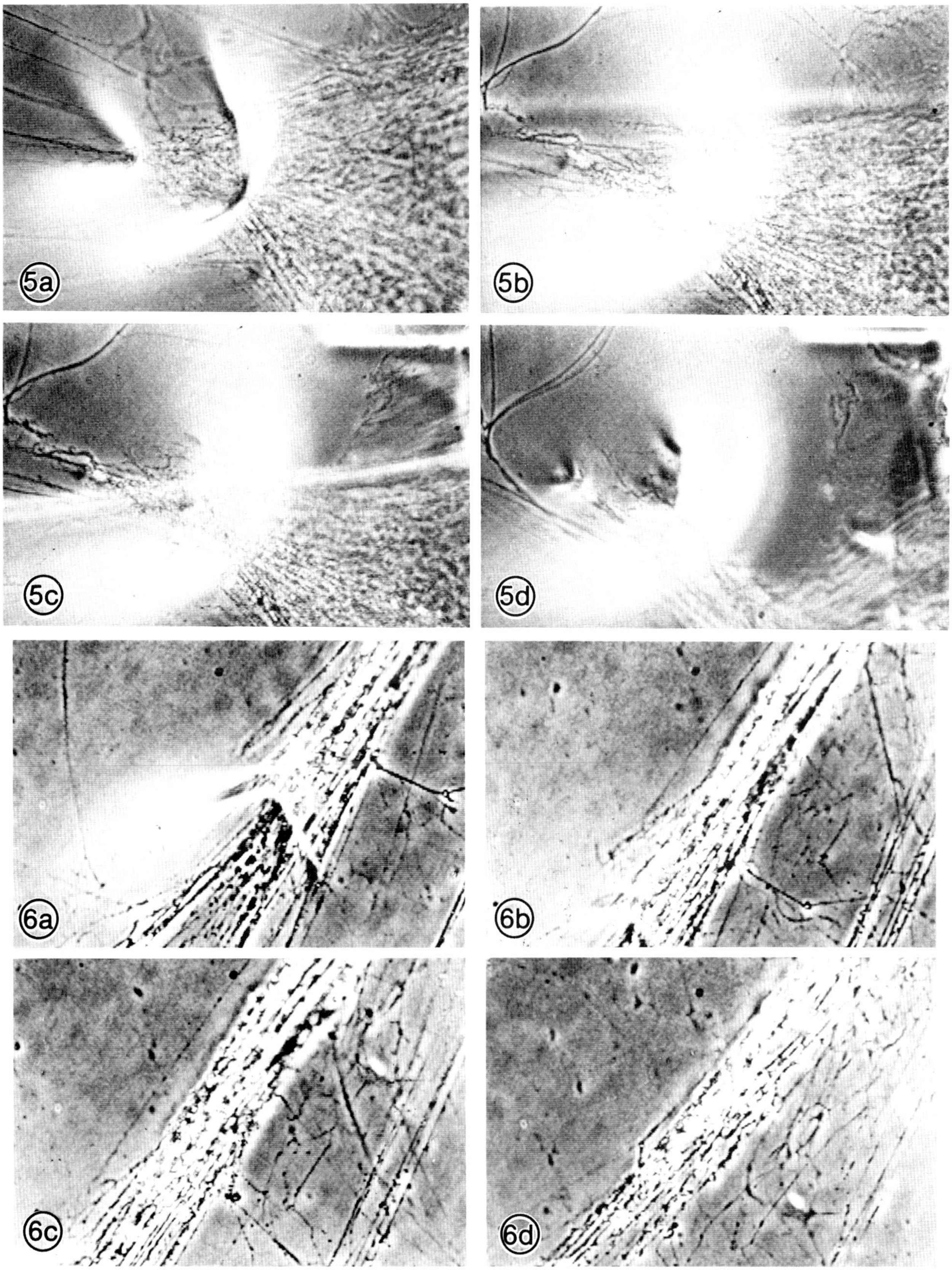

Figs. 5–6

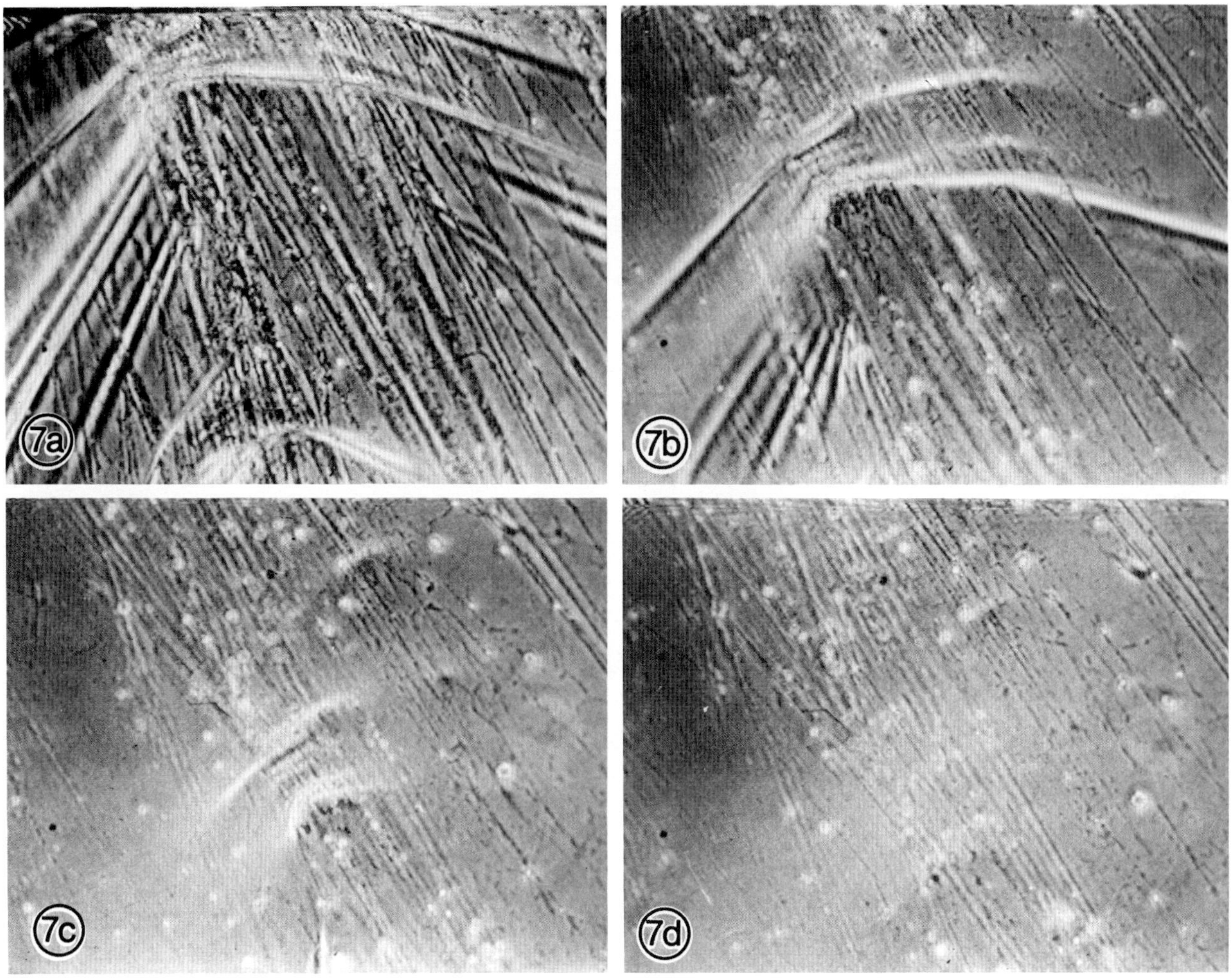

Fig. 7. Treatment of *Amphisorus* reticulopodia with 10 μM cytochalasin B (CB) results in the loss of pseudopodial tension and cell adhesion. (*a*) shows a portion of an *Amphisorus* displaying massive wrinkles. (*b*) By 2 min after the addition of CB, there is a considerable loss of tension, evidenced by the relaxation of the stretched sheet. The loss of tension continues (*c*), and the wrinkles completely disappear by 4 min 40 sec. In (*d*), a characteristic CB poisoned network is evident at an elapsed time of 6 min 30 sec. Magnification: ca. 105 ×

leads to the formation of branch pseudopodia. More recently, TRAVIS and BOWSER (1986 b) suggested that changes in the distribution of pseudopodial tension initiates the dramatic structural rearrangements accompanying the formation of "reticulopodial satellites" from excised portions of the network (ARNOLD 1953, JAHN and RINALDI 1964). However, our present results, and related findings with *Allogromia* (TRAVIS and BOWSER unpublished) have shown that wrinkles in the flexible sheets typically formed where trunk pseudopodia initially contact the substratum, not throughout the pseudopodial network, as might be expected if reticulopodial tension was widespread. These observations do not rule out the possibility that weaker forces, undetected by the rubber sheet method, are present throughout the network.

Although the sites where traction force is applied to the substrate have been identified as proximal pseudopod-substrate adhesion sites, the mechanism of this force production remains in doubt. Our pharmacological studies strongly suggest that pseudopodial tension depends on intact microtubules and actin-containing filament plaques. Electron microscopy provides evidence for a structural interaction between these two filament systems (see also BOWSER et al. 1988). Given these observations, at least two models for the generation of pseudopodial tension can be proposed. In the first, tension is generated by microtubule sliding, presumably through the agency of their lateral crossbridges. (Evidence for sliding microtubules in the reticulopodia has recently been reviewed in TRAVIS and BOWSER 1988.) This force would then be transmitted to the substrate by the subset of microtubules structurally interacting with (*i.e.*, tethered to) the actin-containing adhesion plaques. In this regard, the plaques functionally resemble the z-lines of striated muscle or dense bodies of smooth muscle.

Alternatively, contraction of the actin-containing plaques, in association with an as yet unidentified myosinlike mechanoenzyme, could generate tension along the attached microtubules. In this case microtubules would serve as simple skeletal elements that pull the cell body. An analogous mechanism has been proposed for the locomotion of other cells, *e.g.*, the shelled amoeba *Difflugia* (WOHLMAN and ALLEN 1968, ECKERT and MCGEE-RUSSELL 1973). In this context, one might compare the actin-containing filament plaques of the ventral pseudopodial surface to the stress fibers of vertebrate tissue cells and predict that their contraction would lead to the development of compression wrinkles immediately beneath the adhesion plaques. However, our micromanipulation studies indicate that the comparatively large adhesion plaques themselves do not deform the rubber sheet, but do allow the transmission of an externally applied force against the substratum.

The question of how reticulopodial tension is generated must await the application of methods such as microinjection of specific antibodies to the various mechanoenzymes, to rule out one mechanochemical system or the other. Finally, our studies address the issue of how benthic forams locomote across solid surfaces. In nature, these organisms frequently crawl along or through fine, loose sediment (LEE *et al.* 1969, SEVERIN and ERSKIAN 1981, SEVERIN 1987). It remains to be seen whether a different locomotory mechanism is employed under these conditions.

Acknowledgements

We thank Dr. ALBERT K. HARRIS for demonstrating his rubber sheet method. This work was supported by grants from the National Science Foundation (DCB 85-02875) and the National Institutes of Health (GM-38455) to J.L.T. The Wadsworth Center's high-voltage electron microscope facility is supported by NIH Biotechnology Resource Grant PHS 01219. S.S.B. is an American Heart Association, New York State Affiliate Research Fellow.

References

ALLEN RD (1964) Cytoplasmic streaming and locomotion in marine foraminifera. In: ALLEN RD, KAMIYA N (eds) Primitive motile systems in cell biology. Academic Press, New York, pp 407–431

— ALLEN NS, TRAVIS JL (1981) Video-enhanced contrast differential interference contrast (AVEC-DIC) microscopy: A new method capable of analyzing microtubule related movement in the reticulopodial network of *Allogromia laticollaris*. Cell Motil 1: 291–302

ANDERSON OR, BÉ AWH (1979) Recent advances in foraminiferal fine structure research. In: HEDLEY RH, ADAMS CG (eds) The Foraminifera, vol 2. Academic Press, New York, pp 121–202

ARNOLD ZM (1953) An introduction to the study of movement and dispersal in *Allogromia laticollaris* Arnold. Contrib Cushman Found Foraminif Res 4: 15–21

BOWSER SS (1985) Invasive activity of *Allogromia* pseudopodial networks: Skyllocytosis of a gelatin/agar gel. J Protozool 32: 9–12

— RIEDER CL (1985) Evidence that cell surface motility in *Allogromia* is mediated by cytoplasmic microtubules. Can J Biochem Cell Biol 63: 608–620

— TRAVIS JL, RIEDER CL (1988) Microtubules associate with actin-containing filaments at discrete sites along the ventral surface of *Allogromia* reticulopods. J Cell Sci 89: 297–307

— SPUDICH JA (1981) Mechanism of action of cytochalasin: Evidence that it binds to acin filament ends. J Cell Biol 88: 487–491

ECKERT BS, MCGEE-RUSSELL SM (1973) The patterned organization of thick and thin microfilaments in the contracting pseudopod of *Difflugia*. J Cell Sci 13: 727–739

HARRIS AK, WILD P, STOPAK D (1980) Silicone rubber substrata: A new wrinkle in the study of cell locomotion. Science 208: 177–179

— STOPAK D, WILD P (1981) Fibroblast traction as a mechanism for collagen morphogenesis. Nature 290: 249–251

JAHN TL, RINALDI RA (1959) Protoplasmic movement in the foraminiferan, *Allogromia laticollaris*; and a theory of its mechanism. Biol Bull (Woods Hole) 117: 100–118

KOONCE MP, EUTENEUER U, MCDONALD KL, MENZEL D, SCHLIWA M (1986) Cytoskeletal architecture and motility in a giant fresh water amoeba, *Reticulomyxa*. Cell Motil Cytoskel 6: 521–533

LEE JJ, MULLER WA, STONE RJ, MCENERY ME, ZUCKER W (1969) Standing crop of foraminifera in sublittoral epiphytic communities of a Long Island salt marsh. Marine Biol 4: 44–61

MACCLEAN-FLETCHER S, POLLARD TD (1980) Mechanisms of the action of cytochalasin B on actin. Cell 20: 329–341

SCHLIWA M (1982) Action of cytochalasin D on cytoskeletal networks. J Cell Biol 92: 79–91

SEVERIN KP, ERSKIAN MG (1981) Laboratory experiments on the vertical movement of *Quinqueloculina impressa* Reuss through sand. J Foraminiferal Res 11: 133–136

— (1987) Laboratory observations on the rate of subsurface movement of a small miliolid foraminifer. J Foraminiferal Res 17: 110–116

TRAVIS JL, ALLEN RD (1981) Studies on the motility of the foraminifera. I. Ultrastructure of the reticulopodial network of *Allogromia laticollaris* (Arnold). J Cell Biol 90: 211–221

— KENEALY JFX, ALLEN RD (1983) Studies on the motility of the foraminifera. II. The dynamic microtubular cytoskeleton of the reticulopodial network of *Allogromia laticollaris*. J Cell Biol 97: 1668–1676

— BOWSER SS (1986 a) A new model of reticulopodial motility and shape: evidence for a microtubule-based motor and an actin skeleton. Cell Motil Cytoskel 6: 2–14

— — (1986 b) Microtubule-dependent reticulopodial motility: Is there a role for actin? Cell Motil Cytoskel 6: 146–152

— — (1987) Pseudopodial tension in foraminifera. J Cell Biol 105: 129 a

— — (1988) Optical approaches to the study of foraminiferan motility. Cell Motil Cytoskel (in press)

WOHLMAN A, ALLEN RD (1968) Structural organization associated with pseudopod extension and contraction during cell locomotion in *Difflugia*. J Cell Sci 3: 105–114

Protoplasma (1988) [Suppl 1]: 72–80

Change of Cytosolic Ca-Ion Concentration in the Contraction and Relaxation Cycle of *Physarum* Microplasmodia

R. KURODA[1], S. HATANO[2,]*, Y. HIRAMOTO[3], and H. KURODA[1]

[1] Sugashima Marine Biological Laboratory and [2] Department of Molecular Biology, Faculty of Science, Nagoya University, Nagoya, and [3] National Institute for Basic Biology, Okazaki

Received February 8, 1988
Accepted March 19, 1988

Dedicated to Professor Dr. NOBURO KAMIYA on the occasion of his 75th birthday

Summary

Cytosolic free Ca^{2+} concentration, $[Ca^{2+}]_i$, in microplasmodia of *Physarum polycephalum* showing relatively synchronized contraction and relaxation cycles, was measured by a highly fluorescent Ca^{2+}-sensitive dye, fura-2. Fura-2 dissolved in a solution containing Mg-ATP and sorbitol was directly injected into microplasmodia, and easily diffused into the entire cytosol without being sealed off by the newly formed membrane. $[Ca^{2+}]_i$ oscillated principally around 300 nM with a frequency similar to that of the contraction and relaxation cycle. $[Ca^{2+}]_i$ was revealed to be high in the contracted phase of microplasmodia and low in the relaxed phase, although this clear relation endured for only several cycles.

Artificial elevation of free Ca^{2+} concentration, $[Ca^{2+}]$, of the bathing solution between pCa 8 and 6 brought about the contraction of saponin and EGTA-permeabilized microplasmodia in the presence of ATP. This ensemble of results strongly suggests that in living *Physarum* microplasmodia, the rise in $[Ca^{2+}]_i$ induces the contraction of the actomyosin system.

Keywords: Physarum polycephalum; $[Ca^{2+}]_i$; Fura-2; Contraction and relaxation cycles; Saponin and EGTA-permeabilized microplasmodia.

Abbreviation: EGTA ethylene glycol bis(β-aminoethyl ether)-N,N,N′,N′-tetraacetic acid

1. Introduction

Plasmodia of the slime mold *Physarum polycephalum* exhibit an active shuttle streaming of the endoplasm, of which direction changes every few minutes (KAMIYA 1959). The shuttle streaming is induced by the periodic contraction and relaxation of the actomyosin system residing in the ectoplasm (KAMIYA and KURODA 1958, KAMIYA 1979, 1986, NAGAI *et al.* 1978, YOSHIMOTO and KAMIYA 1978, ISHIGAMI *et al.* 1981, ISHIGAMI 1986, ISHIGAMI *et al.* 1987, WOHLFARTH-BOTTERMANN 1964, ALLERA *et al.* 1971, FLEISCHER and WOHLFARTH-BOTTERMANN 1975, WOHLFARTH-BOTTERMANN and FLEISCHER 1976, NAIB-MAJANI *et al.* 1984, STOCKEM *et al.* 1987). $[Ca^{2+}]_i$, which seems to be one of the important regulatory factors of the actomyosin system, also, has been shown to oscillate with a similar frequency to that of the motile oscillations of plasmodia (RIDGWAY and DURHAM 1976). However, it has not yet been determined whether there is a positive or a negative correlation between the rise in $[Ca^{2+}]_i$ and the contraction. Mutually contradictory results have been reported as follows.

The first observation related to this problem is that the cytoplasmic movements of the plasmodial droplets produced by incubation in a caffeine solution can be induced by about 1 μM Ca^{2+} in the outer medium (HATANO 1970, HATANO and OOSAWA 1971). This was confirmed using droplets permeabilized for Ca^{2+} by a brief treatment with saponin (SATO *et al.* 1981). Electron microscopical visualization of Ca in plasmodia has indicated the Ca release from vacuoles into the cytosol at each contraction and the sequestration into vacuoles during relaxation (KURODA and KURODA 1982). Experiments of the endoplasm replacement with Ca^{2+} buffers (UEDA *et al.* 1978) and of the effects of Ca^{2+} on the torsional movements of Triton-permeabilized plasmodial strands (OGIHARA 1982), also indicate that

* Correspondence and Reprints: Department of Molecular Biology, Faculty of Science, Nagoya University, Furo-cho, Chikusa-ku, Nagoya 464, Japan.

the $[Ca^{2+}]_i$ increment probably causes the contraction of the actomyosin system in *Physarum* plasmodia. On the other hand, the effect of Ca^{2+} on isolated myosin and other regulatory proteins has been shown to relate directly to inhibition of the actin-myosin interaction. Thus, the F-actin severing factor, fragmin, induces the tension reduction of actomyosin threads at 1–10 μM $[Ca^{2+}]$ (SUGINO and MATSUMURA 1983), an inhibitory factor for the myosin-actin-ATP interaction requires Ca^{2+} for its activity (KOHAMA 1981, OGIHARA *et al.* 1983), and Ca^{2+}-binding of the myosin light chain also inhibits the above interaction (KOHAMA *et all* 1985 a, b, 1986), which occurs at the $[Ca^{2+}]$ level of 1 μM.
Another line of experiments has also been interpreted by the inhibitory role of Ca^{2+} for the contraction of plasmodia. Ca^{2+} efflux from the Ca^{2+}-ionophore-treated plasmodial strand falls in the ascending phase of tensile force and rises in the descending phase (YOSHIMOTO *et al.* 1981). Glycerol or dimethylsulfoxide-extracted endoplasmic drops undergo successive cycles of contraction at 100 nM and relaxation at 10 μM of $[Ca^{2+}]$ (ACHENBACH and WOHLFARTH-BOTTERMANN 1986 a, b). Here is an interesting and suggestive experimental result. Optimal $[Ca^{2+}]$ exists for developing the tensile force in saponin-permeabilized plasmodial strands. Hence, Ca^{2+} functions as an activator below 100 nM, and conversely as an inhibitor above that (YOSHIMOTO and KAMIYA 1984).
Almost all of the experiments cited have been performed on a system which could be called an artificial cell model in some respects, or on isolated proteins under an artificial environment. Moreover, none have directly measured $[Ca^{2+}]_i$. For any discussion of the role of Ca^{2+} in the contraction of the actomyosin system of living plasmodia, knowledge of the absolute value of $[Ca^{2+}]_i$ is prerequisite.
Then, we attempted to measure $[Ca^{2+}]_i$ using a fluorescent Ca^{2+}-sensitive dye, fura-2 (GRYNKIEWICZ *et al.* 1985) which could determine $[Ca^{2+}]_i$ from the ratio of its fluorescence intensities at two excitation wavelengths independently of dye concentration, and to clarify the relation between $[Ca^{2+}]_i$ change and the contraction of plasmodia. In the experiments reported below, spherical microplasmodia smaller than 100 μm in diameter were used. The advantage of such microplasmodia is that different regions within one plasmodium perform the relatively synchronized periodic contraction and relaxation, and consequently the interpretation of experimental results could presumably be facilitated. This synchronousness might be attributed to the tiny size and the simple organization of the microfilaments (BRIX *et al.* 1987, ISHIGAMI *et al.* 1987). In order to elucidate a causal relation between $[Ca^{2+}]$ change and the contraction, the effect of artificial elevation of $[Ca^{2+}]$ upon the contractility of microplasmodia permeabilized for Ca^{2+} and ATP with saponin and EGTA was examined in addition to the measurement of $[Ca^{2+}]_i$.

2. Materials and Methods

2.1. *Plasmodia*

For measurement of $[Ca^{2+}]_i$ with fura-2, a white mutant of *Physarum polycephalum* was chosen because it showed less autofluorescence than the wild type strain, and for saponin and EGTA treatments, a colonial strain was employed because of its toughness. Microplasmodia were obtained by vigorous shaking in a semi-defined liquid medium containing 1% bacto-tryptone (DIFCO), 0.15% yeast extract (DIFCO), 1% D-glucose, 50 μg/ml hemin, 11.4 mM Na-citrate, 15 mM KH_2PO_4, 4.1 mM $CaCl_2$, 2.4 mM $MgSO_4$, 0.3 mM $FeSO_4$,

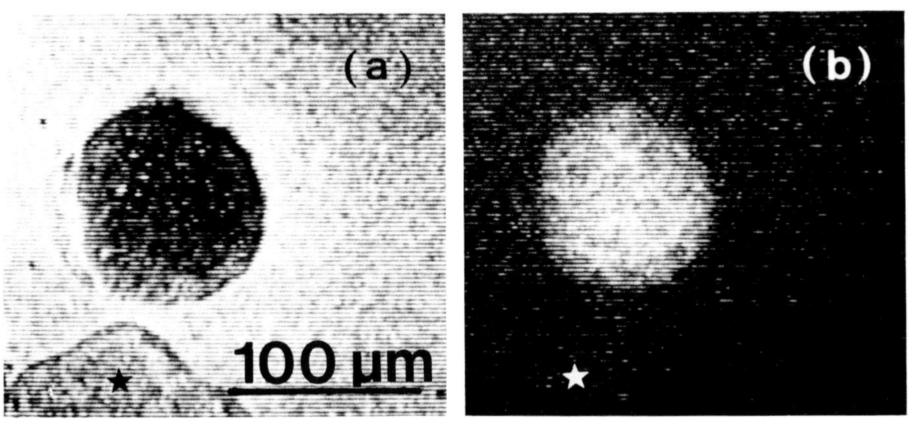

Fig. 1. Fura-2-injected microplasmodium. Images of the ordinary bright field (*a*) and fluorescence (*b*). Fura-2 diffused into the entire region of microplasmodium. Note the autofluorescence of another microplasmodium (★) was negligible

0.42 mM $MnCl_2$, 0.12 mM $ZnSO_4$, pH 4.5 at 18 °C, and used at the log-phase of their proliferation. A few hours before use, microplasmodia were harvested by a brief centrifugation, washed with a solution roughly equiosmolal to the plasmodial cytosol (1 mM KCl, 0.5 mM $CaCl_2$, 2 mM $MgCl_2$, 80 mM mannitol and 5 mM piperazine-N,N'-bis (2-ethane sulfonic acid) (PIPES)-NaOH buffer, pH 7.0), and shaken for about 1 hour. A small aliquot of suspension was placed on a thin agar sheet, covered with a thicker agar sheet, and let stand at room temperature for 1–2 hours until use. During this period microplasmodia became thinned to a thickness of 20–40 µm.

2.2. *Microinjection of Fura-2*

Because *Physarum* plasmodia seemed not to incorporate the acetomethyl derivative of fura-2, we sought to inject fura-2 intracellularly. Plasmodia, however, readily seal off the injected liquid by newly formed membranes, so that injection was maneuvered by dissolving fura-2 pentapotassium salt (Molecular Probes, Inc., Oregon, U.S.A.) in a solution consisting of 1.5 mM ATP (Na_3 salt), 4 mM $MgSO_4$, 200 mM sorbitol and 10 mM N-2 hydroxyethyl piperazine-N'-2-ethanesulfonic acid (HEPES)-KOH, pH 7.0. Thus, fura-2 easily diffused into the entire cytosol (Fig. 1). Microplasmodia were selected with a size smaller than 100 µm in diameter and with a shape as circular as possible. Fura-2 was injected to a final concentration of 40–200 µM in microplasmodia.

2.3. *Measurement of* $[Ca^{2+}]_i$

Fura-2-injected microplasmodia were mounted on a fluoromicroscope (BH-2, Olympus, Tokyo, Japan) equipped with 340 and 365 nm interference filters of 5–10 nm band width (Koshin Kogaku Industry, Kanagawa, Japan) and connected to an ultra-high sensitive camera (C-1966-20, Hamamatsu Photonics, Hamamatsu, Japan). Neutral density filters were used for reduction of the intensity of the 365 nm excitation light to a level comparable to that of the 340 nm light and reduction of the ordinary white light, and a dark green filter was used for elimination of emission light longer than 600 nm. Fura-2-injected microplasmodia were illuminated for 3 seconds each alternately with the 340 nm light, 365 nm light and ordinary white light every 20 seconds. Usually, 10–20 minutes passed from the injection to the beginning of a series of illuminations.

The 365 nm light had to be chosen instead of the 380 or 385 nm light which was recommended in the original papers on fura-2 (POENIE *et al.* 1985, GRYNKIEWICZ *et al.* 1985), because even a white mutant possessed the strong autofluorescence excitable with the light of longer wave length than 360 nM. Under the employed conditions, the autofluorescence of microplasmodia was negligible (Fig. 1). Raw images of fluorescence and of the ordinary bright field were recorded by a Sony U-matic videotape recorder. The total fluorescence which single microplasmodia emitted during four-thirtieth second was measured after background subtraction at each illumination with a two-dimensional photon counting system of Hamamatsu Photonics (VIM C-1966). The ratio of fluorescence from 340 nm excitation to that from 365 nm excitation was determined at each set of illuminations. Under the employed experimental condition except the absence of microplasmodia and a thin agar sheet, the relation between $[Ca^{2+}]$ and the 340 nm/365 nm fluorescence ratio was shown in Fig. 2. $[Ca^{2+}]_i$ was calculated according to the equation of GRYNKIEWICZ *et al.* (1985),

$$[Ca^{2+}] = K_d\left(\frac{R-R_{min}}{R_{max}-R}\right)\left(\frac{Sf_2}{Sb_2}\right)$$

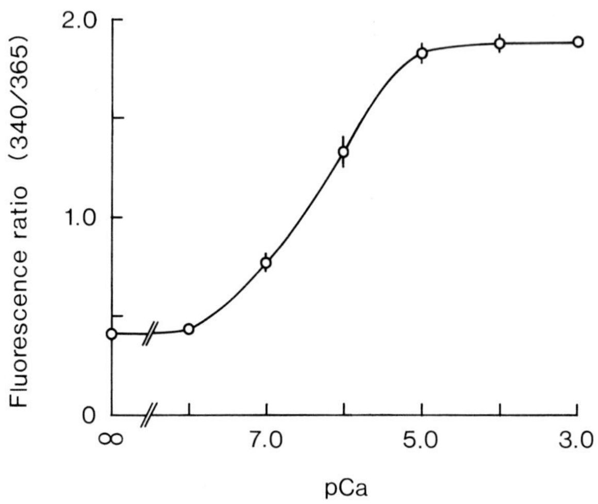

Fig. 2. Relationship between $[Ca^{2+}]$ and fura-2 fluorescence ratio. The relation between $[Ca^{2+}]$ and the 340 nm/365 nm fluorescence ratio of fura-2 was determined with small volumes of calibrating solution containing 20 µM fura-2 (K_4 salt), an appropriate amount of $MgCl_2$ to give 1 mM free Mg^{2+} concentration, an appropriate amount of $CaCl_2$ to give an appropriate pCa in 10 mM EGTA, 100 mM KCl and 10 mM HEPES-KOH, pH 7.0. The apparent Kd of EGTA for Ca^{2+} in this solution was taken as 340 nM. Microscopic conditions and experimental procedures were the same as those in the microplasmodia experiments. Kd of fura-2 for Ca^{2+}, calculated according to the equation of GRYNKIEWICZ *et al.* (1985), was determined to be 210 nM from cited data. Vertical bars represent SD

where R was the measured 340 nm/365 nm ratio, R_{max} and R_{min} were the ratio at saturating $[Ca^{2+}]$ and at zero $[Ca^{2+}]$, respectively, (Sf_2/Sb_2) was the ratio of 365 nm fluorescence at zero $[Ca^{2+}]$ to that at saturating $[Ca^{2+}]$, and K_d was the dissociation constant of fura-2 for Ca^{2+}. The K_d was determined to be 210 nM from the data of Fig. 2.

2.4. *Measurement of Contraction*

Contour of images of microplasmodia under the bright field was traced on a monitor screen with a truckball connected to a microcomputer, and the areas of microplasmodia were computed. Since the apparent area of microplasmodia becomes smaller with the progress of contraction, this area can be the measure of contraction while being overlaid with a coverglass. Microplasmodia which were revealed to show relatively regular oscillations of the area in amplitude and period were selected as data. Microplasmodia with poorly synchronized contraction and relaxation exhibited irregular oscillations of the area.

2.5. *Permeabilization of Microplasmodia*

Microplasmodia adhering to an agar sheet prepared as in 2.1 were incubated in a solution consisting of 30 µg/ml saponin (E. Merck, Darmstadt, Federal Republic of Germany), 40 mM K, 5 mM Mg, 15 mM Cl, 1 mM dithiothreitol, 100 mM sorbitol, 5 mM Ca^{2+}-EGTA buffer, pCa 4, and 10 mM PIPES, pH 7.0 for 30 min at 25 °C, and then washed three times with the above solution lacking saponin. No living microplasmodia were seen after this treatment. EGTA was indispensable for permeabilizing microplasmodia with saponin, and

Ca^{2+} was essential at a concentration higher than 10 µM to prevent disruption of plasmodia during treatments.
Solutions for examining the effect of Ca^{2+} on the contractility contained the same constituents as the above solution except there was no saponin, and 1 mM ATP (Na_3 salt) and an appropriate amount of $CaCl_2$ were added to give appropriate $[Ca^{2+}]$. $[Ca^{2+}]$ was calculated employing a value of 340 nM as Kd of EGTA for Ca^{2+} in this solution at 25 °C. PH was strictly adjusted in a final solution, and pH was ascertained not to change by mixing the solutions of different $[Ca^{2+}]$. Solution change was accomplished by introduction and withdrawal of the test solution of more than 30 volumes of the space formed by a coverglass and an agar sheet, while monitoring under a microscope. The area of permeabilized microplasmodia was measured before the solution change and 5–6 minutes after that as described in 2.4.

3. Results

3.1. Relation Between $[Ca^{2+}]_i$ *and Contraction*

Small spherical microplasmodia of *Physarum polycephalum* obtained by shaking culture in liquid media are thought to be at the early stage of the differentiation into larger microplasmodia of more complex shape according to BRIX *et al.* (1987). For example, the differentiation degree of the cell surface morphology and the cortical microfilament organization has been shown to be generally size- and age-dependent. In fact, round microplasmodia spread on an agar sheet, if left as they were, soon changed to an oval shape and finally to a multi-branched irregular shape. In the round shape stage, if microplasmodia were smaller than 100 µm in diameter, they exhibited almost no endoplasmic streaming, which indicated that the contraction and relaxation cycle was repeated almost synchronously within one microplasmodium, as reported on plasmodial fragments by ISHIGAMI *et al.* (1987). The apparent area of such microplasmodia periodically decreased during contraction and increased during relaxation by an amplitude of 3–10% with a period of 70–120 seconds. The degree of contraction could be measured by this area change. On the other hand, irregular-shaped large microplasmodia showed the endoplasmic streaming and the irregular area change. For this reason, microplasmodia as circular as possible and smaller than 100 µm in diameter were selected in order to investigate the relation in phase between $[Ca^{2+}]_i$ oscillations and the contraction and relaxation cycle.
$[Ca^{2+}]_i$ was measured by a fluorescent Ca^{2+}-sensitive dye, fura-2, injected into microplasmodia. The ratio of fluorescence which one fura-2-injected microplasmodium emitted by excitation with the 340 nm light to that with the 365 nm light and the apparent area at 20-second intervals were determined. Calculated $[Ca^{2+}]_i$ was plotted against time together with the area change of the same microplasmodium in Fig. 3. $[Ca^{2+}]_i$ oscillated around 360 nM by an amplitude of about 50 nM with a period of about 70 seconds similar to that of the area change, although somewhat irregular in waveform and period. $[Ca^{2+}]_i$ changed in the counter direction to the area change except around 6–7 minutes. This result indicates that $[Ca^{2+}]_i$ is high in the contracted phase but low in the relaxed phase of microplasmodia, barring an occasional disorder.
In order to examine critically this tentative conclusion, the correlation was made between the maximal point of $[Ca^{2+}]_i$ and the minimal point of area, and between the minimal point of $[Ca^{2+}]_i$ and the maximal point of area for the initial 5 cycles of contraction and relaxation in each of 5 different microplasmodia. As

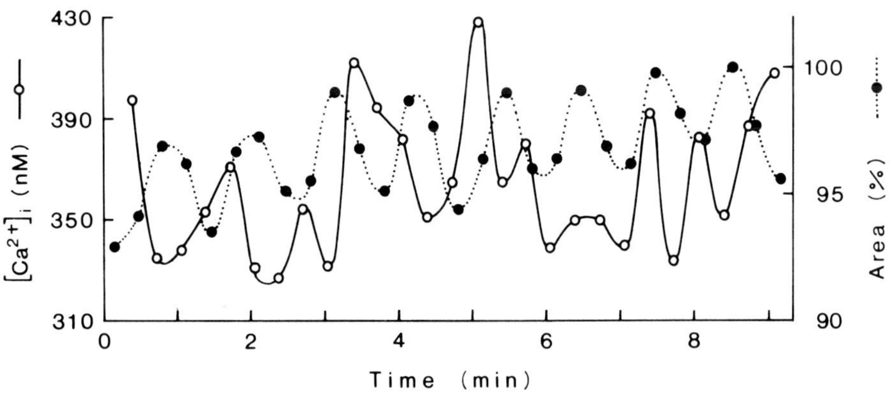

Fig. 3. Change of $[Ca^{2+}]_i$ in the contraction and relaxation cycle of microplasmodia. Fura-2 was injected to a final concentration of 200 µM into a microplasmodium 90 µm in diameter. The 340 nm/365 nm fluorescence ratio and the apparent area at 20-second intervals were determined. $[Ca^{2+}]_i$ was calculated according to the equation of GRYNKIEWICZ *et al.* (1985) and using as Kd 210 nM obtained from Fig. 2. Area is expressed in a percentage of the maximal area, and this represents the degree of contraction. Note high $[Ca^{2+}]_i$ in the phase of small area, and low $[Ca^{2+}]_i$ in that of large area except for 6–7 minutes

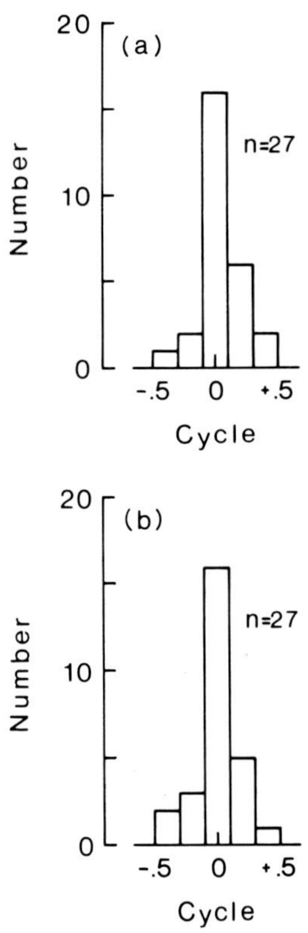

Fig. 4. Correlation between the maximal point of $[Ca^{2+}]_i$ and the minimal point of area (*a*) and between the minimal point of $[Ca^{2+}]_i$ and the maximal point of area (*b*). One cycle of contraction and relaxation is divided into 5 portions, taking the minimal point (*a*) or the maximal point (*b*) of area as zero. Number of maximal points (*a*) or minimal points (*b*) of $[Ca^{2+}]_i$ located in each portion was counted from the initial 5 cycles of contraction and relaxation in each of 5 different microplasmodia. Number of the peak (*a*) or the nadir (*b*) of $[Ca^{2+}]_i$ was larger than that of the contraction and relaxation cycle, because of the irregularity of $[Ca^{2+}]_i$ change. The negative sign represents the peak (*a*) or the nadir (*b*) of $[Ca^{2+}]_i$ ahead of the minimal point (*a*) or the maximal point (*b*) of area, and the positive one represents the peak (*a*) or the nadir (*b*) of $[Ca^{2+}]_i$ behind the minimal point (*a*) or the maximal point (*b*) of area. Mean and SD was 0.03 ± 0.17 (n = 27) in (*a*) and 0.00 ± 0.17 (n = 27) in (*b*)

shown in Fig. 4, a good correlation existed in each combination. This statistical analysis confirms that $[Ca^{2+}]_i$ is high in the contracted phase and consequently low in the relaxed phase of microplasmodia. However, because of the poor time resolution, the above-mentioned experiments could not determine whether or not the rise in $[Ca^{2+}]_i$ was ahead to trigger the contraction. This result, nevertheless, is not contradictory to the hypothesis that the rise in $[Ca^{2+}]_i$ induces the contraction of the actomyosin system in *Physarum* microplasmodia.

One more important result obtained from this series of experiments concerned the absolute value of $[Ca^{2+}]_i$. As for the apparent average $[Ca^{2+}]_i$, the value of 296 ± 65 (SD) nM was obtained form 23 measurements of each of 5 different microplasmodia (n = 115). This value approximated to that from characean algae measured with a Ca^{2+}-selective intracellular microelectrode (MILLER and SANDERS 1987). The magnitude of $[Ca^{2+}]_i$ oscillations was 43 ± 16 (SD) nM, which was measured with 5 waves of each of 5 different microplasmodia (n = 50).

3.2. Irregularity of $[Ca^{2+}]_i$ *Oscillations*

Oscillations in $[Ca^{2+}]_i$ were somewhat irregular in waveform and period as shown in Fig. 3. A similar irregularity of $[Ca^{2+}]_i$ oscillations has been mentioned in studies dealing with mouse B lymphocytes (TSIEN and POENIE 1986), and sheep cardiac cells (EISNER and VALDEOLMILLOS 1986). This phenomenon seems to be due to the spatial and temporal heterogeneity of $[Ca^{2+}]_i$ within single cells.

Then, we measured local $[Ca^{2+}]_i$ in several different regions of one-tenth area of the entire region, each. Figure 5 shows the local $[Ca^{2+}]_i$ change of 3 different regions together with the total $[Ca^{2+}]_i$ change and the area change. $[Ca^{2+}]_i$ oscillations of different regions were relatively well synchronized for the initial 4 minutes in this case, but finally produced differences in the ascending and descending phase at around 5 minutes, which seemed to arise from the difference in their frequencies and some irregular $[Ca^{2+}]$ change in a certain region. These changes coincided with the disorder of the total $[Ca^{2+}]_i$ oscillations and the abrupt increase in area just after that. This irregular change of area might be followed by morphological change of microplasmodia and then the appearance of the endoplasmic streaming as described in 3.1. Thus, the irregularity of $[Ca^{2+}]_i$ oscillations can probably be explained as follows. $[Ca^{2+}]_i$ in different regions within one small microplasmodium might oscillate almost synchronously in the beginning. The period of synchronized oscillations might depend on the conditions of each microplasmodium. In other microplasmodia, synchronized oscillations of local $[Ca^{2+}]_i$ lasted 6–8 minutes. $[Ca^{2+}]_i$ in different regions, however, eventually oscillates at different frequencies with the lapse of time, because of unknown causes. One more cause might be that $[Ca^{2+}]_i$ occasionally changes in an abrupt and irregular fashion because of unknown factors.

As to the magnitude of $[Ca^{2+}]_i$ oscillation, also, there seemed to be some subcellular heterogeneity. Figure 5 shows that the magnitude of $[Ca^{2+}]_i$ oscillations in the

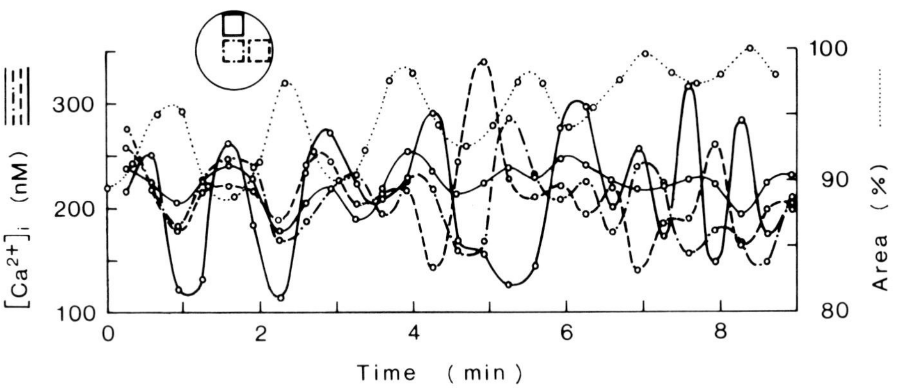

Fig. 5. Changes of local $[Ca^{2+}]_i$ within one microplasmodium. Fura-2 was injected to a final concentration of 50 μM into a microplasmodium 80 μm in diameter. $[Ca^{2+}]_i$ in 3 different local regions, of which area was about one-tenth of the entire region, was determined together with the total $[Ca^{2+}]_i$ in the entire region and the apparent area as in Fig. 3. Each region is indicated by the same lines shown in the insert. The thin solid line represents change of the total $[Ca^{2+}]_i$, and the thin dotted line indicates the area change. $[Ca^{2+}]_i$ oscillations lost their synchronicity in 4 minutes because of the different frequencies and the abrupt irregular change. Note the break in synchronicity of $[Ca^{2+}]_i$ oscillations in advance of the irregular change of area at around 7 minutes

peripheral regions up to 170 nM that was much larger than that in the central region of 10–120 nM, which was still larger than the magnitude of the total $[Ca^{2+}]_i$ oscillations of 10–60 nM. Each apparent average $[Ca^{2+}]_i$, nevertheless, was at almost the same level of 210–220 nM.

The spatial and temporal heterogeneity of $[Ca^{2+}]_i$ within one microplasmodium could be demonstrated also by visualizing the 340 nm/365 nm fluorescence ratio in pseudocolor, which was a relative measure of $[Ca^{2+}]$ (data not shown).

3.3. Effect of $[Ca^{2+}]$ *on Contractility of Saponin and EGTA-Permeabilized Microplasmodia*

Experimental results described in the preceding sections, which clarified the parallelism between the rise in $[Ca^{2+}]_i$ and the contraction of microplasmodia, could claim nothing about the causal relation between them. Then, we examined whether or not the artificial raise in $[Ca^{2+}]$ could induce the increase in contractility of microplasmodia.

Application of a Ca^{2+}-ionophore, A 23187, brought about a remarkable transient contraction, and this effect was impeded by EGTA (data not shown). The absolute level of $[Ca^{2+}]_i$, however, was not easily known in such experiments. Then, permeabilization of microplasmodia for Ca^{2+} and ATP was attempted by treating with saponin and EGTA, and the extent of their contraction was measured by the area change as in living microplasmodia. Contraction of obtained permeabilized microplasmodia was sensitive to ATP and Ca^{2+}, and somewhat reversible. The average area change expressed in a percentage of the original area, given as a function of pCa, was plotted in Fig. 6 for each of 11 microplasmodia of 2 different experiments. Between pCa 8 and 6, the elevation of $[Ca^{2+}]$ in the presence of 1 mM ATP gave rise to the decrease in area, that is, the contraction, whereas the reduction in $[Ca^{2+}]$ induced the increase in area, that is, the relaxation. $[Ca^{2+}]_i$ of living microplasmodia might be between

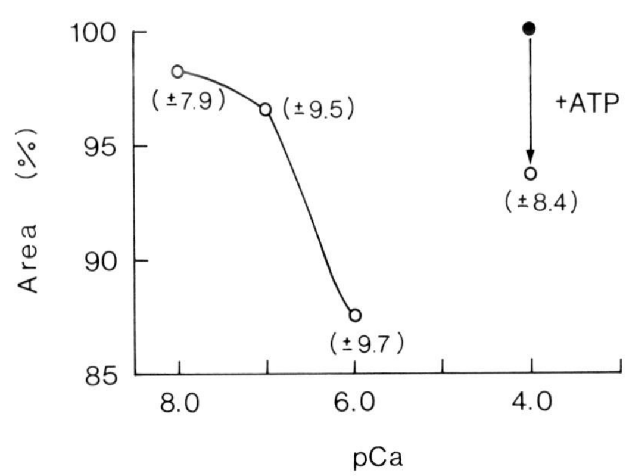

Fig. 6. Contraction of saponin and EGTA-permeabilized microplasmodia as a function of pCa. Permeabilized microplasmodia were prepared at pCa 4 without ATP. At first ATP was added to 1 mM at pCa 4. Then, one series was conducted by sequential shift of pCa from 4 to 7, 6, and 8 under the presence of 1 mM ATP. The other series was shifted from pCa 7 to 8 and then to pCa 6. Permeabilized plasmodia once contracted underwent the relaxation at higher pCa, and vice versa. The average area change expressed in a percentage of the original area is plotted, as a function of pCa, for each of 11 microplasmodia of the 2 different experiments. Note the elevation of $[Ca^{2+}]$ between pCa 8 and 6, where $[Ca^{2+}]_i$ of living microplasmodia ranged, brought about the decrease in area (contraction). Values in parentheses represent SD

pCa 8 and 6. Figure 6 also shows that the addition of 1 mM ATP brought about the clear contraction of permeabilized microplasmodia.

The ensemble of results obtained here and described in the preceding section 3.1 strongly suggests that in living *Physarum* microplasmodia, the ascent of $[Ca^{2+}]_i$ induces more contraction of the actomyosin system, and conversely its descent leads to its relaxation.

4. Discussion

4.1. Phase Relation Between $[Ca^{2+}]_i$ and Contraction

Results from the experiments comparing $[Ca^{2+}]_i$ oscillations with the contraction and relaxation cycles (Figs. 3 and 4) indicate that $[Ca^{2+}]_i$ is high in the contracted phase and conversely low in the relaxed phase in *Physarum* microplasmodia. Since the time resolution of experiments was not enough because of technical problems and being required to protect plasmodia from light, the conclusion that in living microplasmodia the rise in $[Ca^{2+}]_i$ induces the contraction of the actomyosin system hinges on the result of the *in vitro* experiments using microplasmodia permeabilized with saponin and EGTA (Fig. 6). We think this conclusion is applicable to ordinary giant plasmodia.

However, the magnitude of $[Ca^{2+}]_i$ oscillations was much under expected levels, being only 43 ± 16 (SD) nM (n = 50) against the mean $[Ca^{2+}]_i$ of 296 ± 65 (SD) nM (n = 115). Is a $[Ca^{2+}]_i$ rise of only 43 nM able to initiate a series of reactions for the contraction of the actomyosin system? In isolated rat heart cells, $[Ca^{2+}]_i$ of 134 ± 43 (SD) nM (n = 22) at the quiescent stage rises to 270 ± 91 (SD) nM (n = 6) during the spontaneously contracting stage (WIER *et al.* 1987). Considering a $[Ca^{2+}]_i$ rise of only 140 nM in heart cells more active than *Physarum* plasmodia, the value of 43 nM might possibly be not to small. Another reason for the small magnitude might be that, even within one tiny microplasmodium, $[Ca^{2+}]_i$ in different regions oscillates at different frequencies and by different magnitudes (Fig. 5). Thus, at a given moment, $[Ca^{2+}]_i$ in the microplasmodium is spatially heterogeneous (Fig. 5). Regions of high $[Ca^{2+}]_i$ are much smaller than expected. The value of 43 nM was the spatial and temporal average, and probably not too small. In fact, the amplitude of local $[Ca^{2+}]_i$ oscillations observed in 22 regions of 4 different microplasmodia reached 200–300 nM, against the local $[Ca^{2+}]_i$ ranging from 100 nM to 500 nM. This might explain the apparent discrepancy of experiments in which a $[Ca^{2+}]$ rise of 43 nM brought about only 1% contraction of permeabilized microplasmodia, whereas living microplasmodia underwent 3–10% contraction.

As for the subcellular heterogeneity of $[Ca^{2+}]_i$, there exist many reports, for example, on relaxed toad smooth muscle cells (WILLIAMS *et al.* 1985), contracting rat heart cells (WIER *et al.* 1987), and a variety of cells during response to a variety of stimuli (TSIEN and POENIE 1986). Additionally, in a cell of sheep cardiac purkinje fibers, the frequencies of $[Ca^{2+}]_i$ oscillations have been shown to differ from region to region (EISNER and VALDEOLMILLOS 1986).

Experimental results on local $[Ca^{2+}]_i$ oscillations (Fig. 5) lead one to wonder why the period of synchronized local $[Ca^{2+}]_i$ oscillations is only 4 minutes long. A possible explanation against this doubt might be the time span of 10–20 minutes from the fura-2 injection to the beginning of a series of illuminations. Since, at fura-2 injection, microplasmodia as circular as possible were chosen, which were expected to be performing synchronized $[Ca^{2+}]_i$ oscillations, only the last part of synchronized $[Ca^{2+}]_i$ oscillations might be observed. Observations on other microplasmodia, which endured the synchronicity of local $[Ca^{2+}]_i$ oscillations until 6–8 minutes, might support this explanation.

4.2. Absolute Level of $[Ca^{2+}]_i$

We obtained the value of 296 ± 65 (SD) nM (n = 115) as to the apparent average $[Ca^{2+}]_i$ according to the method of GRYNKIEWICZ *et al.* (1985). The K_d of fura-2 for Ca^{2+} was calculated to be 210 nM in a calibrating solution, and the same relation between $[Ca^{2+}]$ and the fluorescence ratio was assumed to be realized within the microplasmodial cytosol. The effect of the intracellular environment of microplasmodia on fura-2 is unknown, but WILLIAMS *et al.* (1985) have estimated the difference in $[Ca^{2+}]$ measured between *in vitro* and *in vivo* to be less than ± 5% in the case of toad stomach cells. Therefore, $[Ca^{2+}]_i$ in microplasmodia might not be so far from the obtained value. This suggests that $[Ca^{2+}]_i$ in microplasmodia might not exceed 1 μM and fluctuate within an unexpectedly narrow range.

Now we think that the optimal $[Ca^{2+}]$ for the contraction of saponin and EGTA-permeabilized microplasmodia might possibly exist, since the extent of contraction at pCa 4 was lower than that at pCa 6. YOSHIMOTO and KAMIYA (1984) have reported that Ca^{2+} exerts a biphasic regulatory action on the generation of tension in saponin-permeabilized plasmodial strand.

We, therefore, have to conclude from the experimental results reported here that, at least in living *Physarum* microplasmodia, the rise in $[Ca^{2+}]_i$ brings about the contraction of the actomyosin system instead of the relaxation which has been preferred in many reports during the past several years. However, some factor which activates the ATPase of *Physarum* actomyosin in dependence on sub-micromolar Ca^{2+} has not yet been isolated. A final conclusion has to be suspended until its isolation.

4.3. Permeabilization of Microplasmodia

The low extent of contraction such as 13% reduction of the area, and some reversibility of the contraction, might be attributed to the firm adherence of permeabilized microplasmodia onto the agar substrate. However, many problems remain to be elucidated. Completely reversible 2 contraction and relaxation cycles have been demonstrated in saponin-skinned isolated smooth muscle cells from the toad stomach (KARGACIN and FAY 1987).

Acknowledgement

This work was supported in part by a Grant-in-Aid (No. 61106009) for Special Project Research of Intracellular Processing of Biomolecules for The Expression of Their Biological Functions from the Ministry of Education of Japan, and carried out under the National Institute for Basic Biology Cooperative Research Program. The authors wish to thank Dr. S. KAWANO for providing microplasmodia of *Physarum polycephalum*, Dr. Y. YOSHIMOTO for his technical assistance and helpful discussion, Dr. M. ISHIGAMI and Dr. S. FUJIME for their valuable suggestions, and Dr. H. SATO for his encouragement throughout this study.

References

ACHENBACH F, WOHLFARTH-BOTTERMANN K-E (1986 a) Reactivation of cell-free models of endoplasmic drops from *Physarum polycephalum* after glycerol extraction at low ionic strength. Eur J Cell Biol 40: 135–138

— — (1986 b) Successive contraction-relaxation cycles experimentally induced in cell-free models of *Physarum polycephalum*. Eur J Cell Biol 42: 111–117

ALLÉRA A, BECK R, WOHLFARTH-BOTTERMANN K-E (1971) Weitreichende fibrilläre Protoplasmadifferenzierungen und ihre Bedeutung für die Protoplasmaströmung. VIII. Identifizierung der Plasmafilamente von *Physarum polycephalum* als F-actin durch Anlagerung von heavy meromysoin *in situ*. Cytobiologie 4: 437–449

BRIX K, KUKULIES J, STOCKEM W (1987) Studies on microplasmodia of *Physarum polycephalum*. V. Correlation of cell surface morphology, microfilament organization and motile activity. Protoplasma 137: 156–167

EISNER DA, VALDEOLMILLOS M (1986) A study of intracellular calcium oscillations in sheep cardiac purkinje fibres measured at the single cell level. J Physiol 372: 539–556

FLEISCHER M, WOHLFARTH-BOTTERMANN K-E (1975) Correlation between tension force generation, fibrillogenesis and ultrastructure of cytoplasmic actomyosin during isometric and istotonic contractions of protoplasmic strands. Cytobiologie 10: 339–365

GRYNKIEWICZ G, POENIE M, TSIEN RY (1985) A new generation of Ca^{2+} indicators with greatly improved fluorescence properties. J Biol Chem 260: 3440–3450

HATANO S (1970) Specific effect of Ca^{2+} on movement of plasmodial fragment obtained by caffeine treatment. Exp Cell Res 61: 199–203

— OOSAWA F (1971) Movement of cytoplasm in plasmodial fragment obtained by caffeine treatment. I. Its Ca^{++} sensitivity. J Physiol Soc Jpn 33: 589–590

ISHIGAMI M, NAGAI R, KURODA K (1981) A polarized light and electron microscopic study of the birefringent fibrils in *Physarum* plasmodia. Protoplasma 109: 91–102

— (1986) Dynamic aspects of the contractile system in *Physarum* plasmodium: I. Changes in spatial organization of the cytoplasmic fibrils according to the contraction-relaxation cycle. Cell Motil 6: 439–447

— KURODA K, HATANO S (1987) Dynamic aspects of the contractile system in *Physarum* plasmodium. III. Cyclic contraction-relaxation of the plasmodial fragment in accordance with the generation-degeneration of cytoplasmic actomyosin fibrils. J Cell Biol 105: 381–386

KAMIYA N, KURODA K (1958) Studies on the velocity distribution of the protoplasmic streaming in the myxomycete plasmodium. Protoplasma 49: 1–4

— (1959) Protoplasmic streaming. Protoplasmatologia 8, 3 a: 1–199

— (1979) Dynamic aspects of movement in the myxomycete plasmodium. In: HATANO S, ISHIKAWA H, SATO H (eds) Cell motility: molecules and organization. Tokyo University Press, Tokyo, pp 399–414

— (1986) Some motility characteristics of living cytoplasm: From observations on *Physarum*. In: ISHIKAWA H, HATANO S, SATO H (eds) Cell motility: mechanism and regulation. Tokyo University Press, Tokyo, pp 577–585

KARGACIN GJ, FAY FS (1987) Physiological and structural properties of saponin-skinned single smooth muscle cells. J Gen Physiol 90: 49–73

KOHAMA K (1981) Ca-dependent inhibitory factor for the myosin-actin-ATP interaction of *Physarum polycephalum*. J Biochem 90: 1829–1832

— UYEDA TQP, TAKANO-OHMURO H, TANAKA T, YAMAGUCHI T, MARUYAMA K, KOHAMA T (1985 a) Ca^{2+}-binding light chain of *Physarum* myosin confers inhibitory Ca^{2+}-sensitivity on actin-myosin-ATP interaction via actin. Proc Jpn Acad 61 B: 501–505

— SHIMMEN T (1985 b) Inhibitory Ca^{2+}-control of movement of beads coated with *Physarum* myosin along actin-cables in *Chara* internodal cells. Protoplasma 129: 88–91

— KENDRICK-JONES J (1986) The inhibitory Ca^{2+}-regulation of the actin-activated Mg-ATPase activity of myosin from *Physarum polycephalum* plasmodia. J Biochem 99: 1433–1446

KURODA R, KURODA H (1982) Relation of cytoplasmic calcium to contractility in *Physarum polycephalum*. J Cell Sci 53: 37–48

MILLER AJ, SANDERS D (1987) Depletion of cytosolic free calcium induced by photosynthesis. Nature 326: 397–400

NAGAI R, YOSHIMOTO Y, KAMIYA N (1978) Cyclic production of tension force in the plasmodial strand of *Physarum polycephalum* and its relation to microfilament morphology. J Cell Sci 33: 205–225

NAIB-MAJANI W, ACHENBACH F, WEBER K, WOHLFARTH-BOTTERMANN K-E, STOCKEM W (1984) Immunocytochemistry of the acellular slime mold *Physarum polycephalum*. IV. Differentiation and dynamics of the polygonal actomyosin system. Differentiation 26: 11–22

OGIHARA S (1982) Calcium and ATP regulation of the oscillatory torsional movement in a Triton model of *Physarum* plasmodial strands. Exp Cell Res 138: 377–384

— IKEBE M, TAKAHASHI K, TONOMURA Y (1983) Requirement of phosphorylation of *Physarum* myosin heavy chain for thick filament formation, actin activation of Mg^{2+}-ATPase activity, and Ca^{2+}-inhibitory superprecipitation. J Biochem 93: 205–223

POENIE M, ALDERTON J, TSIEN RY, STEINHARDT RA (1985) Changes of free calcium levels with stages of the cell division cycle. Nature 315: 147–149

RIDGWAY EB, DURHAM ACH (1976) Oscillations of calcium ion concentrations in *Physarum polycephalum*. J Cell Biol 69: 223–226

SATO H, HATANO S, SATO Y (1981) Contractility and protoplasmic streaming preserved in artificially induced plasmodial fragments, the "caffeine drops". Protoplasma 109: 187–208

STOCKEM W, KUKULIES J, BRIX K (1987) Analysis of cytoplasmic actomyosin functions in *Physarum polycephalum* by fluorescent analog chemistry. In: WOHLFARTH-BOTTERMANN KE (ed) Nature and function of cytoskeletal proteins in motility and transport. Gustav Fischer Verlag, Stuttgart New York (Progress in zoology, vol 34, pp 25–30)

SUGINO H, MATSUMURA F (1983) Fragmin induces tension reduction of actomyosin threads in the presence of micromolar levels of Ca^{2+}. J Cell Biol 96: 199–203

TSIEN RY, POENIE M (1986) Fluorescence ratio imaging: a new window into intracellular ionic signaling. TIBS 11: 450–455

UEDA T, GOTZ VON OLENHUSEN K, WOHLFARTH-BOTTERMANN K-E (1978) Reaction of the contractile apparatus in *Physarum* to injected Ca^{++}, ATP, ADP and 5'AMP. Eur J Cell Biol 18: 76–94

WIER WG, CANNELL MB, BERLIN JR, MARBAN E, LEDERER WJ (1987) Cellular and subcellular heterogeneity of $[Ca^{2+}]_i$ in single heart cells revealed by fura-2. Science 235: 325–328

WILLIAMS DA, FOGARTY KE, TSIEN RY, FAY FS (1985) Calcium gradients in single smooth muscle cells revealed by the digital imaging microscope using fura-2. Nature 318: 558–561

WOHLFARTH-BOTTERMANN K-E (1964) Differentiations of the ground cytoplasm and their significance for the generation of the motive force of amoeboid movement. In: ALLEN RD, KAMIYA N (eds) Primitive motile systems in cell biology. Academic Press, New York London, pp 79–109

— FLEISCHER M (1976) Cyclic aggregation patterns of cytoplasmic F-actin coordinated with oscillating tension force generation. Cell Tissue Res 165: 327–344

YOSHIMOTO Y, KAMIYA N (1978) Studies on contraction rhythm of the plasmodial strand. IV. Site of active oscillation in an advancing plasmodium. Protoplasma 95: 123–133

— MATSUMURA F, KAMIYA N (1981) Simultaneous oscillations of Ca^{2+} efflux and tension generation in the permealized plasmodial strand of *Physarum*. Cell Motil 1: 433–443

— KAMIYA N (1984) ATP- and calcium-controlled contraction in a saponin model of *Physarum polycephalum*. Cell Struct Funct 9: 135–141

Protoplasma (1988) [Suppl 1]: 81–88

Autooscillations in *Physarum* plasmodium
I. Correlation Between Force Generation and Viscoelasticity During Rhythmical Contractions of Protoplasmic Strand

V. A. TEPLOV*

Institute of Biological Physics, USSR Academy of Sciences, Pushchino

Received February 17, 1988
Accepted April 1, 1988

Dedicated to Professor Dr. NOBURO KAMIYA on the occasion of his 75th birthday

Summary

The method of quick release and stretch of the plasmodial strands was used to study their viscoelastic properties at different phases of contraction-relaxation cycle. The rheological behaviour of the strand on a time scale which is much shorter than the period of autooscillations is described in terms of a parallel combination of elastic component and Maxwell element. The stiffness and viscosity oscillate synchronously with cyclic tension production. The 10–20% stepwise stretching of the strand results in a rapid increase of the stiffness. The participation of mechanochemical system in cell oscillator is discussed.

Keywords: Mechanochemical system; Oscillations; *Physarum polycephalum*; Rheological models; Viscosity coefficient; Young's modulus.

1. Introduction

Recently considerable efforts have been aimed at elucidating the physicochemical basis of oscillatory contractile activity in the *Physarum polycephalum* plasmodium which is a large multinuclear mass of protoplasm differentiating into a leading frontal zone and a network of protoplasmic strands (veins). Actomyosin-based motility of plasmodium manifests itself as rhythmic pulsations of plasmodial body and as a fast shuttle streaming of endoplasm, which is superimposed on a slow net movement of the whole organism. The pulsations have a complex spatio-temporal organization (STEWART 1964, BARANOWSKI 1978, CREBECKI and CIESLAWSKA 1978, HEJNOWICZ and WOHLFARTH-BOTTERMANN 1980, BEYLINA *et al.* 1984, TEPLOV and ROMANOVSKY 1987). Typically, their periods are in the 1–3 min range.

Besides, there are simultaneous oscillations of the membrane potential (KAMIYA and ABE 1950, KISHIMOTO 1958), the levels of Ca^{2+} (TEPLOV *et al.* 1973, RIDGWAY and DURHAM 1976, YOSHIMOTO *et al.* 1981 a), H^+ (NAKAMURA and KAMIYA 1985), ATP (YOSHIMOTO *et al.* 1981 b), cAMP (UEDA *et al.* 1986) etc., all occurring with the same frequency as the contractile activity. Because of so many oscillating components it is difficult to single out the group of essential and independent variables which determine the autooscillation as a whole, *i.e.*, to find out the basic instability responsible for generating the rhythm (WOHLFARTH-BOTTERMANN 1979, BERRIDGE and RAPP 1979).

The important peculiarity of electrical and chemical oscillations observed in *Physarum* is that they are accompanied by mechanical oscillations. It might reflect a participation of mechanical variables in the feedback loops responsible for autooscillations. As early as 1952 TURING in his classic paper on the basis of morphogenesis admitted that chemical and mechanical processes in the cell are interdependent. Recently it has been reported that mechanical stresses have an influence on morphogenesis, processes of synthesis, mitosis and on cytoskeletal organization (KOLEGA 1986).

Mechanical system of the cell, the cytoskeleton, is a complex meshwork of filaments attached to the plasma

* Correspondence and Reprints: Institute of Biological Physics, USSR Academy of Sciences, 142292 Pushchino, Moscow region, USSR.

membrane. Such a system alone may play a significant role in the intracellular communication to organize a coordinated motile behaviour (TEPLOV and ROMANOVSKY 1987). The main components of cytoskeleton are contractile proteins. Since enzymatic activity of the protein molecule depends on its conformation, a forced local deformation may change the enzymatic activity of the contractile proteins. This change may result in a change of their active tension which in its turn results in deformation of the next element of the cytoskeleton and so on. Thus, this process seems to be of autowave character and is performed in the given case only by mechanochemical system.

So, it is important to obtain experimental data on the mechanical properties of the cell. In this respect, plasmodial strands which are relatively unharmed by excision give a unique opportunity. In a large plasmodium the lengths of major strands can reach a few dozens of cm and the diameters sometimes attain 2 mm. The strand differentiates into the outer tube-like gel ectoplasm and the more fluid inner endoplasm. The plasmodial strand suspended in moist air starts rhythmic longitudinal contractions 10–20 min after excision (KAMIYA 1970). Thirty minutes later, the local oscillations of length become synchronous and synphase over the entire strand (YOSHIMOTO and KAMIYA 1978 a). Under isometric conditions, the dynamic activity of the strand is revealed in oscillations of longitudinal force. One of the striking characteristics of the strands is their response to stretching (KAMIYA 1970, 1981, KAMIYA and YOSHIMOTO 1972). When the strand is quickly stretched by 10–20% of its length, the amplitude of force oscillations is conspicuously augmented. Under isotonic conditions the augmentation of the load causes an increase in the amplitude of the length oscillations.

In the present work we employed the method widely used for muscle research (HUXLEY 1974) to study *Physarum* protoplasm contractility. The method is to impose a sudden change of the strand length and to follow the time course of the resulting tension change. This tension transient might reflect such processes as elastic or viscoelastic phenomena in the filaments, the attachment and detachment of actin and myosin, mechanical events within the actomyosin complexes.

2. Material and Methods

Plasmodia of the myxomycete *Physarum polycephalum* were maintained by feeding with oatflakes. Prior to experiment the plasmodium migrated on 1% water agar overnight, and thus the formation of smooth strands was provided. The fragments of the plasmodial strands with the diameter of 0.3–0.7 mm and about 4 mm long without any branching were carefully excised from the plasmodium and used for the experiments. The fragments of such a small length were chosen firstly, for the better synchronization of local contractions along the strand and secondly, because the effects arising from propagation of longitudinal oscillations were not detectable. The isolated strand was kept in a horizontal position in water between the arms of a force transducer (resonant frequency 200 Hz, compliance 0.01 mm/g) and an electromechanical device permitting to hold the length of the strand constant, or to make it undergo a predetermined change. The amplified transducer output signal was continuously recorded on a pen recorder and on an oscillograph. All measurements were performed at 20 °C.

3. Results

The tension response to a stepwise change of the strand length could be divided into two phases. The initial tension change occurred simultaneously with the applied length change and hence was an elastic response. After the step had been completed, there was a partial recovery towards the original tension level. This phase of the tension recovery can be well fitted by a single exponential curve with the time constant of about several seconds. The correlation coefficient for a linear regression of function *ln* F(t), describing this fitting, was found to be in the range of 0.992–0.999.

Because the plasmodial strands are characterized by an inherent nonstationary tension, it is difficult to determine exactly the asymptotical value of tension which is approached during the recovery phase of the transient process. Mechanical properties as well as their dependence on deformation should be determined in a short time interval corresponding to a certain phase of autooscillations. For this reason we used short (about 1 s) rectangular impulses of 1% length change in the shortening direction. Series of such impulses with an increasing amplitude was also applied, the maximum amplitude being restricted in most cases to 5% of the strand length. Such deformations did not cause a permanent change in the behaviour of the plasmodial strands.

Figure 1 shows the typical oscillograms obtained by this method. The record of self-oscillatory longitudinal force generated by the strand under isometric conditions is represented in the inset. The same series of deformation impulses with increasing amplitudes were accomplished at the time indicated by 1, 2, 3, and 4. In this case the first series was performed at maximum contraction of the strand, the second was at maximum relaxation, the third and the fourth were in the phases of increasing and decreasing tension, respectively. The induced change of longitudinal force during these four series of the strand deformation are shown on a faster

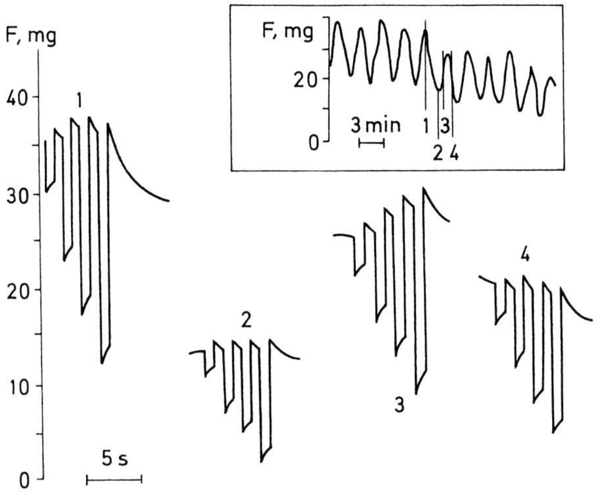

Fig. 1. Typcial force responses of the strand at different phases of the autooscillations to the same series of four short rectangular deformation impulses with an increasing amplitude. The record of spontaneous isometric force and the moments, when the series were carried out, are shown in the inset. The length and diameter of the strand were 4 mm and 0.5 mm, respectively

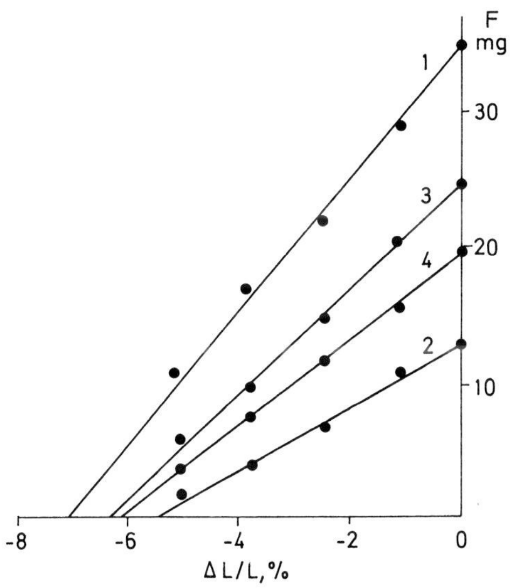

Fig. 2. Instantaneous force-extension relations of the strand at the four phases of the autooscillations shown in Fig. 1

time base on the bottom of Fig. 1. In each force trace at moments of the imposed length changes there are force steps after which the force recovers towards the isometric value.

The extreme forces reached in these steps are plotted against the size of the relative length change in Fig. 2. It can be seen that the instantaneous elasticity responsible for the first phase of the transient is nearly linear. Coefficient of instantaneous stiffness determined by the slope of the straight line depends on the phase of autooscillations but it does not change essentially during

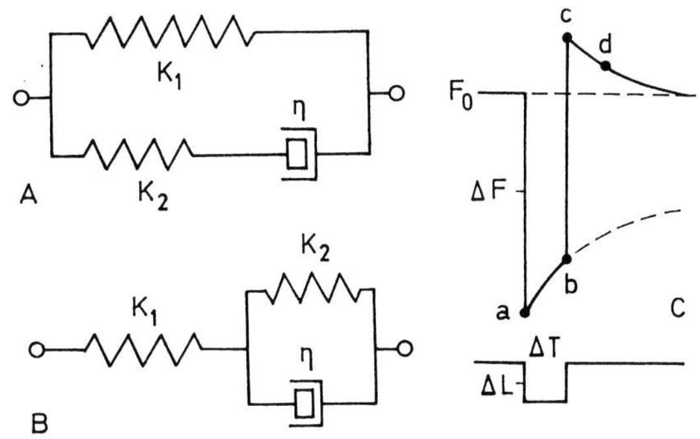

Fig. 3. Equivalent rheological models of the strand and their force response to a brief rectangular impulse of deformation

several seconds. Based on these relationships the important properties of the plasmodial strands such as activation caused by loading or stretching can be explained. It should be noted that when the steps are performed in the lengthening direction, the force-extension relation becomes nonlinear (not shown in Fig. 2). The instantaneous force-length curves approach the length axis at a sharp angle and a negative tension appears at the force transducer when very large steps are applied. These observations suggest that, when isometric force is equal to zero, the ectoplasm responsible for the stiffness of the strand remains rigid, probably, due to the intracellular pressure (LAYRAND *et al.* 1972).

The character of force responses allows depict the viscoelastic properties of the strand on a time scale which is much shorter than the period of autooscillations by either of the arrangements of two elastic components and one viscous component as shown in Fig. 3. If the components have linear characteristics, the both arrangements give the same response (Fig. 3 *C*), provided that the values of the components are appropriately adjusted. Therefore, it would be impossible to say that one arrangement fits a set of experimental data better than the other does. However, as it will be seen below, the K_2 component in the arrangement of Fig. 3 *A* does not change in the course of autooscillations, while all components undergo changes in the arrangement of Fig. 3 *B*. Thus, in contrast to muscle fibers (HUXLEY 1974), the behaviour of plasmodial strands is more simply explained by the arrangement of Fig. 3 *A* than by that of Fig. 3 *B*. Therefore, for an estimation of the viscoelastic characteristics of the strands we used the rheological model shown in Fig. 3 *A*. Solving the simple differential equation describing the behaviour of this model during a rectangular impulse of deformation, it is easy to obtain the formulae for the calculation of unknown parameters (see appendix).

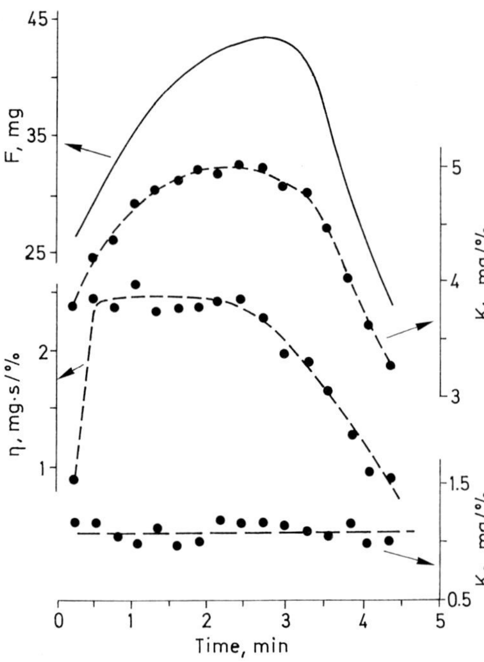

Fig. 4

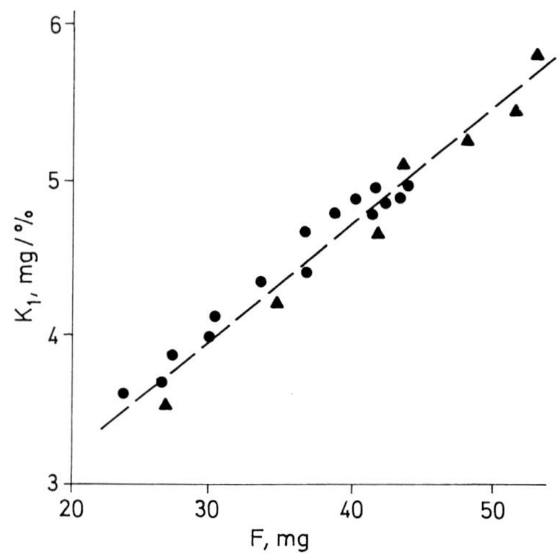

Fig. 5

Fig. 4. Spontaneous changes of isometric tension and of strand viscoelastic parameters calculated on the basis of the model in Fig. 3 *A*, which occurred during one autooscillation period. The length and diameter of the strand were 5 mm and 0.7 mm, respectively

Fig. 5. The stiffness coefficient of the strand in relation to the spontaneous tension generation. The circles correspond to the data in Fig. 4 and the triangles were obtained after the following 10% stepwise stretching of the same strand

Figure 4 shows the time courses of the longitudinal isometric force and the viscoelastic characteristics of the strand during one autooscillation period. The stiffness of the K_1 component and the viscosity of the dashpot change synchronously with the force generation. The force and K_1 curves are similar and quasi-harmonic, but the viscosity oscillation curve is purely relaxational. The stiffness of the K_2 component practically remains constant during the active contraction cycle and seems to reflect passive properties of the strand. Thus, periodical changes of instantaneous elasticity first shown by KAMIYA (KAMIYA and YOSHIMOTO 1972, KAMIYA *et al.* 1972) solely result from the K_1 component, because according to the used model (Fig. 3 *A*), the instantaneous stiffness $K = K_1 + K_2$.
Plotting the stiffness of the K_1 component against isometric force, we see that this relation is linear (Fig. 5). It is noteworthy that after 10–20% stepwise stretching of the strand, the stiffness K_1 increases within several seconds, so that the new values of K_1 and force in the following activated oscillations group again along the same straight line.

4. Discussion

The elasticity of plasmodial strands was first measured by NORRIS (1940) who stretched an excised fragment of the *Physarum* strand by means of a glass needle and estimated the force from the amount of needle bending. NORRIS found that the Young's modulus was of the order of 10^5 dyn/cm^2 for the length changes up to 40%. More recently, KAMIYA determined the elasticity of the same material and elucidated the correlation of Young's modulus with spontaneously oscillating tension (KAMIYA and YOSHIMOTO 1972, KAMIYA *et al.* 1972). Using the method similar to ours, KAMIYA showed that Young's modulus changed by about 30–40% accompanying cyclic tension production. The maximum and the minimum values of Young's moduli were about $5 \cdot 10^5$ and $3 \cdot 10^5$ dyn/cm^2. There was a linear relation between the instantaneous modulus and the generated tension.
Quantitative aspects of viscoelasticity in the slime mold strands were dealt with the same as by ABE (1965). The investigations were made by the two "torsion" methods

without paying attention to the phase of strand autooscillations. By the first method the viscoelastic characteristics were evaluated from the damped torsional oscillations of glass handle with the known moment of inertia, which was attached to the free end of suspended strand. The oscillations were induced by applying a brief torque. By the second method ABE analysed the time course of the change in the twist angle of a strand in response to a long impulse of a certain torque around the longitudinal axis. The rheological behaviour of the strand was found to be described in terms of a parallel combination of viscous and elastic components (Voigt element) in series with a successive combination of elastic and viscous components (Maxwell element). The rigidity moduli obtained by these two methods were of the same order of magnitude and agreed with those obtained by NORRIS and KAMIYA, while the viscosities were exceedingly different. The viscosity coefficient obtained by the second method was of the order of 10^6 poises, that is three orders of magnitude higher than that obtained by the first method.

If we assume that the longitudinal force is determined only by the stress of peripheral gel layer having only 10% thickness of the strand diameter, the stiffness and the viscosity coefficients obtained in the present work can be recalculated in the corresponding average moduli of the ectoplasmic material (see appendix). So the longitudinal Young's modulus calculated according to the data represented in Fig. 4 is about $3 \cdot 10^5$ dyn/cm^2 at the phase of maximum relaxation and about $4.5 \cdot 10^5$ dyn/cm^2 at the phase of maximum contraction. The respective values of the viscosity coefficient at these phases of autooscillations are about $0.5 \cdot 10^5$ and $2 \cdot 10^5$ poises. It should be noted that when studying the migration of plasmodium, the rheological models in Fig. 3 should be supplemented with viscous component in series, the viscosity coefficient determined by the time of the relaxation of the average force being of the order of 10^8 poises. On a time scale comparable to the autooscillation period this component may be neglected. Thus, the values of Young's modulus obtained in the present work are of the same order as those obtained by the above authors (NORRIS 1940, ABE 1965, KAMIYA *et al.* 1972). The values of viscosity coefficient agree with those obtained by the second method of ABE (1965).

At present it is impossible to establish the precise relationship between the macroscopic mechanical properties and the molecular constitution of the *Physarum* cytoplasm. Dynamic activity of the myxomycete plasmodium is closely connected with morphological changes in the ectoplasm. It was first shown by WOHLFARTH-BOTTERMANN (1964) that there are actomyosin-containing fibrils (microfilament bundles) in the ectoplasm, their number correlating with the level of tension (see also KAMIYA 1968). The microfilaments and their assembly undergo a remarkable change in each contraction-relaxation cycle (FLEISCHER and WOHLFARTH-BOTTERMANN 1975, NAGAI *et al.* 1978). The elastic modulus of the strand is apparently proportional to the concentration of cross-linkages in the bundles of microfilaments, and hence the variation of K_1 (see Fig. 4) seems to reveal the changes in the amount of the actomyosin complexes. Thus, the linear correlation between the force and elasticity is quite clear.

More interesting is the fact that the 10–20% stepwise stretching of the strand results in a rapid increase of the stiffness and that the same linear relation between the stiffness and the generated force is restored within several seconds after the stretching (Fig. 5). This mechanical activation might be connected with fibrillogenesis in ectoplasm. A strong stretching results in a *de novo* formation of the fibrils even within endoplasm in 30–40 seconds, these fibrils disappearing during 5 seconds under isotonic conditions (FLEISCHER and WOHLFARTH-BOTTERMANN 1975). The possible mechanisms of the fibrillogenesis and "catch" properties of the plasmodial actomyosin are discussed elsewhere (ROMANOVSKY and TEPLOV 1988).

One of the ways to show the direct involvement of mechanical variables in generating periodic activity is to find out whether they can interfere with the cell oscillator. Moderate stretching of the strand causes neither change in the period nor the shift in the phase of the contraction-relaxation cycles (KAMIYA and YOSHIMOTO 1972, YOSHIMOTO and KAMIYA 1978 b). However, KRÜGER and WOHLFARTH-BOTTERMANN (1978) showed a phase shift of the force oscillations when the strands are stretched by 50% of their original length. The radial deformations of the strands are more essential than the longitudinal ones for such an effect, as it was first demonstrated with the help of the elegant method by KAMIYA (1968). When the endoplasm in a dumbell-shaped plasmodium is pushed away from one half of the plasmodium to the other one by externally applied air pressure, the cortical tension of the former half decreases, while the other half is inflated and the cortex is stretched. KAMIYA clearly demonstrated that the amplitude and the period of the motive force oscillations greatly increase in the inflated plasmodia, but decrease in the shrunken ones. This difference gradually disappeared in about 20 min. The same time is necessary for the intracellular pressure to be equalized into such plasmodia by osmotic process (LAYRAND *et al.* 1972). Using the Kamiya method, we showed that the period and the amplitude of the oscillations of longitudinal force generated by the strand also depend on the intracellular pressure (TEPLOV and ROMANOVSKY 1987).

Any autooscillatory systems interacting with each other are capable of the synchronization. What the synchronization in the plasmodium needs is a normal endoplasm streaming (YOSHIMOTO and KAMIYA 1978c). Compression of the middle part of the plasmodial strand with a glass needle leads to desynchronization of authooscillations in the two halves as soon as the endoplasmic flow can not pass the point of constriction (ACHENBACH and WOHLFARTH-BOTTERMANN 1981). However, when the endoplasm is prevented from flowing between the two halves of the strand by externally applied air pressure, their synphase isotonic oscillations become out of phase with each other, but the same frequency is established again (YOSHIMOTO and KAMIYA 1978c). The similar results were obtained when we studied the oscillations of the membrane potential of the strand in similar experiments (TEPLOV and ROMANOVSKY 1987). Thus, in these experiments the synchrony of plasmodium oscillations seems to be reestablished by the mechanical factor (by means of the air pressure applied periodically) rather than by the diffusion of some chemicals. The essential role of the mechanochemical system in the observed autooscillations is also confirmed by the data of KISHIMOTO (1958) that the plasmodium is capable of frequency pulling of oscillations of the air pressure (external synchronization).

If an instability exists in the mechanochemical system of the cell, the autooscillations might arise from a local positive feedback between the tension and the deformation of the contractile apparatus (BEYLINA *et al.* 1984, OSTER and ODELL 1984). In ameboid cells in contrast to insect flight muscles, the inertial forces are not crucial in the observed mechanical oscillations, and so the positive feedback must be mediated through some regulatory systems of the cell (NAIB-MAJANI *et al.* 1988). Recently we have proposed such mathematical models for mechanochemical autooscillations in the plasmodial strands (TEPLOV and ROMANOVSKY 1987, ROMANOVSKY and TEPLOV 1988) and in the frontal protoplasmic sheet (LATUSHKIN *et al.* 1988). When the mechanical parameters of *Physarum* ectoplasm obtained in the present work are used, the solutions of the mathematical models agree with the available experimental data such as frequency, phase relations, modes of autowaves, velocity of endoplasmic streaming etc. Thus, the autooscillations in *Physarum* plasmodium could really be based on the indirect positive feedback between the tension and the deformation of the contractile apparatus.

In conclusion it should be noted that the rhythm is one of the characteristic features of the movement of ameboid cells (SATOH *et al.* 1985). Periodical processes are apparently not only characteristic but necessary for the cell migration. Small active elastic deformations of the contractile apparatus are the cause for the directional cell movement and it is evident that the contractile machine is not capable of providing the progressive cell movement without working in the oscillatory regime. Therefore, the participation of the mechanochemical system in cell oscillator can be of great functional significance.

Appendix

The behaviour of the rheological model consisting of Maxwell element in parallel with elastic component (Fig. 3 *A*) which well approximates the strand properties is described by the following differential equation:

$$\tau_F \frac{dF}{dt} + F - F_0 = K_1 \cdot (\varepsilon + \tau_\varepsilon \frac{d\varepsilon}{dt}),$$

where $\varepsilon = (L - L_0)/L_0$ is the relative deformation of the strand, F_0 is the force before deformation, $\tau_F = \eta/K_2$ and $\tau_\varepsilon = \eta \cdot (1/K_1 + 1/K_2)$ are the time constants, K_1, K_2 and η are the viscoelastic coefficients according to the rheological model shown in Fig. 3 *A*.

The solution of this equation for the case when short rectangular impulse of deformation is applied yields the formulae for the calculation of the unknown strand parameters. The time constant of the force relaxation is determined:

$$\tau_F = -\triangle T/ln(1 + F_2'/F_1'),$$

where $\triangle T$ is the duration of the deformation impulse, F_1' and F_2' are the derivatives of the force with respect to time at the moments after the beginning and the end of the deformation impulse, respectively. They are practically determined by the slope of the 'straights' ab and cd (see Fig. 3 *C*).

The stiffness coefficients are equal to

$$K_2 = \triangle F \cdot L_0/\triangle L \cdot (1 + exp(-\triangle T/\tau_F));$$
$$K_1 = \triangle F \cdot L_0/\triangle L - K_2,$$

here $\triangle F$ is the force step in response to the length change $\triangle L$, L_0 is the strand length.

Therefore, the viscosity coefficient is

$$\eta = K_2 \cdot \tau_F$$

If the longitudinal force is determined only by the stress of the ectoplasm with the thickness 10% of the strand diameter, the stiffness and the viscosity coefficients can be recalculated in the corresponding average moduli of the ectoplasm material. So the longitudinal instantaneous elasticity modulus E and viscosity one ν are approximately equal to 3.6 $(K_1 + K_2)/D^2$ and 3.6 η/D^2, respectively, where D is the strand diameter.

References

Abe S (1965) The viscoelasticity of slime mold protoplasm. In: Proc Fourth International Congress on Rheology. Wiley, New York, part 4, pp 147–155

Achenbach U, Wohlfarth-Bottermann KE (1981) Synchronization and signal transmission in protoplasmic strands of *Physarum*. Effects of externally applied substances and mechanical influences. Planta 151: 574–583

Baranowski Z (1978) The contraction-relaxation waves in *Physarum polycephalum* plasmodia. Acta Protozool 17: 387–397

Berridge MJ, Rapp PE (1979) A comparative survey of the function, mechanism and control of cellular oscillators. J Exp Biol 81: 217–279

Beylina SI, Matveeva NB, Priezzhev AV, Romanovsky YM, Suchorukov AP, Teplov VA (1984) Plasmodium of the myxomycete *Physarum polycephalum* as an autowave self-organizing system. In: Krinksy VI (ed) Self-organization, autowaves and structures far from equilibrium. Springer, Berlin Heidelberg New York Tokyo, pp 218–221

Crebecki A, Cieslawska M (1978) Plasmodium of *Physarum polycephalum* as a synchronous contractile system. Cytobiologie 17: 335–342

Fleischer M, Wohlfarth-Bottermann KE (1975) Correlation between tension force generation, fibrillogenesis, and ultrastructure of cytoplasmic actomyosin during isometric and isotonic contractions of protoplasmic strands. Cytobiologie 10: 339–365

Hejnowicz Z, Wohlfarth-Bottermann KE (1980) Propagated waves induced by gradients of physiological factors within plasmodia of *Physarum polycephalum*. Planta 150: 144–152

Huxley AF (1974) Muscular contraction (Review lecture). J Physiol 243: 1–43

Kamiya N (1968) The mechanism of cytoplasmic movement in a myxomycete plasmodium. In: Aspects of cell motility (22nd Symp Soc Exp Biol). Cambridge Univ Press, Cambridge, pp 199–214

— (1970) Contractile properties of the plasmodial strand. Proc Japan Acad 46: 1026–1031

— (1981) Physical and chemical basis of cytoplasmic streaming. Ann Rev Plant Physiol 32: 205–236

— Abe S (1950) Bioelectric phenomena in the myxomycete plasmodium and their relation to protoplasmic flow. Colloid Sci 5: 149–163

— Allen RD, Zeh R (1972) Contractile properties of the slime mold strand. Acta Protozool 11: 113–124

— Yoshimoto Y (1972) Dynamic characteristics of the cytoplasm. A study on the plasmodial strand of a myxomycete. In: Hamaguchi K (ed) Aspects of cellular and molecular physiology. Univ Tokyo Press, Tokyo, pp 167–189

Kishimoto U (1958) Rhythmicity in the protoplasmic streaming of slime mold, *Physarum polycephalum* I. A statistical analysis of the electric potential rhythm. J Gen Physiol 41: 1205–1222

Kolega J (1986) Effects of mechanical tension on protrusive activity and microfilament and intermediate filament organization in an epidermal epithelium moving in culture. J Cell Biol 102: 1400–1411

Krüger J, Wohlfarth-Bottermann KE (1978) Oscillating contractions in protoplasmic strands of *Physarum*: stretch-induced phase shifts and their synchronization. J Interdiscip Cycle Res 9: 61–71

Latushkin OA, Netrebko NB, Romanovsky YM, Teplov VA (1988) Mathematical two-dimensional model of autowave processes in living cell. In: Collective dynamics of excitations and structure formation in biological tissues. Inst Appl Phys USSR Acad Sci Press, Gorky (in press; in Russian)

Layrand DB, Matveeva NB, Teplov VA, Beylina SI (1972) The role of elastoosmotic parameters in locomotion of myxomycete plasmodia. Acta Protozool 11: 339–354

Nagai R, Yoshimoto Y, Kamiya N (1978) Cyclic production of tension force in the plasmodial strand of *Physarum polycephalum* and its relation to microfilament morphology. J Cell Sci 33: 205–225

Naib-Majani W, Teplov VA, Baranowsky Z (1988) Morphology and viscoelastic properties of *Physarum* strands during steady-state of their contractile behaviour. Protoplasma [Suppl 1]: 57–63

Nakamura S, Kamiya N (1985) Oscillation of cytoplasmic pH of *Physarum plasmodium* in relation to motility. Cell Struct Funct 10: 133–141

Norris CH (1940) Elasticity studies on the myxomycete, *Physarum polycephalum*. J Cell Comp Physiol 16: 313–322

Oster GF, Odell GM (1984) Mechanics of cytogels I: Oscillations in *Physarum*. Cell Motil 4: 469–503

Ridgway EB, Durham ACH (1976) Oscillations of calcium ion concentrations in *Physarum polycephalum*. J Cell Biol 69: 223–226

Romanovsky YM, Teplov VA (1988) Autowave mechanochemical model for *Physarum* shuttle streaming. In: Lamprecht I, Zotin AI (eds) Thermodynamics and pattern formation in biology. DeGruyter, Berlin (in press)

Satoh H, Ueda T, Kobatake Y (1985) Oscillations in cell shape and size during locomotion and in contractile activities of *Physarum polycephalum, Dictyostelium discoideum, Amoeba proteus* and macrophages. Exp Cell Res 156: 79–90

Stewart PA (1964) The organization of movement in slime mold plasmodia. In: Allen RD, Kamiya N (eds) Primitive motile systems in cell biology. Academic Press, New York, pp 69–76

Teplov VA, Matveeva NB, Zinchenko VP (1973) Free calcium level oscillations in myxomycete plasmodium during protoplasm shuttle streaming. In: Frank GM, Karnaukhov VN (eds) Biophysics of living cell. Biol Res Centre USSR Acad Sci Press, Pushchino 4: 110–115 (in Russian)

— Romanovsky YM (1987) Mechanochemical distributed autooscillations in cell motility. Autowave phenomena in contractile activity of *Physarum polycephalum* plasmodium. Biol Res Centre USSR Acad Sci Press, Pushchino, pp 1–48 (in Russian)

TURING AM (1952) The chemical basis of morphogenesis. Philos Trans R Soc Lond 237 B: 37–72

UEDA T, MATSUMOTO K, AKITAYA T, KOBATAKE Y (1986) Spatial and temporal organization of intracellular adenine nucleotides and cyclic nucleotides in relation to rhythmic motility in *Physarum* plasmodium. Exp Cell Res 162: 486–494

WOHLFARTH-BOTTERMANN KE (1964) Differentiation of the ground cytoplasm and their significance for the generation of the motive force of ameboid movement. In: ALLEN RD, KAMIYA N (eds) Primitive motile systems in cell biology. Academic Press, New York, pp 79–109

— (1979) Oscillatory contraction activity in *Physarum*. J Exp Biol 81: 15–32

YOSHIMOTO Y, KAMIYA N (1978 a) Studies on contraction rhythm of the plasmodial strand I. Sychronization of local rhythms. Protoplasma 95: 89–99

— — (1978 b) Studies on contraction rhythm of the plasmodial strand II. Effect of externally applied forces. Protoplasma 95: 101–109

— — (1978 c) Studies on contraction rhythm of the plasmodial strand III. Role of endoplasmic streaming in synchronization of local rhythms. Protoplasma 95: 111–121

— MATSUMURA F, KAMIYA N (1981 a) Simultaneous oscillations of Ca^{2+} efflux and tension generation in the permealized plasmodial strand of *Physarum*. Cell Motil 1: 433–443

— SAKAI T, KAMIYA N (1982 b) ATP oscillation in *Physarum* plasmodium. Protoplasma 109: 159–168

Protoplasma (1988) [Suppl 1]: 89–95

Motile Activities in Polynuclear Products of Cell Fusion. I. Actin Rearrangements in Spreading on Glass Polycaryons Formed from Ehrlich Ascites Tumour Cells

W. KOROHODA*, and J. KAJSTURA

Department of Cell Biology, The Jan Zurzycki Institute of Molecular Biology, Jagiellonian University, Krakow

Received March 12, 1988
Accepted June 24, 1988

Dedicated to Professor Dr. NOBURO KAMIYA on the occasion of his 75th birthday

Summary

The polycaryons were produced from the Ehrlich ascites tumour (EAT) cells by cell fusion in the presence of polyethylene glycol (PEG). The polycaryons, in contrast to spherical mononucleate EAT cells, were able to flatten, attach to glass and spread in the presence of serum. With direct fluorescent staining with TRITC-phalloidin changes in F-actin organization were followed within these polycaryons during various stages of their spreading on glass. The organization of actin filaments to bundles and finally to stress fibres was not needed for cell spreading but it gradually developed in already attached and spread polycaryons.

It is hypothesized that in tissue culture cells, similarly as in lower organisms, nucleus influences cortical contractions and polarization of cell motile activities. These nuclear functions are more or less impaired in neoplastic cells. Polynucleation of these cells restores more efficient nuclear control over cell shape, adhesiveness, and cytoskeleton organization.

Keywords: EAT cells; Cell fusion; Cell spreading; Nucleus; Cytoskeleton.

Abbreviations: DIC differential interference contrast; EAT Ehrlich ascites tumour; EMEM Eagle's minimal essential medium; Hepes N-2-hydroxyethylpiperazine N′-2-ethane sulfonic acid; PEG polyethylene glycol; TRITC-phalloidin tetramethylrhodamine isothiocyanate labeled phalloidin.

1. Introduction

In 1981 BERSHADSKY *et al.* reported that transformed mouse L line fibroblasts and cells of their highly tumorigenic TLSF subline, when made polycaryotic, increased spreading on glass, developed bundles of filaments in their cytoplasm and focal contacts to the substratum. These effects of polynucleation upon organization of cytoskeleton are just opposite to those occurring during the onset of neoplastic transformation (AMBROSE *et al.* 1970 a, POLLACK *et al.* 1975, BOSCHEK *et al.* 1981, BEN ZE'EV 1985).

Ascitic tumours represent almost extreme forms of neoplasia. In their cells all features which commonly accompany neoplastic transformation are visible in an exaggerated form. Whereas transformed cells are usually less adherent and less spread on glass or plastic than non-transformed cells (ABERCROMBIE and AMBROSE 1962, CASTOR 1970, CLARKE *et al.* 1970, SHIELDS and POLLOCK 1974, WEBER *et al.* 1977, FOLKMAN and MOSCONA 1978), ascitic cells remain spherical as well *in vivo* as *in vitro* and in culture grow in an anchorage-independent manner. In their cytoplasm the bundles of actin filaments organized to stress fibres are absent (CIEŚLAK and KOROHODA 1979). The experiments reported below were therefore undertaken to examine how polynucleation affects cell morphology and actin organization in ascitic EAT cells.

2. Material and Methods

EAT cells were grown by serial intraperitoneal transplantation in Swiss mice (Camura Breeding Labs., Cracow, Poland) and harvested after 9–10 days (CIEŚLAK and KOROHODA 1978). The cells were washed and fused by means of polyethylene glycol (PEG, M.W. 6000), as described elsewhere (KAJSTURA and KOROHODA 1988).

* Correspondence and Reprints: Department of Cell Biology, The Jan Zurzycki Institute of Molecular Biology, Jagiellonian University, al. Mickiewicza 3, PL-31-120 Krakow, Poland.

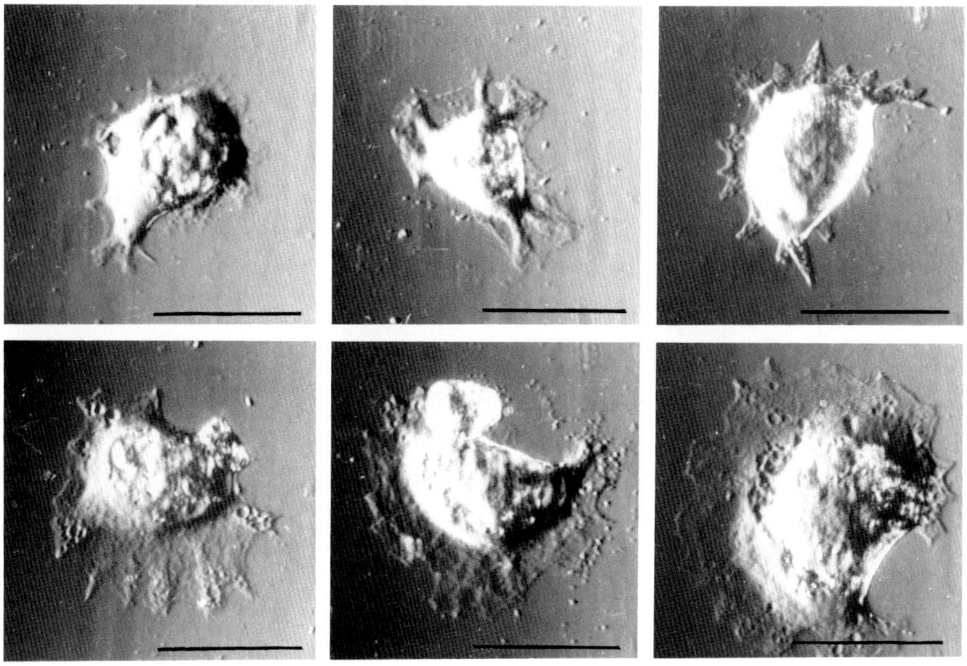

Fig. 1. Examples of morphological changes during the spreading of products of EAT cell fusion. Cells were photographed in differential interference contrast 2 days after fusion. Bar = 50 μm

After 60–90 s treatment with 60% PEG in Eagle's minimal essential medium (EMEM) the cells were immediately diluted with EMEM, left for 15 minutes to be sedimented and resuspended in a new medium several times, to remove traces of PEG and enrich the cell population in fusion products. Then the cells were incubated at 37 °C in Leighton tubes with inserted coverglasses. The medium used was EMEM supplemented with 5% calf serum, 20 mM Hepes, and antibiotics.

For the photographs in differential interference contrast (DIC), taken with Biolar (PZO, Poland) microscope, cells were fixed in 3% glutaraldehyde. For photographs in phase contrast and epifluorescence, taken with the Leitz Orthoplan (Wetzlar, Federal Republic of Germany) microscope, cells were fixed in 4% formaldehyde, permeabilized with 0.1% Triton X-100 instead of Nonidet 40, and stained with TRITC-phalloidin (Bereiter-Hahn and Kajstura 1988). The TRITC-phalloidin was a generous gift of Prof. Dr. H. Faulstich (Heidelberg, Federal Republic of Germany).

3. Results

Already our first observations showed that multinucleated products of fusion of EAT cells differ in adhesion from non-treated cells and from those EAT cells which remained mononucleate after the fusion experiment. Polycaryons containing four to some dozens of nuclei, and 50 to 300 μm in diameter, after 24–48 h in culture were attached to the glass and showed the first stages of spreading. At the onset of spreading only in certain regions of their surface were cell surface protrusions formed and adhered to the glass. The remaining cell surface appeared to be under tension, as in the spherical cells.

The spreading in normal macrophages and fibroblasts usually lasts 10 to 40 minutes, respectively, but in the polycaryons formed from EAT cells was prolonged up to 20–40 h. The time needed for spreading of polycaryons which contained four to eight nuclei was usually shorter than in those with dozens of nuclei. The largest, giant multicaryons which exceeded sometimes 1 mm in diameter, did not spread, though showed active changes in their shape. In the polycaryons nuclei coalesced only 2–3 days after the fusion experiment and formed great nuclei of size comparable with that of the whole mononucleate cells.

For further experiments presented here the intermediate size polycaryons, which adhered and well spread on glass, were selected. Figure 1 shows a few examples of the morphology of such polycaryons as observed with DIC light microscope. Frequently only a part of the polycaryon cytoplasm spread forming large, rather flat hyaline lamellipodium, while the nuclei and granular cytoplasm remained in a thick, centrally situated "cell body". Some polycaryons formed narrower and longer surface protrusions. These attached at their tips to the glass and flattened in the region of adhesion. Gradually extending, they pulled and stretched the polycaryon body.

With direct fluorescent staining with TRITC-phalloidin changes in F-actin were followed in the polycaryons spreading on the glass. Mononucleate spherical

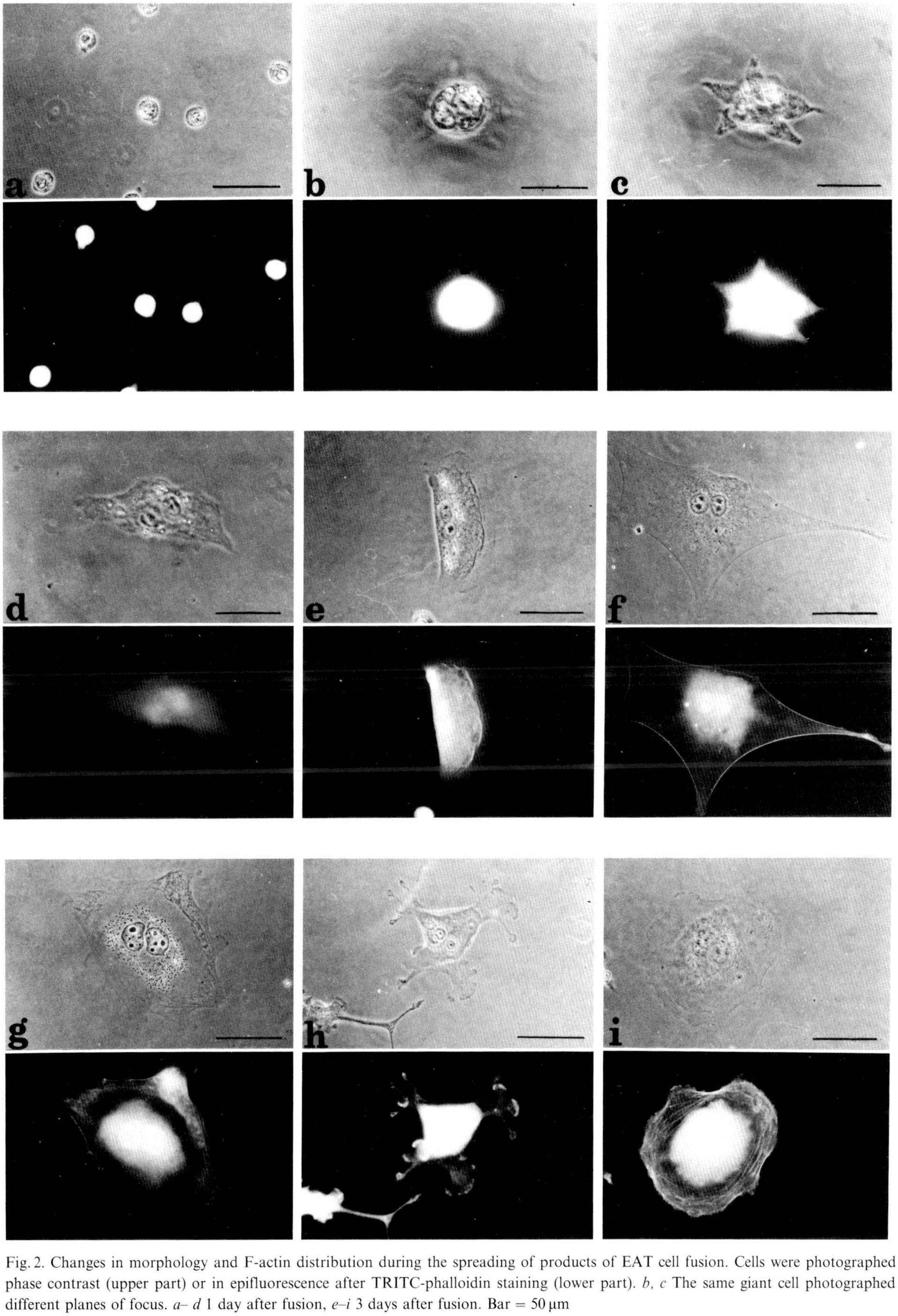

Fig. 2. Changes in morphology and F-actin distribution during the spreading of products of EAT cell fusion. Cells were photographed in phase contrast (upper part) or in epifluorescence after TRITC-phalloidin staining (lower part). *b, c* The same giant cell photographed at different planes of focus. *a– d* 1 day after fusion, *e–i* 3 days after fusion. Bar = 50 μm

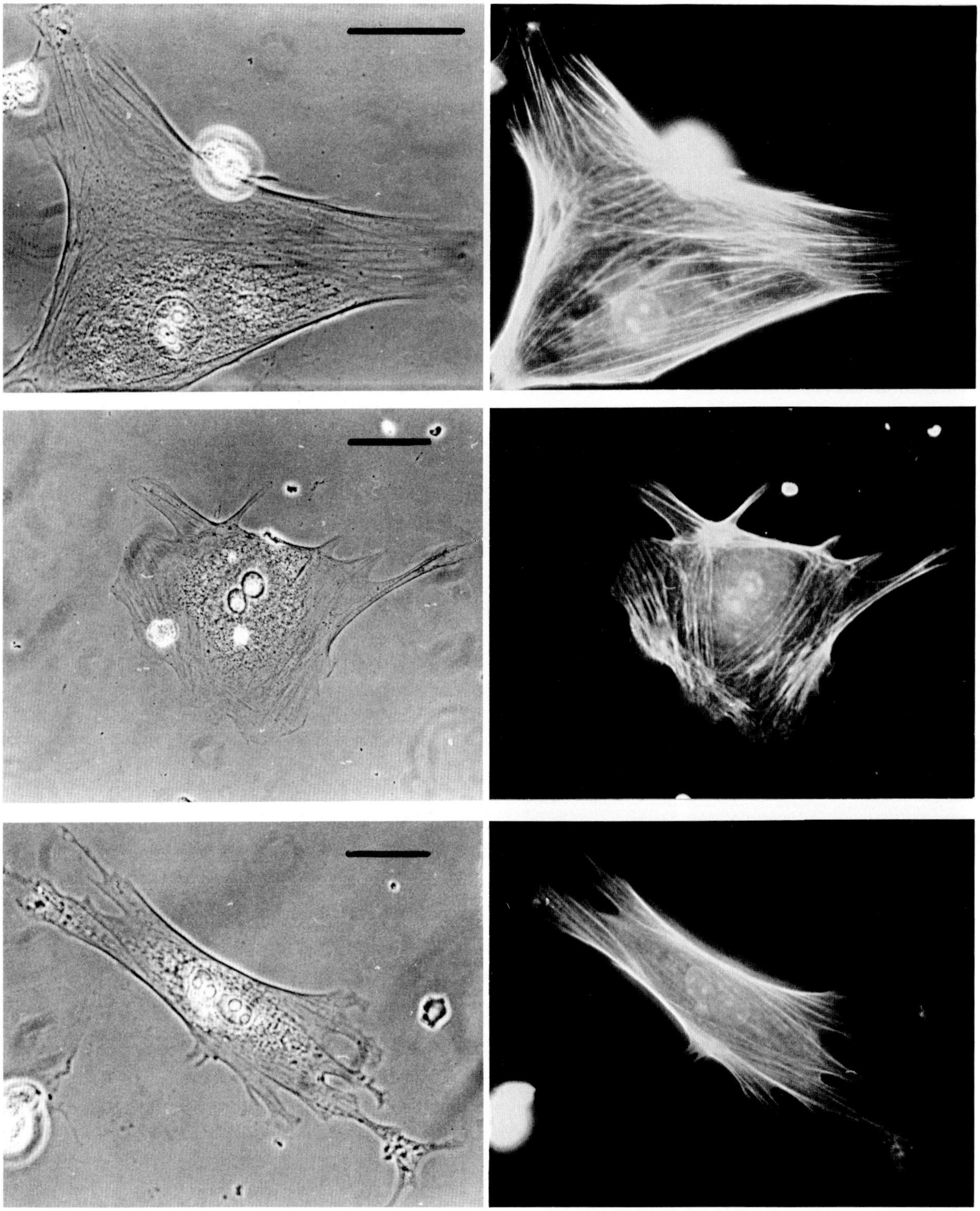

Fig. 3. Formation of stress fibres in the spread products of EAT cell fusion. Cells were photographed in phase contrast (left) or in epifluorescence after TRITC-phalloidin staining (right) 4 days after fusion. Bar = 50 μm

cells (Fig. 2 *a*) showed bright, diffuse fluorescence. Similar fluorescence was observed in the thicker pernuclear body of polycaryons, but here the cortical localization of the microfilament system could be observed more clearly. This was easy to see under the microscope but the thickness of the object makes it difficult to record it in the microphotographs. An additional line of bright fluorescence was often present at the border between the centrally situated thicker granuloplasm and the more hyaline lamelliposdium (Fig. 2 *a*). In spread polycaryons their borders, if remaining under visible tension, showed strong fluorescence in the region beneath the plasma membrane (Fig. 2 *f, g*). The formation of cell surface protrusions occurred where more diffuse distribution of the phalloidin-stained network of the F-actin filaments in the cortical cytoplasm was observed (Fig. 2 *b–d*). Frequently intensively-stained filaments accumulated at the flattened tips of fully spread surface protrusions (Fig. 2 *h*). More organized bundles of actin appeared only in polycaryons remaining 3 days in culture (Fig. 2 *i*). In the following days the polycaryons fully spread and within their cytoplasm typcial stress fibres developed (Fig. 3).

The above described F-actin rearrangements corresponded to various stages of flattening and spreading of the polycaryons and occurred sequentially.

4. Discussion

Malignant-malignant cell hybrids usually retain their malignant character (STANBRIDGE 1976, HARRIS 1985). Such homocaryons contain, however, 2 or 3 nuclei and differ significantly from the polycaryons formed by neoplastic cell fusion or multinucleation with cytochalasin B, as described by BERSHADSKY *et al.* (1981). Here we confirmed that also in EAT cells polynucleation results in changes in many cell features, in particular in those commonly assumed to be associated with cell neoplastic character.

As described in Results the polycaryons formed from the EAT cells flattened, attached to glass and spread. This behaviour was qualitatively different from that of mononucleate EAT cells in the same culture conditions. The changes in morphology probably did not stem from the greater amount of membrane material in polycaryons than in mononucleate cells, as suggested by BERSHADSKY *et al.* (1981). The mononucleate EAT cells under some experimental conditions efficiently spread within 40–60 minutes (CIEŚLAK and KOROHODA 1978) which shows that it is not the insufficient amount of the membrane material that normally prevents their flattening and attachment to the substratum.

The polynucleate EAT cells when spreading did not show any presence of the F-actin bundles in their cytoplasm. Formation of stress fibres in flattened and spread polycaryons was always observed only 4 and more days after their attachment to glass. This clearly shows that the organization of stress fibres is not needed for cell spreading (*cf.* KOROHODA and KAJSTURA 1982, TOSNEY and WESSELLS 1983, ALBRECHT-BUEHLER 1987).

The formation of stress fibres in anchorage-dependent, non-transformed cells requires not only the presence of all necessary proteins building these fibres, but also (i) the attachment of the actin bundles to the cytoplasmic face of the plasma membrane by means of specific proteins (JACOBSON 1983, BRAY *et al.* 1986) and (ii) the creation of tension which permits isometric contraction. Actin bundles usually disintegrate after isotonic contraction (FLEISCHER and WOHLFARTH-BOTTERMANN 1975, KOROHODA and WOHLFARTH-BOTTERMANN 1976, KREIS and BIRCHMEIER 1980, PIES and WOHLFARTH-BOTTERMANN 1985). The condition of isometric contraction requires fairly strong adhesion of cells to the solid substratum and/or neighbouring cells (KOROHODA and KAJSTURA 1982, TOSNEY and WESSELLS 1983, BEREITER-HAHN 1985, OPAS 1987).

Polynucleate products of EAT cell fusion 4 days after spreading developed typical stress fibres. This shows that nuclei of the EAT cells contain all the information for all proteins needed for development of typical cytoskeletal organization of spread cells with the stress fibres present in their cytoplasm, *i.e.* the cytoskeleton organization normally found in non-transformed cells (*cf.* POLLACK *et al.* 1975, BEN ZE'EV 1985). It might be therefore deduced that in the mononucleate EAT cells there is quantitatively insufficient control of cytoplasm by the nucleus.

The cell spreading process is connected with adhesiveness to glass and with dynamic activity of the cell surface complex consisting of cell surface coat, plasma membrane and membrane-associated cytoskeleton forming the cell cortex (AMBROSE *et al.* 1970 b, AMBROSE 1972). The results here presented show that in the polycaryons surface protrusions were formed where in cell cortex filaments were in a diffuse arrangement. One may expect that in these regions of the cell surface complex its deformability locally increased as a result of its transient relaxation. This suggestion corresponds to similar conclusions based on observations of amoebae and microplasmodia (KUKULIES and STOCKEM 1985, KUKULIES *et al.* 1987, BRIX *et al.* 1987). As pointed out by KOLEGA (1987) the tissue cells also cannot easily extend protrusions in the direction perpendicular to tension.

The ascitic cells have spherical shape for G 1, S, and G 2 phases of the cell cycle. Normal, non-transformed cells usually assume spherical shape only during mitosis. They shrink just before metaphase and after telophase they again flatten, attach to the substratum of neighbouring cells and spread. After daughter nuclei formation the cell surface changes rapidly within a few minutes. Simultaneously, after attachment to solid substratum or to other cells their cytoskeleton organizes. During mitosis their cytoplasm shows absence of any stress fibres and diffuse organization of F-actin filaments but after spreading normal organization of their cytoplasm reasumes. In transformed cells the cell surface properties remain similar to those observed during mitosis for the whole G 1 phase. The intermitotic cells remain less strongly attached to the substratum and less spread than non-transformed cells. In extreme cases, as in ascitic or leukaemia cancers, the cells remain spherical throughout the entire cell cycle (see ABERCROMBIE 1962, FOX *et al.* 1971, SMETS and DELEY 1974, LAZARIDES 1975, SANGER and SANGER 1983, ELVIN and EVANS 1983).

The results discussed herein indicate that the polynuclear products of the EAT cell fusion show extending with time relaxation of their surface complex which enables them to produce cell surface processes, to attach to glass, to flatten and spread. After 4 days in culture fully spread polycaryons show typically developed stress fibres. Our working hypothesis would be that ascitic mononucleate cells are in the stage of overall tension in their surface complex due to an insufficient control of the cell cortex contractions by the nucleus. In normal, non-transformed cells only temporally such state arises during mitosis, when the nuclear envelope disintregrates. During other phases of the cell cycle (and in enucleated cytoplasts made from interphase cells) cytoplasm is under efficient control of the nucleus. Cortical contractions can be affected either by secretion of a relaxing substance from the interphase nucleus or by sequestration in it of factors causing unrestrained cortical contractions.

The above formulated working hypothesis need further experimental verification. It should be however pointed out that it is consistent with numerous observations and conclusions concerning othert cells than tissue culture ones. It was for example demonstrated that motionless, enucleated, spherical fragments of amoeba produced fronts and extended pseudopodia at the site of local applications of general anaesthetic (benzene, ether, chloroform) when it was allowed to diffuse from the orifice of a micropipette. Accordingly, it was concluded that in the intact amoeba the nucleus controls the extent and polarization of ectoplasmic contractions (KOROHODA 1977, GRĘBECKI *et al.* 1981). In line with this suggestion polynucleate forms of Amoeba proteus produced microsurgically were unable to locomote and showed generalized relaxation of their cytoplasm. Their oriented movements appeared only when the light was applied in the manner to induce a stronger contraction in one part of the polycaryon (GRĘBECKI *et al.* 1978).

Acknowledgement

This work was partly supported by the grant CPBP 04.01 from the Polish Academy of Sciences. Grateful thanks are extended to the Alexander von Humboldt Foundation (Bonn, Federal Republic of Germany) for donation of the Leitz Orthoplan microscope. TRITC-phalloidin was a kind gift from Prof. H. FAULSTICH (Heidelberg, Federal Republic of Germany).

References

ABERCROMBIE M (1962) Contact-dependent behaviour of normal cells and the possible significance of surface changes in virus-induced transformation. Cold Spring Harbor Symp Quant Biol 27: 427–431

— AMBROSE EJ (1962) The surface properties of cancer cells: a review. Cancer Res 22: 525–548

ALBRECHT-BUEHLER G (1987) Role of cortical tension in fibroblast shape and movement. Cell Motil Cytoskel 7: 54–67

AMBROSE EJ (1972) Cell shapes and cell contacts. Acta Protozool 11: 9–20

— BATZDORF U, OSBORN JS, STUART PR (1970 a) Subsurface structures in normal and malignant cells. Nature 227: 397–398

— OSBORN JS, STUART PR (1970 b) Structure and properties of cell surface complex. In: PORTER G (ed) Liquid crystals and grounded fluids. Plenum Press, Oxford, pp 83–96

BEN ZE'EV A (1985) The cytoskeleton in cancer cells. Biochim Biophys Acta 780: 197–212

BEREITER-HAHN J (1985) Architecture of tissue cells. The structural basis which determines shape and locomotion of cells. Acta Biotheoretica 34: 139–148

— KAJSTURA J (1988) Scanning microfluorimetric measurement of TRITC-phalloidin labelled F-actin: dependence of F-actin content on density of normal and transformed cells. Histochemistry, in press

BERHSHADSKY AD, GELFAND VI, VASILIEV JM (1981) Multinucleation of transformed cells normalizes their spreading on the substratum and their cytoskeleton structure. Cell Biol Int Rep 5: 143–150

BOSCHEK CB, JOKUSCH BM, FRIIS RR, BACK R, GRUNDMANN E, BAUER H (1981) Early changes in the distribution and organization of microfilament proteins during cell transformation. Cell 24: 175–184

BRAY D, HEATH J, MOSS D (1986) The membrane-associated "cortex" of animal cells: its structure and mechanical properties. J Cell Sci [Suppl 4]: 71–88

BRIX K, KUKULIES J, STOCKEM W (1987) Studies on microplasmodia of *Physarum polycephalum*. V. Correlation of cell surface morphology, microfilament organization and motile activity. Protoplasma 137: 156–167

CASTOR LN (1970) Flattening, movement and control of division of epithelial-like cells. J Cell Physiol 75: 57–64

CIEŚLAK J, KOROHODA W (1978) Dextran T-500 induction of spreading in Ehrlich ascites tumour cells on glass surface. Cytobiologie 16: 381–392

— — (1979) Preliminary morphological study on the Ehrlich ascites tumour cells induced to spread on glass at the presence of Dextran T-500. Acta Protozool 18: 123–124

CLARKE GD, STOKER MPG, LUDLOW A, THORTON M (1970) Requirement of serum for DNA synthesis in BHK-21 cells: effects of density, suspension and virus transformation. Nature 218: 438–441

ELVIN P, EVANS CW (1983) Cell adhesiveness and the cell cycle: Correlation in synchronized Balb/c 3T3 cells. Biol Cell 48: 1–10

FLEISCHER M, WOHLFARTH-BOTTERMANN KE (1975) Correlation between tension force generation, fibrillogenesis, and ultrastructure of cytoplasmic actomyosin during isometric and isotonic contractions of cytoplasmic strands. Cytobiologie 10: 339–365

FOLKMAN J, MOSCONA A (1978) Role of cell shape in growth control. Nature 273: 345–349

FOX TO, SHEPPARD JR, BURGER MM (1971) Cyclic membrane changes in animal cells: Transformed cells permanently display a surface architecture detected in normal cells only during mitosis. Proc Natl Acad Sci USA 68: 244–247

GRĘBECKI A, KALININA LV, GRĘBECKA L (1978) Response to light-shade difference in anucleate and polynucleate specimens of Amoeba proteus. Cytobiologie 17: 323–353

— GRĘBECKA L, KŁOPOCKA W (1981) Testing steering functions of the frontal zone in the locomotion of *Amoeba proteus*. Cell Biol Int Rep 5: 595–600

HARRIS H (1985) Suppression of malignancy in hybrid cells: the mechanism. J Cell Sci 79: 83–94

JACOBSON BS (1983) Interaction of the plasma membrane with the cytoskeleton: an overview. Tissue Cell 15: 829–852

KAJSTURA J, KOROHODA W (1988) The improvement of polyethylene glycol-induced cell fusion by incubation of cells in a medium with increased NaCl concentration. Folia Histochem Cytobiol 26: 77–82

KOLEGA J (1986) Effects of mechanical tension on protrusive activity and microfilament and intermediate filament organization in an epidermal epithelium moving in culture. J Cell Biol 102: 1400–1411

KOROHODA W (1977) Experimental induction of locomotion in enucleated fragments of *Amoeba proteus* and its bearing on the theories of ameboid movement. Cytobiologie 14: 338–349

— KAJSTURA J (1982) Demonstration of the significance of isometric contraction for the formation of stress fibers in chick embryo fibroblasts. Folia Histochem Cytobiol 20: 153–156

— WOHLFARTH-BOTTERMANN KE (1976) Effects of relaxation and contraction stimulating solutions on the structure of cytoplasmic fibrils in plasmodia of *Physarum polycephalum*. Acta Protozool 15: 195–202

KREIS TE, BIRCHMEIER W (1980) Stress fiber sarcomeres of fibroblasts are contractile. Cell 22: 555–561

KUKULIES J, BRIX K, STOCKEM W (1987) Studies on microplasmodia of *Physarum polycephalum*. VI. Functional analysis of a cortical and fibrillar actin system by use of fluorescent-analog cytochemistry. Cell Tissue Res 250: 125–134

— STOCKEM W (1985) Function of the microfilament system in living cell fragments of *Physarum polycephalum* as revealed by microinjection of fluorescent analogs. Cell Tissue Res 242: 323–332

LAZARIDES E (1975) Immunofluorescence studies on the structure of actin filaments in tissue culture cells. J Histochem Cytochem 23: 570–628

OPAS M (1987) The transmission of forces between cells and their environment. In: BEREITER-HAHN J, ANDERSON OR (eds) Cytomechanics. Springer, Berlin Heidelberg New York Tokyo, pp 273–285

PIES NJ, WOHLFARTH-BOTTERMANN KE (1985) Enrichment of fibrillar cytoplasmic actomyosin in protoplasmic strands of *Physarum polycephalum* for the production of cell-free models. Cell Tissue Res 239: 365–374

POLLACK R, OSBORN N, WEBER K (1975) Patterns of organization of actin and myosin in normal and transformed cultured cells. Proc Natl Acad Sci USA 73: 544–548

SANGER JW, SANGER JM (1980) Surface and shape changes during cell division. Cell Tissue Res 209: 177–186

STANBRIDGE EJ (1976) Suppression of malignancy in human cells. Nature 260: 17–20

SHIELDS R, POLLOCK K (1974) The adhesion of BHK and PyBHK cells to the substratum. Cell 3: 31–38

TOSNEY KW, WESSELLS NK (1983) Neuronal motility: the ultrastructure of veils and microspikes correlates with their motile activities. J Cell Sci 61: 389–411

WEBER MJ, HALE H, LOSASSO L (1977) Decreased adherence to the substrate in Rous sarcoma virus-transformed chicken embryo fibroblasts. Cell 10: 45–51

Cell and Organelle Division

Protoplasma (1988) [Suppl 1]: 99–112

Relation of F-actin Organization to Microtubules in Drug Treated *Haemanthus* Mitosis

JADWIGA MOLÈ-BAJER and A. S. BAJER*

Department of Biology, University of Oregon, Eugene, Oregon

Received December 21, 1987
Accepted April 4, 1988

Dedicated to Professor Dr. NOBURO KAMIYA on the occasion of his 75th birthday

Summary

We report observations on the effects of cytochalasin D, pressure injection of anti-actin antibody, and of a microtubule-disrupting drug (amiprophos methyl), on the distribution of F-actin bundles in endosperm of the higher plant, *Haemanthus*. The course of mitosis and cell plate formation *in vivo* was recorded with video time-lapse. Cells recorded *in vivo* were fixed and processed with non-fading immunogold and/or immunogold silver-enhanced stain to follow the distribution of actin bundles or microtubules. Double stain (actin and microtubules in the same cell) was applied extensively. Three-dimensional spatial relation of actin to microtubules was analyzed with serial optical sectioning in video microscopy.

In most of the control cells, actin bundles are located in the cytoplasm and only sparse actin strands occur within the mitotic spindles. Cytoplasmic actin reorganizes very little in the control or in the experimentally treated cells. A pronounced reorganization of F-actin, resulting in the distribution characteristic for each stage of mitosis, is restricted to the cell domains where actin bundles are in close contact with microtubules. Actin bundles appear to be gradually eliminated from the spindle. Cytochalasin D does not affect chromosome movements and cell plate formation, but it does change the properties of actin bundles. They are fragmentated more extensively, than those in controls and are nearly totally eliminated from the spindle, phragmoplast and cell plate. The most extensive fragmentation occurs in the phragmoplast. Injection of anti-actin antibody also prevents actin reorganization during mitosis. Shortly after the disassembly of microtubules by amiprophos methyl, the actin network invariably collapses. The presence of microtubules is therefore required for actin rearrangement, but the experimental disturbance of actin has no detectable effect on mitosis and cell plate formation. We conclude, therefore, that actin is not involved in the intrinsic molecular mechanism of mitosis and cell plate formation in the higher plant *Haemanthus*.

* Correspondence and Reprints: Department of Biology, University of Oregon, Eugene, OR 97403, U.S.A.

Keywords: Mitosis; Cell plate; F-actin; Microtubules; Immunogold-silver.

Abbreviations: APM amiprophos methyl; CD cytochalasin D; 3-D three dimensional; IGS immuno-gold stain; IGSS immuno-gold stain silver-enhanced; MT(s) microtubule(s); OMDR optical memory disc recorder.

1. Introduction

Recent advances in visualization of the cytoskeleton with nonfading, high-contrast immunogold (INOUÉ *et al.* 1985) combined with more recently developed serial ultrathin optical sectioning with video microscopy (INOUÉ 1987 b, 1988) provide a new approach to the understanding of the 3-D organization and structural relation of the two omnipresent components of the eukaryotic cytoskeleton, MTs and F-actin (MOLÈ-BAJER *et al.* 1988).

Studies on actin in higher plants were initially concentrated on cytoplasmic streaming (reviewed by KAMIYA 1986) without implication of MTs in this process. It has been documented only recently that in some motile processes in living cells, F-actin filaments and MTs may be very closely associated (KOONCE *et al.* 1986). There is also compelling evidence that actin and tubulin interact *in vitro* (POLLARD *et al.* 1984). It seems unlikely, therefore, that F-actin and MTs do not influence or interfere with each other, merely because of their close association. Both the exact determination of the degree of the functional interaction and the association itself deserve closer attention.

The process where the collective contribution of MTs/

actin filaments to the molecular mechanism has been extensively studied and discussed is mitosis (reviewed by Forer 1985). However, not only all available data did not supply any compelling evidence for functional interaction in mitosis (Aubin *et al.* 1979, Kiehart *et al.* 1984) and other processes (Euteneuer and Schliwa 1985), but all critical data point to the contrary.

F-actin has been located in the mitotic spindle of *Haemanthus* with HMM probes (Forer *et al.* 1979) and in the cell plate (Schmit and Lambert 1985, 1988) with immunofluorescence or with rhodamine-labelled phalloidine. It has been suggested that actin is involved in the motive force production for chromosome movements (Forer *et al.* 1979), and more recently, that it plays a functional role during the cell plate formation (Clayton and Lloyd 1985, Gunning 1982, Gunning and Wick 1985, Schmit and Lambert 1985, 1988). The final steps of actin bundles organization during mitosis in *Haemanthus* were described by Schmit and Lambert (1988). More detailed 3-D information concerning the relations of F-actin to MTs in normal endosperm cells of *Haemanthus* (Molè-Bajer *et al.* 1988) was obtained by use of a new technique of optical ultrathin sectioning (Inoué 1987 a, b) of non-fading immunogold labelled actin and MTs in *Haemanthus*. It appeared that the rearrangement of actin during mitosis and cell plate formation mimics the random distribution of other inert bodies in the spindle. These studies (Molè-Bajer *et al.* 1988) clarify some aspects of actin bundle redistribution and argue on the structural grounds against its intrinsic contribution to the molecular mechanism of mitosis and cell plate formation. Similarly, critical analysis of phragmoplast and cell plate formation of *Allium* root tips led Palevitz (1987) to conclude that MTs are responsible for actin reorganization and not *vice versa*.

This report demonstrates that the experimental perturbance of MTs has a pronounced effect on actin distribution and that changes of actin properties and its behaviour do not influence the course of mitosis and cell plate formation in *Haemanthus*. Thus the present data further strengthen our previous conclusions on the role of actin during mitosis and cell plate formation in *Haemanthus* (Molè-Bajer *et al.* 1988).

2. Material and Methods

Endosperm cells of the African globe (blood) lily, *Haemanthus katherinae* Bak., were used as material for experiments which were carried out at room temperature (21°–24 °C). Cells extruded first on agar-coated cover slips, were flattened in a moisture chamber by surface tension and then covered with a thin layer of Gelrite (Kelco Co., San Diego, CA). Cells thus prepared are shaped like fried eggs, with the flat sides facing the agar and the nuclei protruding towards the Gelrite side. The methods of observations *in vivo* are described previously (Bajer *et al.* 1986 a).

The fixation/permeabilization procedure for *Haemanthus* (De May *et al.* 1982) was followed. It appeared also that more than 80% for 5 nm and 100% for 1 nm immunogold, of cells do not require the presence of detergent during fixation for single immunostain IGS or by IGSS.

Cells were stained either for single (F-actin) or double antigen (F-actin and MTs). The micrographs of simultaneous visualization of F-actin and MTs with double stain during normal mitosis in *Haemanthus* are presented elsewhere (Molè-Bajer 1988). The principle of double stain for video microscopy is based on the differences of the light signal of IGS and IGSS, applied for different antigens. For this report MTs were stained with IGS using 10 nm gold particles and F-actin with 5 nm or 1 nm IGS subsequently silver-enhanced. The two stains can be distinguished by differences in dichroism (Inoué *et al.* 1981) in a custom-designed (Inoué 1986) super-sensitive polarizing microscope, or in mixed epipolarized and transmitted light with the Antiflex method (this report). Video microscopy is used in these two methods to visualize and distinguish two antigens in a single video frame.

Double immuno-stain with permanent label. The best results were obtained with a modified Tapia *et al.* (1983) method, using two primary antibodies applied simultaneously (mouse-anti-actin, rabbit-anti-tubulin). The indirect immunogold stain with two corresponding secondary antibodies, coupled with two non-overlapping gold sizes (GAM IgM 1 or 5 nm for actin and GAR IgG 10 or 15 nm for MTs), followed. Since 1 and 5 nm gold was later enhanced by silver (IGSS), the second step was sequential and not concurrent as in the Tapia *et al.* (1983) method. Application of fish oil (Birrell *et al.* 1987) improved specific binding of immunogold and minimized the background.

The immunostaining procedure was as follows: After standard procedure for IGS with or without prepermeabilization (De May *et al.* 1982) the preparations were incubated with the mixture of both antibodies overnight at room temperature and were processed first for actin. The stain which is not detectable in conventional light microscope (1 and 5 nm IGS), was enhanced with silver (Holgate *et al.* 1983). The procedure described in Janssen Intense Kit II was followed except that, in the last step, the final reduction of silver was done with the fixative from Intense Kit I. Buffering of the fixative (pH 6.9–7) improved the preservation of the second antigen (MTs). Washing in TBS, equilibration in 20 mM BSA buffer (pH 8.2), and incubation with the second secondary antibody coupled with gold (GAR 10 and 15 nm) followed. The development of the label was monitored under the conventional light microscope. On some preparations, chromosomes were counterstained with toluidine blue.

The general principles of video microscopy and ultrathin optical sectioning are outlined by Inoué (1986, 1987 a, b). The course of mitosis *in vivo* (CD, APM-treated and injected cells) was recorded with a Sony 3/4-inch time-lapse recorder model TVO-9000 on an inverted Nikon microscope equipped with X 40 LDW DL N. A. 0.55 and Nikon X 20 DL N. A. 0.4 objectives. The microscope objectives for video microscopy [Nikon X 40 Plan Apo N. A. 0.95, Olympus X 60 Plan Apo N. A. 1.4 (strain-free-factory-selected) and Zeiss X 63 Plan Apo N. A. 1.4] were used in conjunction with a Nikon rectified condenser N. A. 1.35. Zeiss Antiflex Neofluar objective X 63 N. A. 1.25 was used extensively for the analysis in epi-illumination. This objective appeared to be far superior for selectively distinguishing

IGSS F-actin and 10–15 nm immunogold-stained MTs. Similarly, as is conventional light microscopy (HOEFSMIT *et al.* 1986), IGSS-enhanced structures observed with the Antiflex appear several times brighter in video microscopy than do other cellular components.

Optical sectioning of immunogold stained cells was facilitated by the permanent attachment of a mechanical reduction gear to the microscope fine-adjustment shaft. A clutch worm gear assembly connected to a leaf detent stop corresponded approximately to 0.25 μm of axial displacement of fine focus. The slip clutch on the output shaft of the reduction gear permitted manual overriding during routine use of the microscope. High magnification on the TV monitor (5000 X or higher) is required for optical serial sectioning. Estimated thickness of optical sections done with our modified microscope (BAJER *et al.* 1986) and Antiflex Neofluar objective is not more than 0.17 μm, and surpasses measured lateral resolution (0.195 μm) of our system.

The video image captured by a Newvicon camera, (Dage-MTI Model 67 M, Michigan City, IND), was processed digitally as described in the figure legends, using an Image-1 processor (Universal Imaging Corp. Ithaca, NY). Video-processes sequential images at different levels of focus were recorded with OMDR (Panasonic Model TQ-2025 F). Through-focus-time-lapse projection of such sequentially recorded images allowed us to follow and compare the 3-D distribution of F-actin and MTs within a large number of cells. We consider such projection absolutely necessary for evaluation of precise 3-D relations of actin to MTs. For publication, images stored in the OMDR were reproduced by photographing their display on a nine-inch video monitor (Conrac SNA 9-C) on Kodak Panatomic-X film used and processed in Kodak High Contrast developer (110, formula B).

CD was obtained from Sigma (St. Louis, MO) and used at concentrations from 5 μg/ml to 10 μg/ml, *i.e.*, 10 to 20 μM. APM [methyl-0-(4-methyl-6-nitrophenyl)-N-isopropyl-phosphorothioamidate] was obtained from Dr. C. A. Anderson, Chemagro, Mobay Chemical Corp., Kansas City, MO. Monoclonal mouse antibody against actin from chicken gizzard N. 350, which reacts with different isoforms of actin including that in *Haemanthus*, as demonstrated by nitrocellulose blots (SCHMIT and LAMBERT 1988), was obtained from Amersham (Arlington Heights, IL). All the immunogold (GAM IgM) reacting to the mouse antiactin was obtained from Janssen Pharmaceutica, Beerse, Belgium. The immunogold staining was enhanced with a Janssen Pharmaceutica silver-staining kit (Intense II). Cells stained with IGS and IGSS were embedded in Permount (Fisher Sci. Co. Fair Lawn, NJ).

Injected cells were processed in two ways: Injected antibody used as a primary antiactin antibody, or antiactin antibody used as in the standard IGS/IGSS procedure.

The methods of pressure injection of antiactin antibody and APM treatment were the same as those described previously (BAJER *et al.* 1986, BAJER and MOLÈ-BAJER 1986 b). The average size of the injected drop was 1.4×10^{-6} μl. The amount of injected antibody varied from 1.4×10^{-8} μg to 1.4×10^{-6} μg.

3. Results

In interphase cells of endosperm, a spontaneous reorganization of MTs occurs after the removal of cells from young seeds. Thus MT organization just after preparation, and for example, 1 hour later is very different (BAJER and MOLÈ-BAJER 1986 a). Comparable studies on the organization of F-actin were not done. Preliminary observations indicate, however, that actin strands and MT arrays gradually separate and that the former tend to accumulate at the cell periphery. In this report we restrict the analysis to the mitotic cells from endosperm which have been extracted from young seeds 0.5–1 hour before the drug was administered. These data concern mainly anaphase and telophase, when the spatial relation of F-actin to MTs is very characteristic and uniform in control cells.

3.1. Distribution of F-actin and MTs in Control Cells

Rearrangement of MTs during mitosis and interphase of *Haemanthus* endosperm has been studied extensively (summarized BAJER and MOLÈ-BAJER 1986 a). The behaviour of actin bundles has been followed to a much lesser extent (SCHMIT and LAMBERT 1985, 1988, MOLÈ-BAJER *et al.* 1988). During typical mitosis of *Haemanthus* endosperm cell, all MTs are incorporated into the mitotic apparatus, and nearly all actin is located in the cytoplasm. Actin strands are organized as a loose anastomosing 3-D network. The pronounced actin rearrangement during mitosis is restricted to the thin zone between the spindle and the cytoplasm, *i.e.*, to the close proximity of MTs. This initial reorganization results in the decrease of branching of actin strands at the spindle surface and in the tendency to arrange along MT arrays.

Actin has not been detected inside the nucleus, but a variable quantity of actin strands occurs in the prometaphase spindle. It presumably gets there during the abrupt rearrangement of MTs after the breaking of the nuclear envelope. Its amount decreases progressively during prometaphase-metaphase. The irregular network of actin strands in the cytoplasm, except those strands located at the polar regions, does not reorganize significantly during mitosis. Actin bundles at the polar regions of the spindle, often arranged perpendicular to the spindle axis, are usually increasingly compressed during the progress of anaphase (Fig. 1 *a*). Up to mid-anaphase, actin cables at the surface of the spindle and at the rapidly elongating interzone between the separating chromosome groups follow the long axis of the spindle and tend to arrange approximately parallel to the MT arrays (Fig. 1 *a, b*).

During later stages of anaphase, abundant formation of polar MTs takes place within the rapidly disorganizing sister half-spindles. They grow in all directions and may form pronounced aster-like arrays. Most of these MTs elongate, toward the equator and surround the interzone like two domes, which finally fuse and interdigitate at the equator. The ring-shaped cell plate forms in the zone of the interdigitation.

Simultaneously with the elongation of polar MTs, the parallel arrangement of actin cables around the interzone, especially at the vicinity of the equator, is progressively disturbed, leading to their irregular arrangement (Fig. 1 *c*). Increasing numbers of actin filaments accumulate at the equatorial region. These strands rapidly rearrange and become parallel to the phragmoplast axis during the cell plate formation. Finally, some of them are incorporated into the cell plate (Fig. 1 *d, e*). The quantity and average length of actin strands in the cell plate decreases, suggesting that they are gradually broken and the fragments eliminated toward the sister nuclei, as the cell plate ages.

3.2. *Effects of APM*

Depending on the stage of mitosis in which APM is applied, all or most of *Haemanthus* MTs are disassembled (BAJER and MOLÈ-BAJER 1986 b). In interphase MTs are completely depolymerized and during mitosis chromosome movements are arrested instantly. In cells in mitosis prolonged action of APM (1–2 hours at 1–5 μM) results in disassembly of all MTs except short stubs in the old cell plates and occasionally traces at the kinetochores. Shorter treatment or lower concentrations of APM leave only a few kinetochore MTs. The core MTs of kinetochore fibers, which comprise not more than 25 percent of the kinetochore fiber complex, and phragmoplast MTs close to the cell plate, are most resistant and are not disassembled.

The general effect of APM is change of the cell shape, which occurs after disassembly of the MTs. The thickness of the cells determines the degree of change and in most preparations 50 to 80 percent of the cells become rounded to a different extent. Very flat cells retain their shape after the treatment. The change in cell shape causes the sister chromosomes and telophase nuclei often to move back 5–10 μm, somewhat compressing these actin cables located in the interzone between them.

In all cells, the effect of APM on the organization of actin cytoskeleton is not immediate, but finally the actin meshwork collapses. The change of the cell shape always results in some redistribution of actin in the cell cortex. The most typical effect is invasion of the spindle domains, previously occupied by MTs, by the actin network (Figs. 1 *f* and 2). The actin network appears to be much thicker in APM-treated mitotic cells than in controls. The final arrangement of actin strands depends on two major factors, primarily, their organization at the onset of the drug action (stage of mitosis) and secondarily, by the change of the cell shape. Thus, the mitotic stage-dependent organization of actin bundles at the spindle surface is, as a rule, reflected in the final actin arrangement after MT disassembly. In APM-treated very flat cells in which change of shape does not occur, the arrangement of actin bundles is indistinguishable from that in control cells in the same stage, for at least 10 minutes *i.e.*, the time sufficient to disassemble most MTs. Later, however, the arrangement of the actin meshwork changes and appears to be collapsed (Figs. 1 *f* and 2).

3.3. *Effects of CD*

Preparations were perfused with CD, and selected mitotic cells were followed in CD *in vivo* up to 3 hours. The course of mitosis and cell plate formation in about 50 cells was recorded with a video time-lapse tape re-

Fig. 1. F-actin distribution in control (*a–e*) and (*f–g*) APM-treated anaphase. *a* Mid-anaphase. Long actin strands, approximately parallel to the spindle axis, stretch between the polar regions. Compressed irregular bundles, predominantly perpendicular to the spindle axis in the polar area (above the nucleus) occur at the upper pole. At the equatorial region of the spindle only a few actin strands are arranged across the spindle. *b* General view of the same cell (DIC – Normarski). *c* Onset of telophase, illustrating the early stages of the cell plate formation at the surface of the phragmoplast. It is a later stage than is *a*. Parallel organization of actin strands characteristic of mid-anaphase is already disturbed; short pieces are arranged in all directions and accumulate predominantly at the equator of the interzone. Some strands are trapped in a more advanced (older) cell plate, which is at the edges of the phragmoplast (arrow with black core). In the center of the phragmoplast, strands are still longer and more regularly arranged (parallel to the phragmoplast axis). The irregular actin meshwork in the cytoplasm marked by thick arrows does not reorganize during mitosis. Chromosomes are arranged in half-circles; kinetochore regions are marked by hollow arrows. *d, e* Late telophase; cell plate in the same cell at two different levels of focus: *d* close to the surface and *e* within the cell plate. Axial distance between *d* and *e* is about 2 μm. Only short pieces of F-actin strands are present in the cell plate. Several longer strands are near and around the nuclei (marked by *N*). *f* Effects of 10^{-6} M APM on F-actin distribution in anaphase. MTs are depolymerized and the spindle is collapsed. Actin strands invade the area previously occupied by the spindle. Actin distribution differs drastically from that of control cells. *g* General view of the same cell (phase contrast). Image processing: DIC Nomarski, analog enhancement only, for *a, b, d,* and *e*; *c* original image in transmitted polarized light (crossed polarizers) was sharpened through convolution; *f* transmitted polarized light after background subtraction. Microscope objectives: *a, d, e, f* × 60; *b, g* × 40. Scales: 10 μm intervals

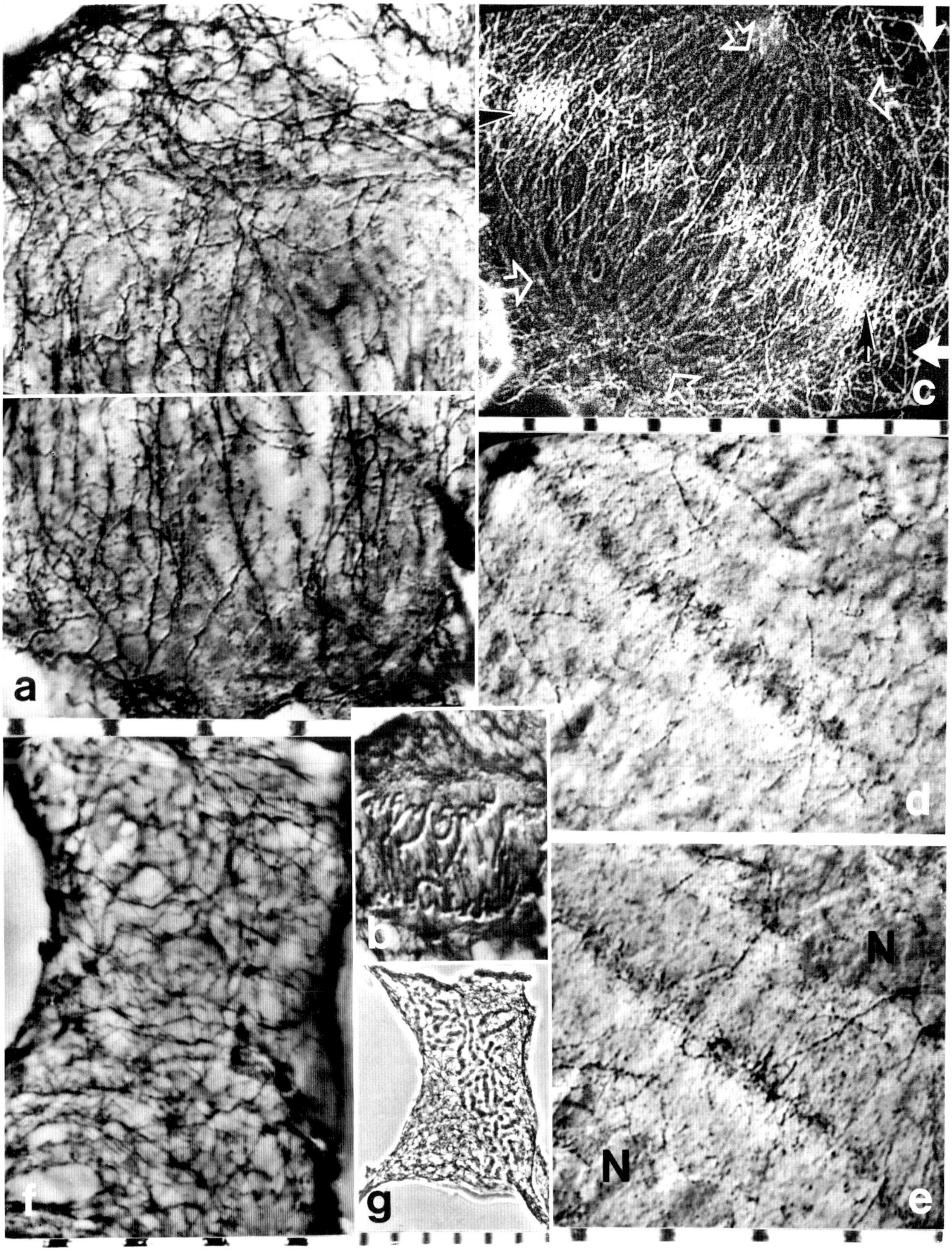

Fig. 1

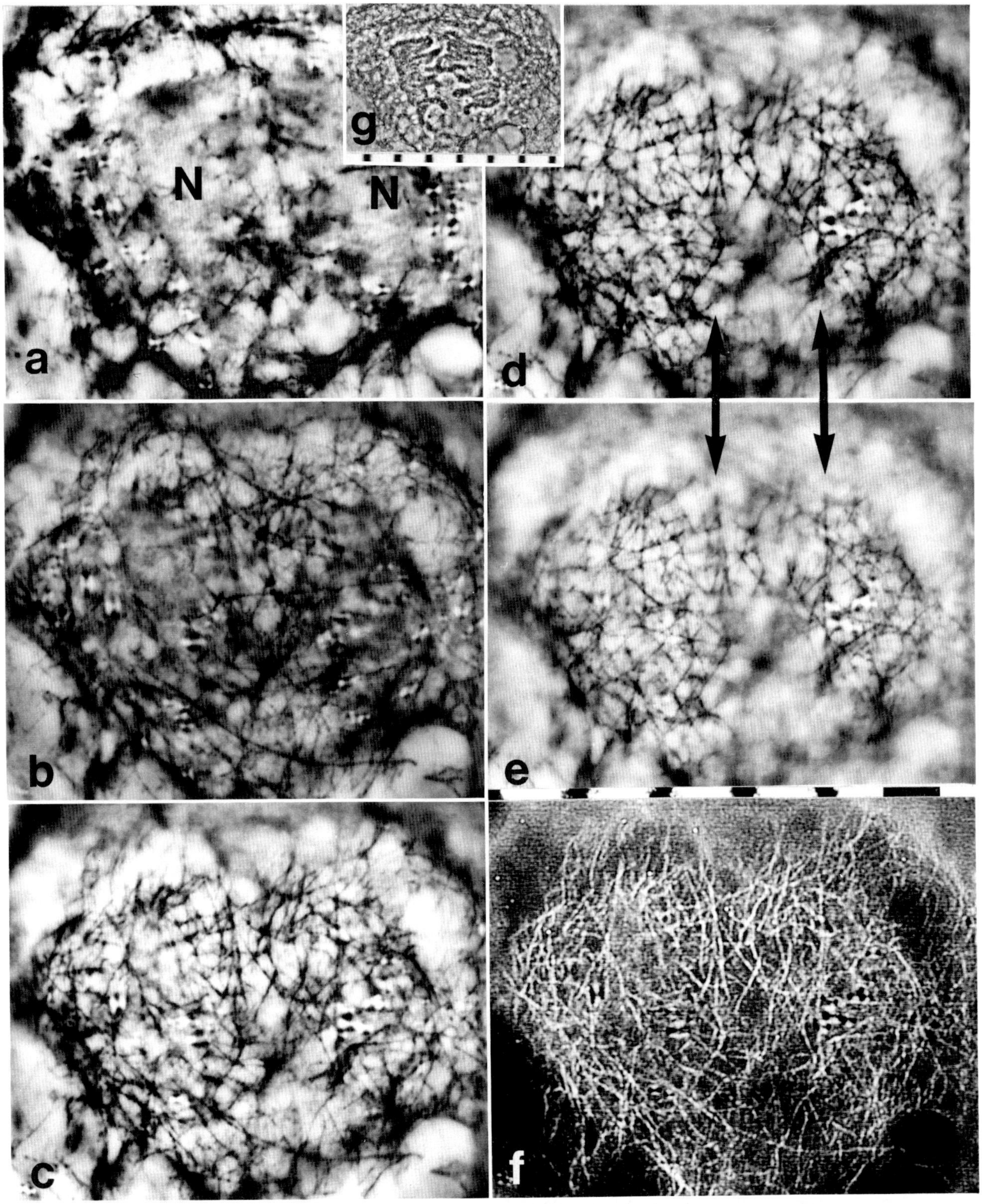

Fig. 2. Effects of 45 minutes action of 10^{-6} M APM on F-actin distribution on a telophase cell. The cell is not flat, and consecutive "sections" show F-actin distribution in the cytoplasm on the surface of the nuclei and the area previously occupied by the phragmoplast (*cf.*, Fig. 1). The latter, which in control cells is almost entirely deprived of actin, is occupied by a thick meshwork of actin strands. *N* nuclei in *a*. The double-headed arrows between *d* and *e* point to compressed meshworks of actin. Such compression takes place during slight translocation of sister nuclei toward the equator, after disassembly of polar MTs in the interzone. *Cf.*, the arrangement of actin strands in the "groove" at the equator in *b* and *c*. *g* General view of the same cell (phase contrast). Image processing: a–*e* transmitted polarized light (crossed polarizers) after background subtraction, *f* original image in bright field with crossed polarizing prisms was sharpened through convolution, and contrast of convoluted image was inverted. In this mode chromosomes are barely visible. Optical sectioning of the same cell. Approximate thickness between sections is: *a–b* 2 µm, *b–c* 0.5 µm, *c–d* 0.5 µm, *d–e* 0.5 µm. *f* the same optical section as *c*. Microscope objectives: *a–f* 60 ×; *g* × 40. Scale 10 µm intervals

corder. Long-term effects (more than 3 hours) of CD were not studied.

MT reorganization during the course of mitosis and cell plate formation was not affected by CD for at least 2 hours after its application. Similarly CD had no effect on the duration of anaphase and cell plate formation. The appearance of the cell plate at our level of resolution was indistinguishable from that in control cells (Fig. 3 *a, b, c*). The effects on actin distribution were time-dependent and were not detected for the first few minutes after the perfusion with the drug solution. Short-term effects (15–30 minutes) were much less pronounced than longer ones (1–2 hours). The effects of short-term action of CD will not be discussed further. The overall result was less actin in the spindle; the actin strands were shorter and often appeared only as fragments arranged along MT arrays. Regardless of the stage of mitosis when the action of CD began, the further course of mitosis was normal.

The global effects of CD on actin distribution were perhaps most striking when cells were observed with low-power (X 20) microscope objectives. The actin network in the cytoplasm and on the surface of the spindle, appeared to be well developed and resembled control cells, while the spindle interior seemed to be completely devoid of actin. The details, however, could not be resolved at low magnifications. Image processing at high resolution appeared necessary to follow the details. Actin strands in the cytoplasm often seemed to be somewhat thicker than those in control cells, and the 3-D meshwork appears somewhat coarser (Fig. 4 *a—f*). Our technique is not adequate, however, to determine that single strands contain more filaments. Similarly, as in control cells, the cytoplasmic actin meshwork did not reorganize appreciably during mitosis.

Video microscopy with high N. A. objectives demonstrates that, on the spindle surface and inside the spindle, F-actin does not form long strands as in control cells, but appears usually as short pieces (Fig. 3 *d, f*). The amount of actin depends on the duration of cytochalasin action and the stage of mitosis. The longer the action of CD, the less F-actin is present and the shorter the pieces. Thus, if the CD was applied in prometaphase, only a negligible number of actin strands is found in the anaphase half-spindles; the strands appeared to be literally pulverized in the phragmoplast, after the cell plate is formed. Little effect was observed if the cells were fixed a short time (less than 15 minutes) after onset of CD action. Thus the overall effect appeared to be considerable depletion of actin in CD-treated spindles/phragmoplasts. The strands were also much shorter, often seen as discontinuous short fragments arranged along MT arrays. Since the amount of F-actin in the spindles/phragmoplasts in control cells was already very small in comparison to cytoplasm, the number in CD-treated cells can be considered negligible. CD seems to leave the fully functional mitotic apparatus with only traces of F-actin. Our methods did not allow us to estimate quantitatively and to compare the amount of actin in control and CD-treated cells.

3.4. Injection of Antiactin Antibody

Fifty-five cells in various mitotic stages were injected with antiactin antibody. Six cells did not survive injection. All other cells completed mitosis, followed by the formation of a phragmoplast and cell plate (Fig. 5 *a—f*). The number of occasional minor disturbances, such as chromosome bridges or defective cell plates, did not exceed those in control cells recorded under similar conditions, with strong, continuous uninterrupted illumination during the recording with a video time-lapse tape recorder.

Injection of antiactin has a pronounced, concentration-dependent effect on the distribution of actin strands. This effect seems to be restricted only to the spindle and phragmoplast, *i.e.*, cell regions containing MT arrays. The injected area and its vicinity retain characteristic features of mechanical interference which is often detected in video microscopy even a few hours after the injection. The injected region of the cell frequently remains bulged and is surrounded by concentric actin cables. Actin strands are detected only in the region of the cell close to the injected spot, if the injected antibody is used as the only primary antibody. The organization of these actin strands is, however, changed. They appear shorter and crumbled, and their distribution appears to be no longer related to the MTs cycle which is so characteristic in control cells. Thus, actin bundles seem to be immobilized by the antibody and their anaphase arrangement may persist through telophase. Actin strands look somewhat thinner and the stain is less compact than in untreated cells.

4. Discussion

New optical and immunostaining techniques—serial ultrathin optical sectioning combined with video microscopy (Inoué 1987 a, b) and double permanent immuno-staining (Molè-Bajer *et al.* 1988, Molè-Bajer 1988)—allowed us to relate the 3-D organization of actin strands to that of MTs during normal mitosis. Extensive illustrations of organization of actin strands and MTs visualized with double stain in control cells has been also presented (Molè-Bajer 1988). The same technique employed in this report was combined with observations in living cells. This experimental approach provided the information on actin behaviour in conditions when either MTs or actin were experimentally disturbed. It appears that the 3-D organization of F-actin strands during mitosis is determined by MTs, while experimental disturbance of actin is not reflected in the MTs organization of the spindle. If MTs are disrupted, then, the final 3-D organization of the actin network depends on the manner MTs are disassembled and on the stage of mitosis. These conclusions are consistent with observations of Palevitz (1987, 1988 b).

We have used APM, one of the most potent drugs known, which inhibits the assembly of higher plants MTs *in vitro* (Morejohn and Fosket 1984) and *in vivo* (Bajer and Molè-Bajer 1986 b). This drug disrupts all except kinetochore (core only) and phragmoplast ring MTs at μM concentration. Vinblastine (Schmit and Lambert 1986) and colchicine (Molè-Bajer 1958), drugs very effective for animal MTs, disassemble *Haemanthus* MTs in living cells slowly and only at non physiological levels, at about 100 times higher concentrations. They have also several adverse and impossible to-characterize side effects, and the redistribution of actin is not as drastic (Schmit and Lambert 1987) as in APM. The disassembly of MTs by APM has a delayed but inevitable effect on F-actin redistribution in *Haemanthus*. The general appearance of actin bundles does not seem to change although the branching decreases and the number of sharp kinks increases. Gradually, the actin network invades the space previously occupied by MTs, suggesting that the general trend during mitosis is the progressive decrease of intermingling of actin strands with MT arrays and that rare close intertwining has no functional significance. The final 3-D organization of actin strands depends on the duration of APM action and stage of mitosis, and may be affected by the change of the cell shape. The latter alone cannot, however, account for the pronounced redistribution of actin cables after MT disassembly. The time delay of this effect suggests that MTs are instrumental in maintaining or producing the tension which keeps actin strands stretched or anchored to MTs. Thus, although MT arrays and the actin framework form separate entities during mitosis, MTs have a pronounced influence on the 3-D redistribution of actin strands. The effect of MTs is limited to the narrow region of the unavoidable close contact between MTs and actin bundles, primarily the border between the spindle and the cytoplasm.

Similarly, the experimental disturbance of actin by CD is only expressed in close proximity to MT arrays and does not have a noticeable effect on the MT cycle in mitosis or cell plate formation in *Haemanthus*. Thus the organization of actin strands, well defined and different for each mitotic stage in control cells, is therefore not a prerequisite either for the MT reorganization in the successive mitotic stages or for regular progress of mitosis and cell plate formation, both in *Haemanthus* endosperm and in epidermal cells of another monocot, *Allium* (Palevitz 1980).

These conclusions concern comparatively short (less than 2 hours) action of CD. Long-range, indirect effects may not be not detected with our techniques and cannot be excluded. The structure and further development of the cell plate is unknown, because only early stages of cell plate formation can be followed in *Haemanthus*. We do not know whether at the EM level the cell plate in CD-treated cells is identical to the controls and whether it would develop a normal cell wall.

We assume that molecular principles of actin assembly (Korn *et al.* 1987) for all actin isoforms are similar *in vitro*, but the regulation of actin turnover *in vivo* may be different in various cell types, especially in higher plant cells. The action of cytochalasins on *Haemanthus* endosperm and other higher plant cells seems to vary. We cannot distinguish among several possible explanations of cytochalasin action. One possibility, which

Fig. 3. Effect of CD on mitosis and cell plate formation. *a, b, c* Mitosis in a living cell (phase contrast). The action of CD (5 μg/ml) started 35 minutes before *a*. *a* Prometap hase, *b* anaphase, *c* telophase with a well developed cell plate. The course of mitosis and cell plate formation are normal and the duration of mitosis is within the range for control cells. *d, e, f* The same cell after immunostain (15 nm IGS for actin only). IGS gives much weaker signal than does IGSS, and actin strands are visualized in epipolarization (Antiflex) as very thin strands (*f*). Fibers seen in d with DIC Nomarski represent most likely bundles of MTs visualized non-selectively owing to image processing. Only small fragments of actin are left in the interzone (*e*). Actin strands are less affected at the polar areas between the nucleus and the cell boundary, where MT network is often poorly developed. Note that the signal of 15 nm gold is less intense than that of silver: *i.e.*, in cells processed with IGSS (*cf.*, Fig. 4). *c, d* Black arrows mark the same nucleus. Black arrow with white margins in *d* and white arrows in *e* and *f* mark the sister nucleus. Image processing: *d* DIC Nomarski with background subtraction; *e, f* epi-illumination (Antiflex) with background subtraction. Microscope objectives: *a–c* × *40*, *d–f* × 63 Antiflex. Scale 10 μm intervals

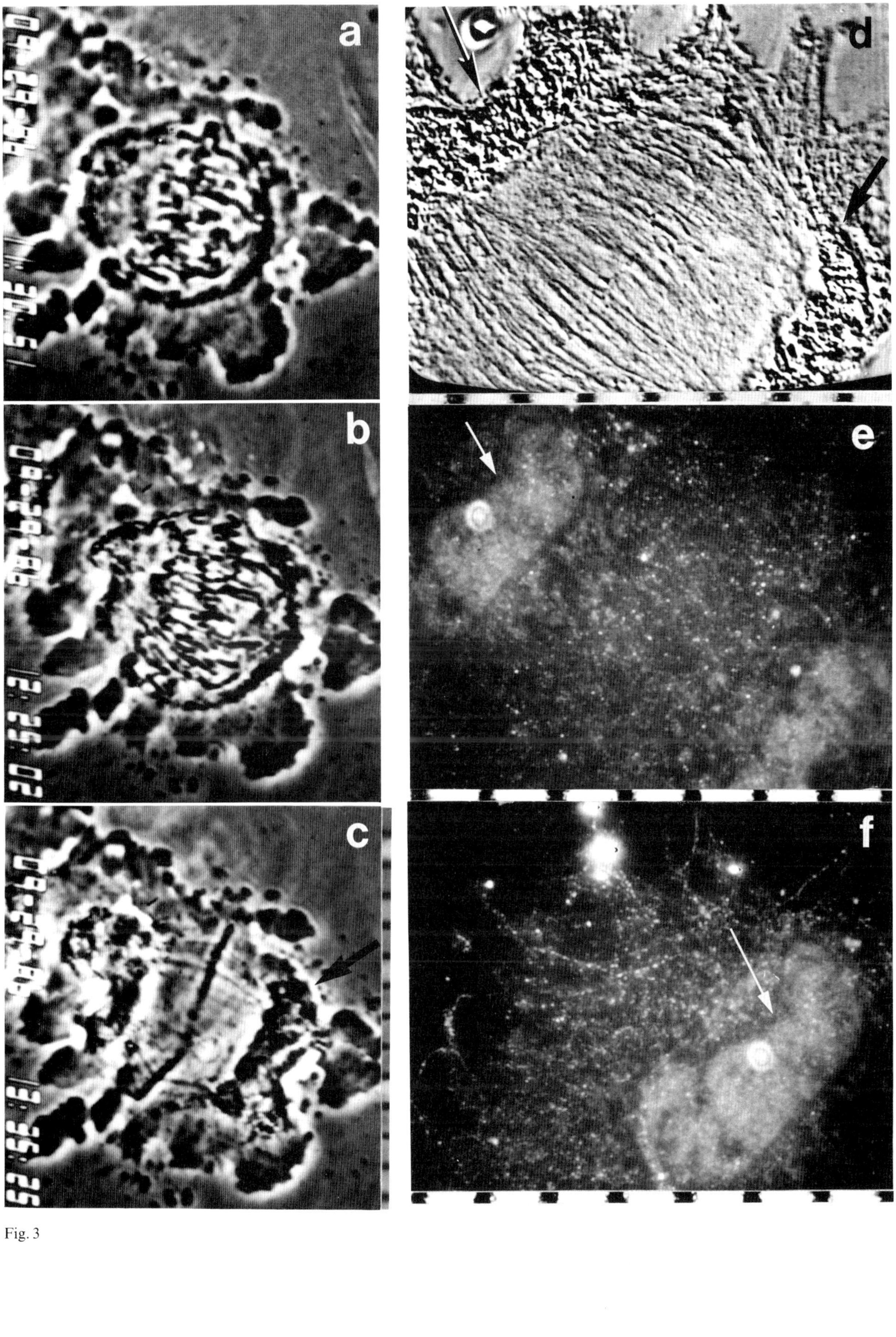

Fig. 3

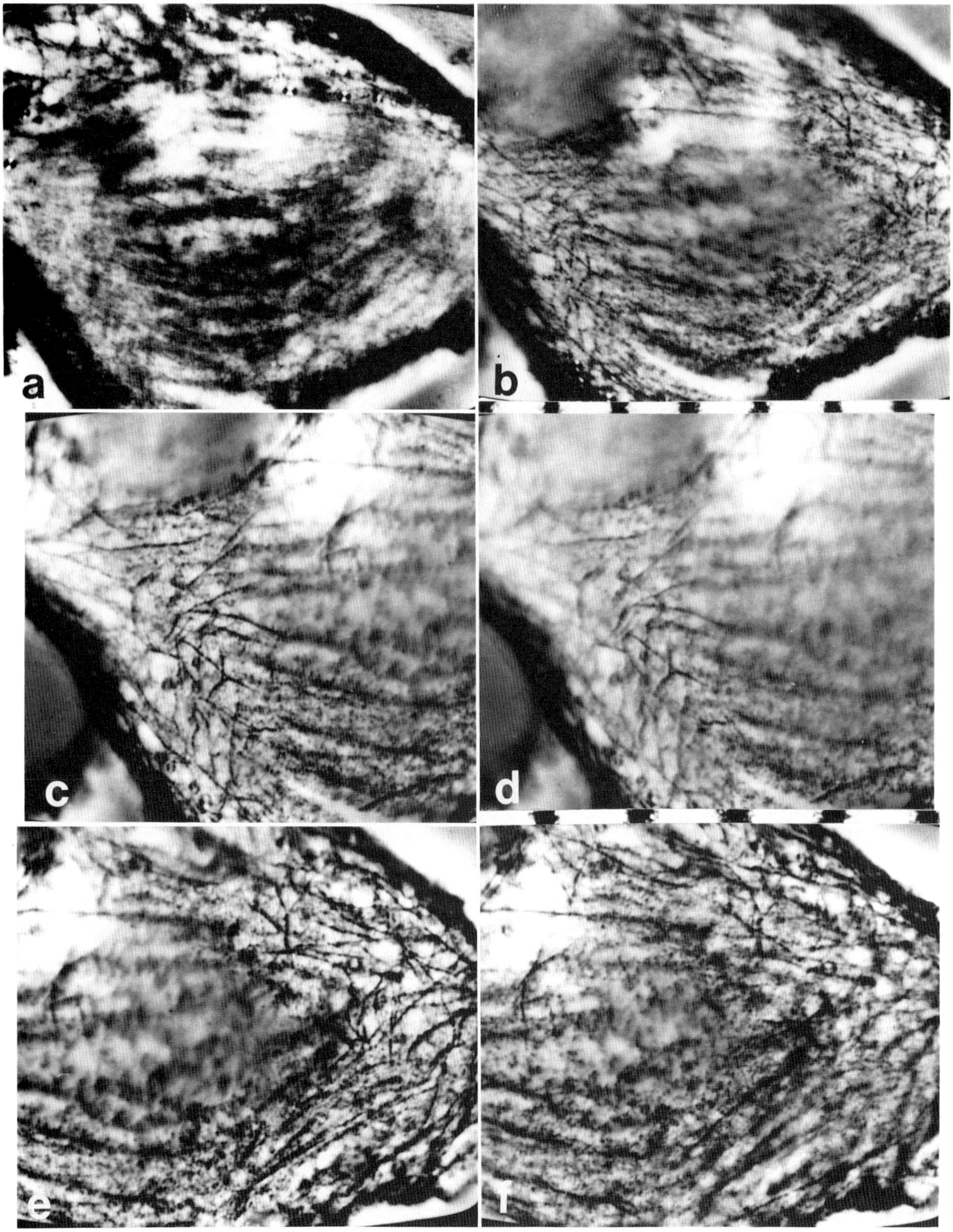

Fig. 4. Distribution of F-actin strands in anaphase under the influence of CD (5 µg/ml). The action of CD (5 µg/ml) started 45 minutes before *a*. Only traces of F-actin are present inside the spindle and on its surface while it remains well developed in the surrounding cytoplasm. *a*, *b* Two optical sections of the interzone (the center) of the same cell 4.5 µm apart; *a* chromosome level and *b* surface of the spindle. *c*–*f* Details of actin strands organization in two sister half-spindles under high magnification. *c*, *d* Two optical sections through one half-spindle and cytoplasm and *e*, *f* through the second. The distance between *c*–*d* and *e*–*f* is 1 µm. Image processing: transmitted polarized light (crossed polarizers) after background subtraction. Microscope objective: × 63 Antiflex. Scale 10 µm intervals

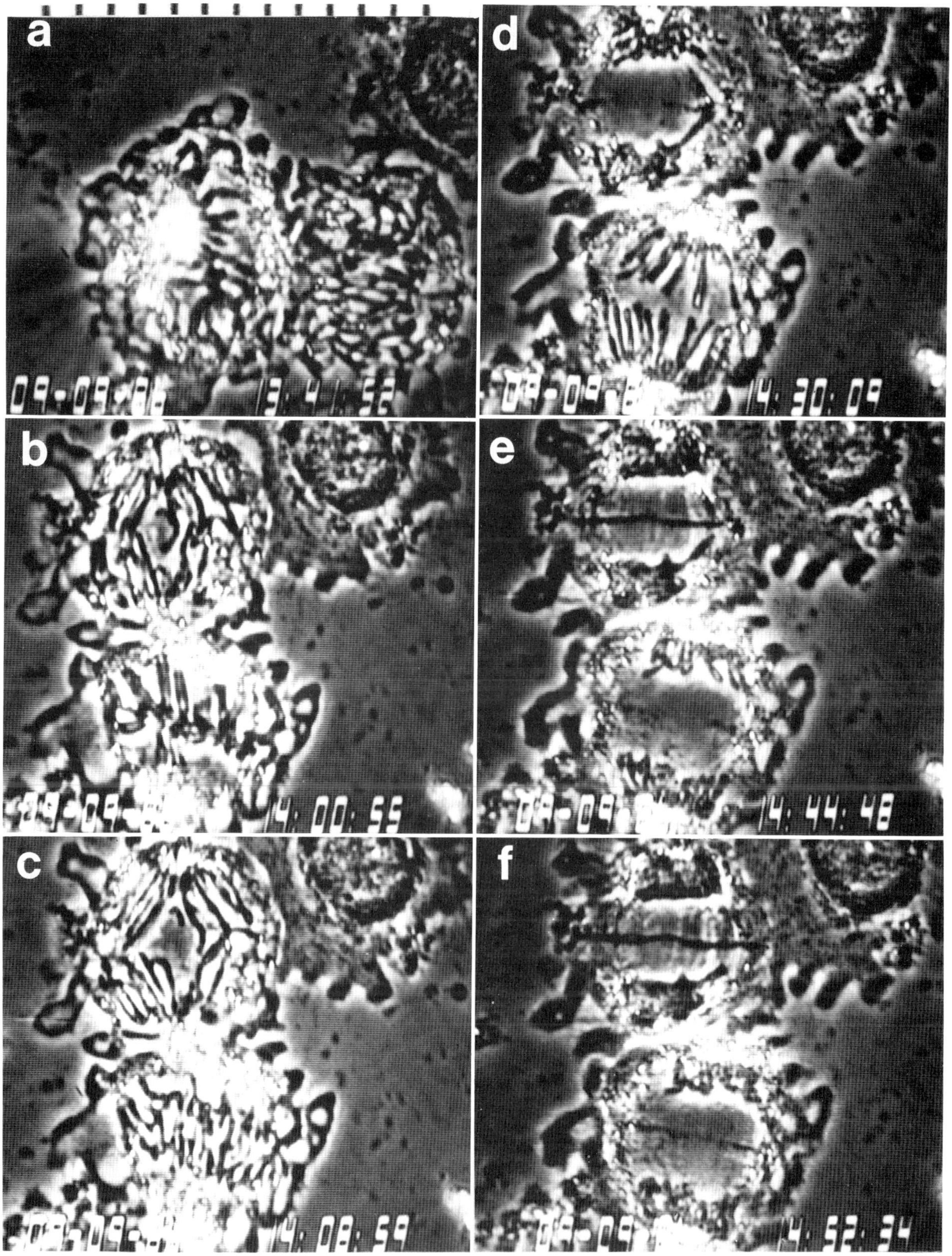

Fig. 5. The course of mitosis after injection of antiactin (concentration 1 mg/ml). Both anaphase and late prometaphase cell (*a*) were injected. The actual injection of prometaphase cell shown in *a*. In both cells the course of mitosis and the cell plate formation are normal. The duration of the mitotic stages is within the range for control cells. The position of cells in *b*–*f* is at 90° in comparison to *a*. Phase contrast. Microscope objective: × 40. Scale 10 μm intervals

seems rather unlikely, is that this variance reflects the differences between immuno- vs. phalloidin derivative stain, used extensively in severel earlier studies. Cytochalasins are small rapidly penetrating molecules with high binding affinity and capping properties for the barbed end of actin *in vitro* (reviewed by Cooper 1987). We used CD, which is 10 times more efficient than the often previously applied cytochalasin B. The latter disrupts other processes in cells and the results may not be fully comparable with CD and Cooper (1987) provides convincing evidence against its use. Thus very high concentrations (1×10^{-4} mM) of cytochalasin B were needed for reproducible results (Parthasarathy 1985); lower concentrations resulted in unclear data (Gunning and Wick 1985). Cytochalasins may (Dancker and Low 1979, Hertwig and Stossel 1979) or may not (MacLean-Fletcher and Pollard 1980) shorten actin filaments, but in general they do not appear to have severing properties in several animal cell types. The data on *Haemanthus in vivo* and those on other monocots, *Avena* coleoptile (Parthasarathy 1985) and *Allium* root tips (Palevitz 1988 a), suggest, however, that cytochalasins always have pronounced, variable and often severing effects on actin in higher plant cells. They produce characteristic spots or foci in *Allium*, similar to those previously reported in different cell types (Weber *et al.* 1976, Pardue and Parthasarathy 1985, Menzel and Schliwa 1986, and others). Such spots were not observed in *Haemanthus*, where the changes of some physical properties of actin bundles seem to appear *in vivo*. The nature of such changes cannot be closely analyzed with our techniques. It is possible that plant actin, perhaps due to different and still unknown associated proteins, fragments more easily than animal actin. It can be fragmented not only by CD but by increase of free Ca^{2+} (2×10^{-6} M) in pollen tubes in the presence of ionophore A23187 (Kono and Shimmen 1984).

It remains unexplained why both the immunostaining and the pattern of redistribution of actin bundles in *Haemanthus* after CD treatment differs considerably from control cells. Actin strands in the spindle, but not in the cytoplasm, have a punctate appearance. They appear to fragment only if they are in close contact with MTs. Cytochalasins inhibit either polymerization, filament/filament interaction, or both (Brown and Spudich 1979, MacLean-Fletcher and Pollard 1980). It is therefore plausible that, in *Haemanthus*, CD changes the mechanical properties of actin, and the strands are subsequently broken by MT-related transport. The result of prolonged CD action is that only traces of actin are left in the spindle and phragmoplast. The mitosis and the cell plate formation proceed normally, and it seems unlikely that actin contributes to the basic molecular mechanism of these processes, unless it is needed only in traces. Our data support experiments on sea urchin mitosis (Kiehart *et al.* 1982).

The effects of antiactin antibody injection, although not conclusive, due to the lack of information on biochemical properties of plant actin and associated proteins, are fully consistent with these conclusions. We believe that antiactin antibody diffuses in the cell because identically produced injected antitubulin antibody arrests instantly mitosis in *Haemanthus* (Molè-Bajer unpublished). It seems unlikely that antiactin antibody would behave differently, but we do not know why there is an apparent change in immunostaining properties of actin after injection.

We cannot explain why actin in *Haemanthus* appears to behave differently from that in other plant species (Palevitz 1988 a, b, Clayton and Lloyd 1985), nor can we distinguish among several alternatives. Well-defined actin bundles in the cytoplasm are not detected during mitosis and within the cell plate of *Allium* (Palevitz 1988 a, b) or in several other plant species (Clayton and Lloyd 1985) with the immunofluorescence technique. One possible explanation is the differences in techniques, both procedures and visualization combined with specific response of material. Kakimoto and Shibaoka (1987) could not simultaneously preserve actin and MTs with paraformaldehyde fixation. Glutar aldehyde fragments actin *in vitro* (Lehrer 1981) while in *Haemanthus* both F-actin and MTs are well preserved, as demonstrated by Schmit and Lambert (1987). We used a high-contrast nonfading label (IGS/IGSS) combined with digital video microscopy and a modified microscope (Bajer *et al.* 1986) while a majority of other studies are based on immunofluorescence and/or fluorescent phalloidin derivatives. Our initial magnification in video microscopy and resolution are much higher than those used with fluorescent methods. It seems possible, therefore, that with the higher magnification and resolution applied to other much smaller higher plant cells, diffuse immunofluorescence would be visualized as thin bundles.

It is not surprising that strands are not resolved in the cell plate with fluorescent methods. The latter may provide a weak signal which may be difficult to distinguish from the nonspecific background (see Figs. 47–60 in Gunning and Wick 1985). Image processing is required in *Haemanthus* to resolve the dense aggregation of actin bundles in the cell plate, and details cannot be resolved in conventional light microscopy. Conversely, actin may disassemble during the rather harsh (as compared to *Haemanthus*) immuno-procedures required for cells with cellulose cell walls. Still another interpretation (an additional assembly in *Haemanthus* endosperm during our procedures) cannot be excluded, although it is not supported by the following observation: in more then 80 percent of *Haemanthus* cells, actin bundles stain without any permeabilization when we use standard EM glutaraldehyde fixation. Since the distribution in prepermeabilized and non-prepermeabilized cells is identical, the latter explanation seems implausible. The latter observations suggest that the lack of diffuse fluorescence, which perhaps indicates the presence of monomeric actin, cannot be attributed to the leaking during prepermeabilization. The problem may be more complex, because there are always some silver particles in the cytoplasm. It is unknown at present whether they are comparable to diffuse immunostain or represent nonspecific background.

Our major conclusions are that the redistribution and the behaviour of F-actin cables during mitosis is related to the presence of MTs and that the experimental disturbance of actin is not reflected in the global MT organization and cell plate formation. We assume, therefore that, because of the close association in the spindle and on its surface, MT-related transport redistributes actin in mitosis. It is not a negligible transport and it may break chromosome arms and eliminate them from the spindle (BAJER and VANTARD 1988). We believe that underestimation of the role of this transport and implication of actin in the mechanism to cell plate formation (GUNNING 1982) contributed to the present confusion. The final pattern of distribution of actin in *Allium* (PALEVITZ 1988 a), although in a different form from that in *Haemanthus*, indicates that it is redistributed by the same type of MT-related transport (MOLÈ-BAJER *et al.* 1988). Thus our data demonstrate MT/actin interaction but argue against the molecular functional involvement of F-actin in the cell plate formation and mitosis in higher plants (in two processes where MTs play an essential but not clearly understood role).

Acknowledgements

We thank Dr. SHINYA INOUÉ (MBL, Woods Hole, MA), Dr. BARRY PALEVITZ (U. of GA, Athens, GA), Ms. CAROL COGSWELL (U. of O), and MILDRED DETLING for critical reading of the MSS, constructive comments and for sharing with us their latest results; to Mr. HARRISON HOWARD for printing of the micrographs, and Mr. STEVE CROWE for editorial help. Supported by NSF (DCB 850-1264) to ASB.

References

AUBIN J, WEBER K, OSBORN M (1979) Analysis of actin and microfilament-associated proteins in the mitotic spindle and cleavage furrow of PtK_2 cells by immunofluorescence microscopy. A critical note. Exp Cell Res 124: 93–109

BAJER AS, SATO H, MOLÈ-BAJER J (1986) Video microscopy of colloidal gold particles and immuno-gold labelled microtubules in improved rectified DIC and epi-illumination. Cell Struct Funct 11: 317–330

— MOLÈ-BAJER J (1986 a) Reorganization of microtubules in endosperm cells and cell fragments of the higher plant *Haemanthus in vivo*. J Cell Biol 102: 263–281

— — (1986 b) Drugs with colchicine-like effects that specifically disassemble plant but not animal microtubules. Ann NY Acad Sci 466: 767–784

— VANTARD M (1988) Microtubule dynamics determine chromosome transport of acentric fragments. Mut Res (in press)

BIRRELL BG, HEDBERG KK, GRIFFITH OH (1987) Pitfalls of immunogold labelling: analysis by light microscopy, transmission electron microscopy, and photoelectron microscopy. J Histochem Cytochem 35: 843–853

BROWN SS, SPUDICH JA (1979) Cytochalasin inhibits the rate of elongation of actin filament fragments. J Cell Biol 83: 657–662

CLAYTON L, LLOYD CW (1985) Actin organization during the cell cycle in meristematic plant cells. Exp Cell Res 156: 231–238

COOPER JA (1987) Effects of cytochalasin and phalloidin on actin. J Cell Biol 105: 1473–1478

DANCKER P, LOW I (1979) Complex influence of cytochalasin B on actin polymerization. Z Naturforsch Sec B 34 c: 555–557

DE BRABANDER M, GEUENS G, NUYDENS R, MOEREMANS M, DE MEY J (1985) Probing microtubule-dependent intracellular motitility with nanometre particle video ultramicroscopy (nanovid ultramicroscopy). Cytobios 43: 273–283

DE MEY J (1983) Colloidal gold probes in immunocytochemistry. In: POLAK JM, VAN NOORDEN SW (eds) Immunocytochemistry. Wright-PGS, London, pp 92–110

EUTENEUER U, SCHLIWA M (1985) Evidence for an involvement of actin in the positioning and motility of centrosomes. J Cell Biol 101: 96–103

FORER A, JACKSON WT, ENGBERG A (1979) Actin in spindles of *Haemanthus katherinae* endosperm. II. J Cell Sci 37: 349–371

— (1985) Does actin produce the force that moves a chromosome to the pole during anaphase? Can J Biochem Cell Biol 63: 585–598

GUNNING BES (1982) The cytokinetic apparatus: its development and spatial regulation. In: LLOYD CW (ed) The cytoskeleton in plant growth and development. Academic Press, New York, pp 229–299

— WICK SM (1985) Preprophase bands, phragmoplasts, and spatial controls of cytokinesis. J Cell Sci [Suppl] 2: 157–179

HERTWIG JH, STOSSEL TP (1979) Cytochalasin B and the structure of actin gels. J Mol Biol 134: 539–553

HOEFSMIT ECM, KORN C, BLIJLEVEN N, PLOEM JS (1986) Light microscopical detection of single 5 and 20 nm gold particles used for immunolabelling of plasma membrane antigens with silver enhancement and reflection contrast. J Microsc 143: 161–169

INOUÉ S, MOLÈ-BAJER J, BAJER AS (1981) Dichroism of mitotic microtubules displayed by gold conjugated anti-tubulin. J Cell Biol 91: 321 a

— (1986) Video microscopy. Plenum Press, New York, pp 584

— (1987 a) Video microscopy of living cells and dynamic molecular assemblies. Appl Optics 26: 3219–3225

— (1987 b) Ultrathin optical sectioning tomography achieved witn the light microscope. Biol Bull 173: 419

— (1988) Imaging of unresolved objects, superresolution, and precision of distance measurement, with video microscopy. In: Taylor DL, Wang Y-L (eds) Fluorescence microscopy of living cells in culture: Quantitative fluorescence microscopy: Imaging and spectroscopy. Academic Press, New York (Methods in cell biology, vol 30, in press)

Kamiya N (1986) Cytoplasmic streaming in giant algal cells: a historical survey of experimental approaches. Bot Mag 99: 441–467

Kiehart DP, Mabuchi I, Inoue S (1982) Evidence that myosin does not contribute to force production in chromosome movement. J Cell Biol 94: 165–178

Kakimoto T, Shibaoka H (1987) Actin filaments and microtubules in the preprophase band and phragmoplast of tobacco cells. Protoplasma 140: 151–156

Kono T, Shimmen T (1984) Ca^{2+}-induced fragmentation of actin filaments in pollen tubes. Protoplasma 122: 145–152

Koonce MP, Euteneuer U, McDonald KL, Menzel D, Schliwa M (1986) Cytoskeletal architecture and motility in a giant freshwater amoeba, *Reticulomyxa*. Cell Motil Cytoskel 6: 521–533

Korn ED, Carlier MF, Pantaloni D (1987) Actin polymerization and ATP hydrolysis. Science 238: 638–644

Lehrer SS (1981) Damage to actin filaments by glutaraldehyde: protection by tropomyosins. J Cell Biol 90: 459–466

MacLean-Fletcher S, Pollard TD (1980) Mechanism of action of cytochalasin B on actin. Cell 20: 329–341

Menzel D, Schliwa M (1986) Motility in siphonaceous green alga Bryopsis. II. Eur J Cell Biol 40: 286–295

Molè-Bajer J (1958) Cine-micrographic analysis of c-mitosis in endosperm. Chromosoma 9: 332–358

— (1988) F-actin redistribution during mitosis and cell plate formation in higher plant *Haemanthus*. Simultaneous visualization of F-actin and microtubules with non-fading immunomarkers in conventional and video microscopy. (Submitted)

— Bajer AS, Inoué S (1988) 3-D localization and redistribution of F-actin in higher plant mitosis and cell plate formation. Cell Motil Cytoskeleton 9 (in press)

Morejohn LC, Fosket DE (1984) Inhibition of plant microtubule polymerization *in vitro* by the phosphoric amide herbicide amiprophos-methyl. Science 224: 874–876

Palevitz BA (1980) Comparative effects phalloidin and cytochalasin B on motility and morphogenesis in *Allium*. Can J Bot 58: 773–785

— (1987) Actin in the preprophase band of *Allium cepa*. J Cell Biol 104: 1515–1519

— (1988 a) Cytochalasin induced reorganization of actin in *Allium* root cells. Cell Motil Cytoskeleton, in press

— (1988 b) Accumulation of F-actin during cytokinesis in *Allium*. Correlation with microtubule distribution and the effects of drugs. Protoplasma 141: 24–32

Parthasarathy MV (1985) F-actin architecture in coleoptile epidermal cells. Eur J Cell Biol 39: 1–12

Perdue TD, Parthasarathy MV (1985) *In situ* localization of F-actin in pollen tubes. Eur J Cell Biol 39: 13–20

Pollard TD, Selden SC, Maupin P (1984) Interaction of actin filaments with microtubules. J Cell Biol 99: 33 s–37 s

Schmit AC, Lambert AM (1985) F-actin dynamics is associated to microtubule function during cytokinesis in higher plants. A revised concept. In: De Brabander M, De Mey J (eds) Microtubules and microtubule inhibitors. Elsevier, Amsterdam, Ser 8, pp 243–252

— — (1988) Characterization and dynamics of F-actin in higher plant endosperm cells during interphase, mitosis and cytokinesis. J Cell Biol 105: 2157–2166

Tapia F, Varndell JM, Probert L, De Mey J, Polak JM (1983) Double immunogold staining method for the simultaneous ultrastructural localization of regulatory peptides. J Histochem Cytochem 31: 977–981

Weber K, Rathke PC, Osborn M, Franke WW (1976) Distribution of actin and tubulin in cells and glycerinated cell models after treatment with cytochalasin B (CB). Exp Cell Res 102: 285–297

Protoplasma (1988) [Suppl 1]: 113–121

Characterization of Monoclonal Antibodies Against the Mitotic Apparatus Associated 51-kD Protein and the Effect of the Microinjection on Cell Division in Sand Dollar Eggs

K. OHTA*,[1], M. TORIYAMA[1], Y. HAMAGUCHI[2], S. MAEKAWA[1], S. ENDO[1], and H. SAKAI[1]

[1] Department of Biophysics and Biochemistry, Faculty of Science, University of Tokyo, and [2] Biological Laboratory, Tokyo Institute of Technology, Tokyo

Received December 28, 1987
Accepted April 1, 1988

Dedicated to Professor Dr. NOBURO KAMIYA on the occasion of his 75th birthday

Summary

We prepared monoclonal antibodies against the protein fraction capable of reassembling into microtubule organizing granules (MTOGs) from eggs of sea urchin, *Hemicentrotus pulcherrimus*. Immunoblot analysis presented evidence that the antigen to the monoclonal antibodies corresponded to the mitotic apparatus associated 51-kD protein which has been shown to be involved in the aster forming activity *in vitro*. Indirect immunofluorescence staining of mitotic eggs showed that the monoclonal antibodies stained the centrosomal region, basal regions of asters and half spindles in the same fluorescence pattern stained by a polyclonal antibody to the 51-kD protein. Some of the monoclonal antibodies cross-reacted with mitotic apparatus (MA) proteins of several other kinds of sea urchins. The immunoreactive proteins had molecular weights close to that of the 51-kD protein of *Hemicentrotus pulcherrimus*. Localization of the immunoreactive 52-kD antigen in *Clypeaster japonicus* MAs was generally the same as that of the 51-kD protein in *Hemicentrotus pulcherrimus* MAs. When one of monoclonal antibodies was injected into *Clypeaster japonicus* eggs before prophase, the formation of the MA was not observed, thereby no cleavage occurred. When injected at prometaphase, the spindle formed was shorter than the control, causing failure of nuclear as well as cell division. These results suggested that the 52-kD antigen in *Clypeaster japonicus* eggs is prerequisite for the formation of a functional spindle.

Keywords: 51-kD protein; Sea urchin egg; Mitotic apparatus; Monoclonal antibody; Microinjection; Tubulin; Microtubules; Microtubule organizing granules; Spindle matrix.

* Correspondence and Reprints: Department of Biophysics and Biochemistry, Faculty of Science, University of Tokyo, Bunkyo-ku, Tokyo 113, Japan.

Abbreviations: MA mitotic apparatus; MTOG microtubule organizing granule; PC-fraction phosphocellulose column 0.5 M KCl eluate; TBS Tris-buffered saline; PBS-II phosphate-buffered saline-II; SDS-PAGE sodium dodecyl sulphate-polyacrylamido gel electrophoresis; DAPI 4'-6-diamidino-2-phenylindole dihydrochloride; CHO chinese hamster ovary; DMSO dimethylsulfoxide; Ig immunoglobulin; CBB Coomasie brilliant blue; FITC fluorescein isothiocyanate.

1. Introduction

Microtubule organizing centers have been shown to govern the spatial arrangement of astral and spindle microtubules in mitosis (PICKETT-HEAPS 1969, MAZIA 1984, BRINKLEY 1985). In sea urchin eggs, clusters of granular materials with unit granules of 90 nm in diameter have been detected in the centrosome regions of the MA (ENDO 1979 and 1980). They are considered to nucleate microtubules, and are called microtubule organizing granules (MTOGs) (ENDO *et al.* 1985). The MTOGs were isolated by the homogenization of isolated MAs at low temperature. They could initiate microtubule assembly *in vitro*, thus forming asters (TORIYAMA *et al.* 1984). MTOGs were solubilized in a buffer solution containing 0.5 M KCl-50% glycerol, and proteins capable of reassembling into granules were fractionated by phospho-cellulose column chromatography. The 0.5 M KCl eluate (PC-fraction) formed granules upon dialysis against a solution of low ionic

strength. When incubated with tubulin, the granules (reconstituted MTOGs) initiated microtubule assembly to form asters in a manner plus end distal to the center. The PC-fraction contained a major basic protein of 51,000 molecular weight which was different from tubulin (51-kD protein). Using polyclonal antibodies raised in rabbits against the 51-kD protein and affinity purified, indirect immunofluorescence staining revealed that the 51-kD protein was localized in the centers of asters formed *in vitro* (TORIYAMA et al. 1988). The antibodies also stained sperm asters, centrosome regions, basal regions of astral microtubules, and half spindles (OHTA et al. submitted). The localization of the 51-kD protein except in the spindle suggested that the 51-kD protein may play an important role in initiating microtubule assembly *in vivo*. Spindle microtubules are surrounded by a structural matrix (spindle matrix) which plays roles in the formation of the spindles as well as the chromosome movement (PICKETT-HEAPS et al. 1984). The 51-kD protein in spindle was supposed to play a role in stabilizing spindle microtubules which lay through the spindle matrix.
Though affinity purified polyclonal antibodies showed high specificity, the antibody concentration was not high enough for some purposes. We prepared monoclonal antibodies against the 51-kD protein that allows us to perform experiments that require higher concentration of the antibody, such as microinjection and affinity chromatographic preparation of a large amount of purified 51-kD protein. This paper deals with properties of the monoclonal antibodies and the influence of one of the antibodies on cell division which was observed after microinjection into sand dollar eggs.

2. Materials and Methods

2.1. Immunization and Cell Fusion

Female BALB/c mice were immunized with the PC-fraction prepared from a whole egg homogenate as described before (TORIYAMA et al. 1988). About 60 µg of protein in a Rivi Adjuvant System (+ trehalose dimycolate and monophosphoryl lipid A) emulsion (RIBI Immunochem Research, Inc.) were injected intraperitoneally. The mice were further immunized with 60 µg of protein twice. Thereafter they received a last boost injection of 240 µg proteins. On the 4th day after the last boost, they were killed and the spleens were removed for cell fusion. Hybridomas (pai) were prepared according to the method of KÖHLER and MILSTEIN (1975). After fusion of spleen cells and pai myeloma cells in 50% polyethylenglycol 4000 (Merck), cells were plated into 24-well plates.

2.2. Screening

Hybridomas were selected in a medium containing hypoxanthine, aminopterin, and thymidine, and screened by a immunoblotting method. The fraction one step before the phosphocellulose column chromatography (TORIYAMA et al. 1988) was electrophoresed and transferred to Durapore membranes (GVHP, 0.22 µm, Millipore) by the method of TOWBIN et al. (1979). After blocking by 0.3% gelatin in Tris-buffered saline (TBS; 20 mM Tris-HCl (pH 7.4), 150 mM NaCl), vertical strips were sliced, dried up, and stored at room temperature in a dark box. The strips were incubated in culture media overnight at room temperature. The following immunoblot procedure was performed according to the method of TOWBIN et al. (1979). A mixture of anti-mouse IgA, G, and M (Cappel) was used for peroxidase-linked secondary antibody. Selected hybridomas were subcloned twice on 94-well plates with feeder layers of mouse spleen cells by means of limiting dilution.

2.3. Preparation of Gamma Globulin Fractions of Monoclonal Antibodies and Antibody Classification

Positive clones were then grown to larger volumes and injected into mice intraperitoneally. Ascites fluids were obtained from the mice, and gamma globulin fractions were prepared by precipitation with ammonium sulphate at 50% saturation. Antibodies were finally dialyzed against phosphate-buffered saline II (PBS II; 250 mM NaCl and 50 mM KH_2PO_4, pH 7.4). SDS-polyacrylamide gel electrophoresis (SDS-PAGE) of the gamma globulin fraction showed that they contained gamma globulin dominantly.
Antibody classes of monoclonal antibodies were analyzed by immunoblottings using subclass specific secondary antibodies (Zymed Laboratories).

2.4. Electrophoresis and Immunoblottings

SDS-PAGE was performed on 10% polyacrylamide gel according to LAEMMLI (1970). The final concentration of SDS in the gel was 0.1%. SDS was from Nakarai Chemicals, LTD. The pH of the separation gel was 8.8, and that of the stacking gel was 6.8. Two-dimensional electrophoresis was performed according to the method of nonequilibrium pH gradient electrophoresis (O'FARREL et al. 1977). Gels were stained with Coomasie brilliant blue R-250. Immunoblotting was performed by the method of TOWBIN et al. (1979) with 1/100 dilutions of the gamma globulin fractions of the HP 1 and HP 2 antibodies and 1/250 dilution of anti-tubulin antibody. The mouse monoclonal anti-α-tubulin antibody was from Amersham, Inc. (U.K.).

2.5. Immunofluorescence Microscopy

Indirect immunofluorescence staining against sections of paraffin embedded eggs was carried out as described elsewhere (OHTA et al. in press). Culture media containing antibodies were directly applied on egg sections. In the case of the gamma globulin fractions, 1/100–1/1000 dilutions were used. We used 1/250 dilution of the anti-tubulin antibody for immunostaining. For control experiments, culture medium from a negative clone was reacted with egg sections. For detection of DNA, 4'-6-diamidino-2-phenyl-indole dihydrochloride (DAPI, Sigma, 10 µg/ml) was added to the secondary antibody solution.

2.6. Antibody Blocking Experiments of Aster Forming Activity

Freshly prepared MTOG fractions without glycerol from the isolated MAs of *Pseudocentrotus depressus* were mixed with antibodies and goat IgG. Final dilution were 1/4 (AP17-1 and 1 mg/ml goat IgG, 3% BSA in 150 mM NaCl, 50 mM KH_2PO_4, pH 7.4) and 1/10 (gamma globulin fractions of HP 1 and HP 2 antibodies in PBS II).

The mixture incubated for 10–30 min at room temperature. The aster forming activity was measured as described (TORIYAMA *et al.* 1988).

2.7. Preparation of Eggs

We obtained gamates of sea urchins, *Hemicentrotus pulcherrimus, Anthocidaris crassispina, Pseudocentrotus depressus*, and *Temnoplurus hardwicki*, by injecting 0.5 M KCl into the coelomic cavity. Gametes of the sand dollar, *Clypeaster japonicus* were obtained by the injection of 1 mM acetylcholine in sea water into the coelomic cavity. The fertilization membranes were removed by treatment of the fertilized eggs with 1 mM 3-amino-1,2,4-triazole (SHOWMAN and FOERDER 1982) or 1 M urea.

2.8. Isolation of MAs from Sea Urchin Eggs and Centrosomes from Chinese Hamster Ovary (CHO) Cells

MAs were isolated from eggs at metaphase according to the glycerol-DMSO method (ENDO *et al.* 1983).
Isolation of centrosomes from CHO cells was performed according to the methods of MITCHISON and KIRSCHNER (1984), GOSTI-TETSU *et al.* (1986) and BORNENS *et al.* (1987) with some modifications. In the method of MITCHISON and KIRSCHNER, centrosomes were concentrated on a Ficoll cushion by centrifugation. This step was eliminated. GOSTI-TETSU *et al.* and BORNENS *et al.* used a 40, 50, and 70% discontinuous sucrose gradient set. We employed a 40 and 60% alternative discontinuous sucrose gradient set.

2.9. Microinjection and Microscopy

Microinjection was carried out as described by HIRAMOTO (1974) and HAMAGUCHI *et al.* (1985). The gamma globulin fraction of the HP 1 antibody (20 mg/ml) was injected into eggs of *Clypeaster japonicus* 30 min (10 pl), and 45–46 min (24–25 pl) after insemination. Final concentration of the HP 1 antibody in the eggs was 0.68 mg/ml. For the control, we injected with 24–25 pl of 20 mg/ml rabbit IgG in PBS II 45–46 min after insemination. Observations by polarization and differential interference microscopy were previously described (HAMAGUCHI *et al.*, 1985). The microscopic images were recorded on Kodak Tri-X film (400ASA).

2.10. Others

Protein concentration was determined by the method of BRADFORD (1976) and LOWRY *et al.* (1951). Whole egg sample of *Hemicentrotus pulcherrimus* and *Clypeaster japonicus* for SDS-PAGE were prepared as described below. Eggs at metaphase were homogenized in 10 vol of 100 mM PIPES, 0.5 mM $MgCl_2$, 10 mM EGTA, at pH 6.8 and 0 °C. SDS was added to the samples to make a final concentration of 6%, followed by immediate boiling for 10 min. The boiled samples were centrifuged at 350,000 × **g** and 25 °C for 20 min. The supernatant was used for the whole egg sample. Myeloma cell line pai was a gift from Dr. TAKAO ARAI (Institute of Basic Medical Science, University of Tsukuba).

3. Results

3.1. Detection of Antigens to Monoclonal Antibodies by Immunoblotting

We obtained eight antibody generating clones. Their antigens were analyzed by immunoblottings. Figure 1 shows the antigens in the fraction used for the screenings. Figure 1, lane *sc*, shows the CBB staining of the fraction used for the screenings. Eight clones were categorized into four groups (Fig. 1). Four hybridomas (HP 1, 2, 5 and 6) generated monoclonal antibodies against a 51,000 molecular weight protein which existed dominantly in the fraction used for the screenings (Fig. 1, lane *a*, arrow). We also obtained antibodies against minor components of the PC-fraction. Antibodies from two hybridomas (HP 3 and 8) reacted with

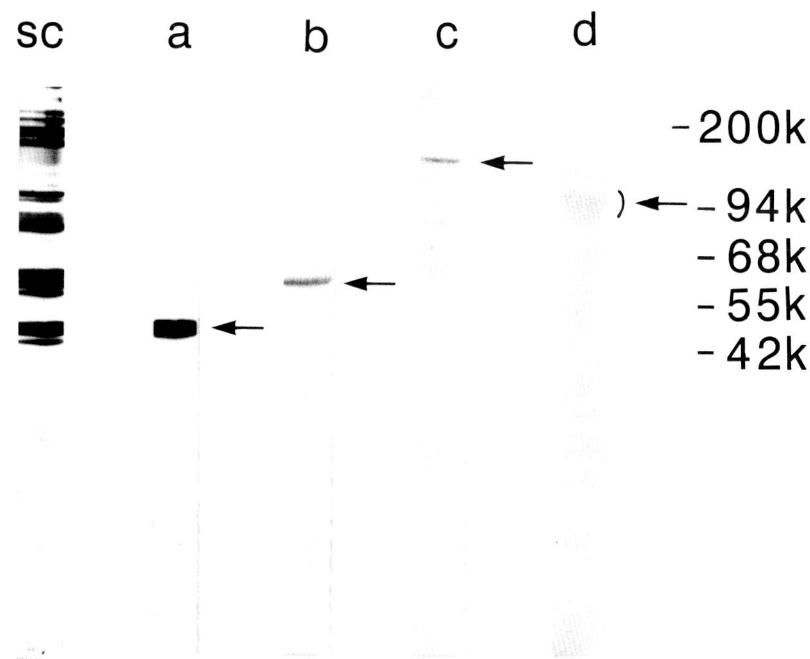

Fig. 1. Immunoblotting analysis using monoclonal antibodies. Antigens of monoclonal antibodies in the fraction used for the screenings were shown. Lane *sc* shows the CBB staining of the fraction used for the screening. Monoclonal antibodies were categorized into four groups. Lane *a*, immunoblot of the HP 1 antibody. The same results were obtained in cases of the HP 2, 5, and 6 antibodies. These antibodies reacted with a 51,000 molecular weight protein (arrow). Lane *b*, immunoblot of the HP 3 antibody. The HP 8 antibody also gave the same results. The arrow indicated a 63,000 molecular weight antigen. Lane *c*, immunoblot of the HP 4 antibody. The antibody recognized a 140,000 molecular weight protein. Lane *d*, immunoblot of the HP 7 antibody. The antibody detected multiple polypeptides whose molecular weights were in a range from 110,000 to 1000,000. The apparent molecular masses of the standard proteins (in daltons) are indicated to the right of lane *d*. Proteins in the fraction used for screening were separated by SDS-PAGE in 10% polyacrylamide gel

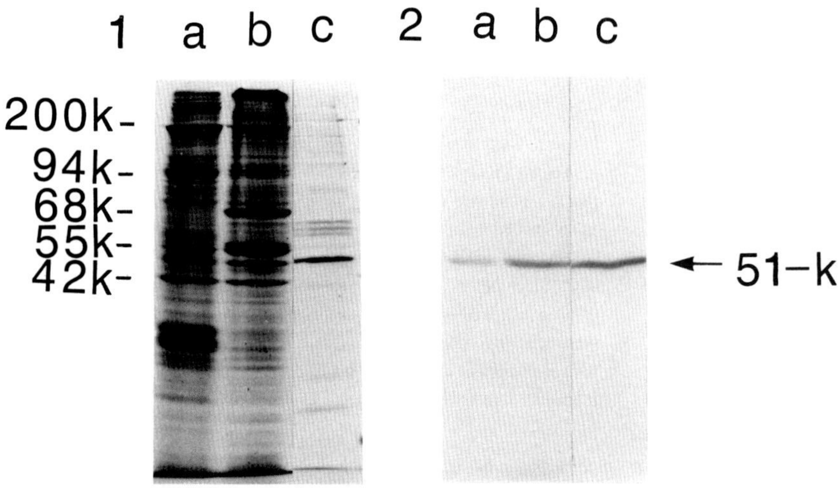

Fig. 2. Distribution of the 51,000 molecular weight antigen detected by the HP 1, 2, 5, and 6 antibodies during the fractionation of MTOGs. Panel *1*, CBB staining, panel *2*, immunoblot of the HP 1 antibody. *a* Whole metaphase egg homogenate, *b* isolated MAs, *c* PC-fraction from whole metaphase egg homogenates. The HP 2, 5, and 6 antibodies presented the same results. Arrow indicates the 51,000 molecular weight antigen. The apparent molecular masses of the standard protein (in daltons) are indicated to the left of panel *1*

a 63,000 molecular weight protein (Fig. 1, lane *b*, arrow). Two other hybridomas secreted antibodies against a 140,000 molecular weight protein (HP 4) (Fig. 1, lane *c*, arrow) and 110,000–100,000 molecular weight proteins (HP 7) (Fig. 1, lane *d*, arrow). The anti-tubulin antibody did not reacted with any protein in the fraction used for the screenings. HP 1, 2, 3, and 6 antibodies were all of the IgG class, and HP 4, 5, 7, and 8 were all IgM.

3.2. Distribution of the 51,000 Molecular Weight Antigen Detected by Monoclonal Antibodies HP 1, 2, 5, and 6 During the Fractionation of MTOGs and Its Localization in MA of Hemicentrotus pulcherrimus

The distribution of the 51,000 molecular weight antigen detected by the HP 1, 2, 5 and 6 antibodies during the fractionation of MTOGs was examined by immunoblotting. Figure 2 shows the results of an immunoblotting with the HP 1 antibody against fractions from *Hemicentrotus pulcherrimus* eggs. The antigen of the HP 1 antibody existed in all fractions examined, that is, the whole metaphase egg homogenate (Fig. 2, lane *a*), the isolated MAs (lane *b*), and the PC-fraction (lane *c*). The HP 1 antibody specifically reacted with the 51,000 molecular weight protein (arrow), which was enriched in the PC-fraction. The same results were obtained with the HP 2, 5 and 6 antibodies. Relative amounts of the 51,000 molecular weight protein in these fractions corresponded to those of the 51-kD protein (TORIYAMA *et al.* 1988). The anti-tubulin antibody reacted with a protein band just above the 51-kD protein in the whole metaphase egg homogenate and the isolated MAs, but not in the PC-fraction. Further, when immunoblotting by the HP 1 protein was carried out after 2-dimensional electrophoresis of the isolated MAs, we could not detect any protein which had the same molecular weight as the 51-kD protein other than the 51-kD protein.

The localization of the 51,000 molecular weight antigen in *Hemicentrotus pulcherrimus* eggs was studied by postembedding immunofluorescence labeling of paraffin embedded anaphase eggs. Figure 3 shows the staining patterns of the anti-tubulin antibody (Fig. 3 *A*) and the HP 1 antibody (Fig. 3 *B*). The antigen of the HP 1 antibodies is localized in the centrosomes, the spindle and the basal regions of the astral fibers. As a control, the culture medium from a negative clone was applied instead of HP 1. It revealed no staining. This localization was quite identical to that of the 51-kD protein, which was detected by polyclonal antibodies (OHTA *et al.* in press). The same results were obtained using the HP 2, 5 and 6 antibodies. From these results, we concluded that the 51,000 molecular weight antigen detected by the HP 1, 2, 5 and 6 antibodies was the 51-kD protein which was previously reported (TORIYAMA *et al.* 1988).

3.3. Cross-Reactivity of the HP 1 and 2 Antibodies with Proteins in MAs from Several Kinds of Sea Urchin Eggs

We examined the cross-reactivity of the HP 1 and 2 antibodies to proteins of MAs from *Anthocidaris crassispina, Pseudocentrotus depressus, Temnoplurus hardwicki*, and *Clypeaster japonicus*. Figure 4 shows that the HP 1 antibody reacted with 52-kD proteins in MAs of *Anthocidaris crassispina, Temnoplurus hardwicki*, and *Clypeaster japonicus* (Fig. 4, lanes 2 *b*, 4 *b*, and 5 *b*, respectively), and a 51-kD protein of *Pseudocentrotus depressus* (Fig. 4, lane 3 *b*). For the blotting of MA

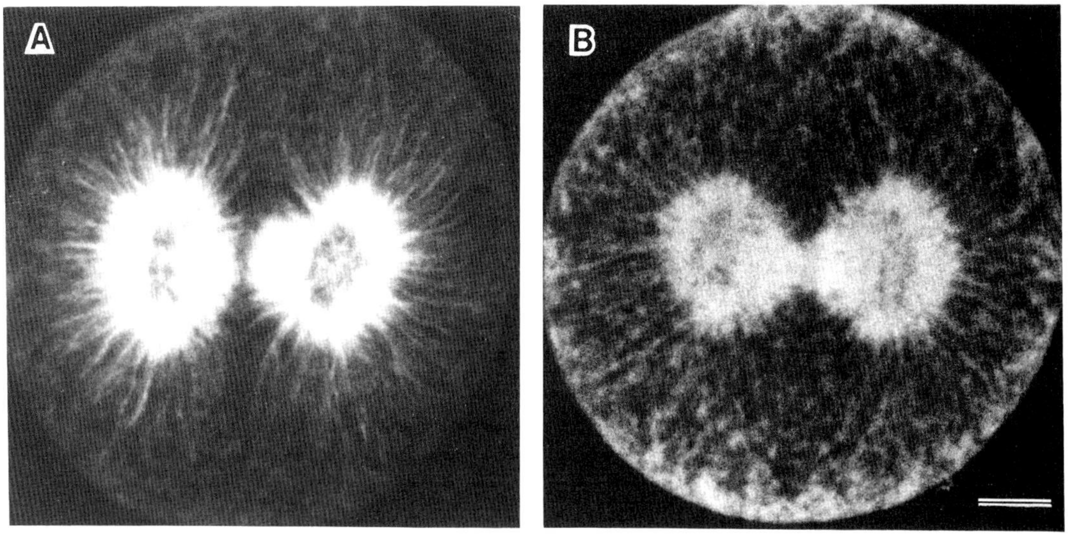

Fig. 3. Localization of the HP 1 antigen in *Hemicentrotus pulcherrimus* egg. The anti-tubulin (*A*) and the HP 1 antibody (*B*) was reacted with sections of paraffin embedded *Hemicentrotus pulcherrimus* eggs. The anti-tubulin antibody clearly visualized mitotic apparatus (*A*). The antigen of the HP 1 antibody was shown to be localized in basal regions of astral microtubules, centrosomes, and half spindles (*B*). The same results were obtained using the HP 2, 5, and 6 antibodies. The localization corresponded to that of the 51-kD protein detected by polyclonal anti-51-kD protein antibodies (OHTA *et al.* submitted). Bar, 10 μm

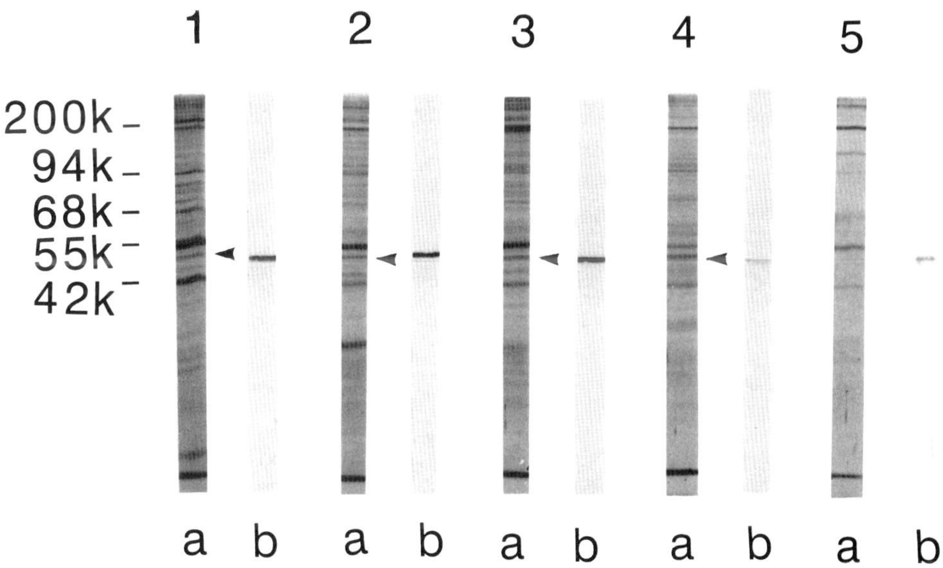

Fig. 4. Cross-reactivity of the HP 1 antibody with proteins of MAs from several kinds of sea urchin. Lanes *a*, CBB staining; lanes *b*, immunoblotting. *1* MA proteins of *Hemicentrotus pulcherrimus*, *2 Anthocidaris crassispina*, *3 Pseudocentrotus depressus*, *4 Temnoplurus hardwicki*, *5 Clypeaster japonicus*. Arrowheads indicated the immunoreactive proteins detected in CBB staining. The same results were obtained using the HP 2 antibody

proteins from *Clypeaster japonicus*, approximately 10 times more protein was loaded. The HP 2 antibody presented the same results. The molecular weights of these immunoreactive proteins are close to the 51-kD protein of *Hemicentrotus pulcherrimus* (Fig. 4, lane 1 *b*). Except for the 52-kD protein of *Clypeaster japonicus*, they are recognizable in CBB staining as protein bands just below tubulin (Fig. 4, lanes 1 *a*, 2 *a*, 3 *a*, and 4 *a*, arrowheads). The 52-kD antigen of *Clypeaster japonicus* was present in low concentrations compared with its presence in other species. The reactivity of the HP 1 and 2 antibodies with *Hemicentrotus pulcherrimus*, *An-*

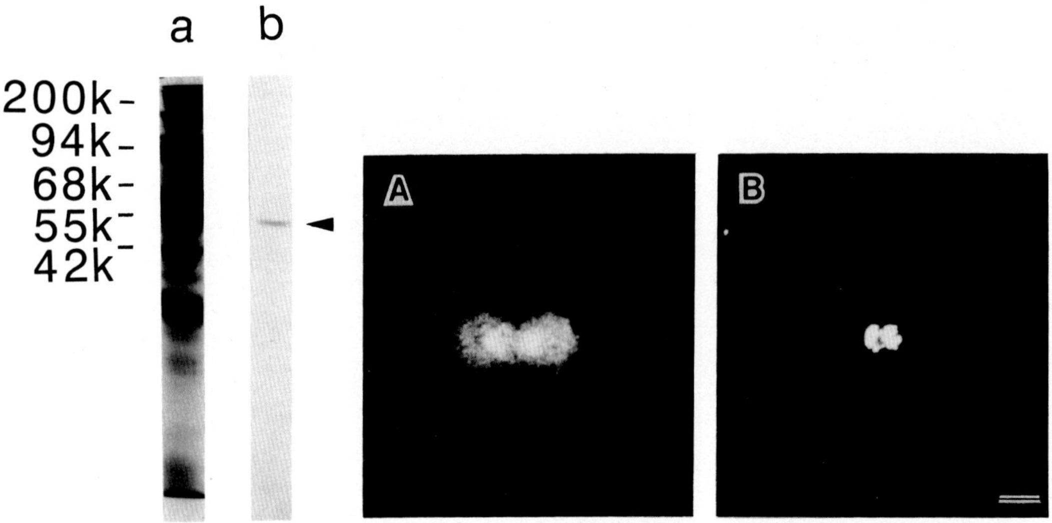

Fig. 5. Localization of the 51-kD antigen to the HP 1 antibody in *Clypeaster japonicus* MA. The HP 1 antibody was applied to sections of paraffin embedded eggs, and FITC-labeled secondary antibody to mouse IgG (*A*) and DAPI (*B*) were reacted with the sections. The HP 1 antigen was shown to be localized in the half spindle and the basal region of asters. DAPI counter staining indicated that the egg was at early anaphase. Lanes *a* (CBB staining) and *b* (immunoblotting) demonstrated the specificity of the HP 1 antibody to a 52-kD protein in the whole metaphase egg homogenate of *Clypeaster japonicus* (arrowhead). Bar, 10 μm

thocidaris crassispina, and *Pseudocentrotus depressus* seemed to be stronger than that with *Clypeaster japonicus* and *Temnoplurus hardwicki*.

The HP 1 and 2 antibodies did not cross-react with any protein in the centrosome fraction from CHO cell line. The HP 1 and 2 antibodies were supposed to recognize epitopes which were conserved in several kinds of sea urchin species.

3.4. Localization of the 52-kD Protein Which Was Recognized by the HP 1 Antibody

Localization of the 52-kD protein, which was the antigen to the HP 1 antibody in *Clypeaster japonicus* MA, was studied using immunofluorescence labeling with sections of paraffin embedded *Clypeaster japonicus* eggs (Fig. 5). DAPI counter staining (Fig. 5 *B*) showed that the egg was in early anaphase. The 52-kD protein was shown to be localized in the basal regions of the asters and the spindle (Fig. 5 *A*). Because of the poor preservation of *Clypeaster japonicus* eggs by fixation, staining of the centrosomes was not clear in contrast to the staining of *Hemicentrotus pulcherrimus* eggs (Fig. 3). However, the localization of the 52-kD protein in *Clypeaster japonicus* MA was generally similar to that of the 51-kD protein in *Hemicentrotus pulcherrimus* MA.

Immunoblotting (Fig. 5, lane *b*) shows the specificity of the HP 1 antibody against the whole metaphase egg homogenate of *Clypeaster japonicus*. The HP 1 antibody recognized specifically the 52-kD protein band (arrowhead).

3.5. The Effect of the Antibodies on Aster Forming Activity of MTOGs

We have previously shown that when MTOGs were preincubated with an affinity purified polyclonal antibody to the 51-kD protein, the initiation of astral microtubules from MTOGs of *Hemicentrotus pulcherrimus* was suppressed up to 65% even in the presence of 50% glycerol (TORIYAMA *et al.* 1988). It was suggested that glycerol inhibits the binding of the antibody to the antigen on the MTOGs in this study.

To test the effect of antibodies on the aster forming activity of MTOGs, monoclonal antibodies HP 1 and 2 as well as the polyclonal AP 17-1 as a reference were incubated with freshly prepared MTOG fraction in the absence of glycerol. Thereafter, the aster forming activity of the fraction was examined after incubation with tubulin.

Under the same experimental conditions where the polyclonal AP 17-1 antibody largely suppressed (up to 88–94%) the aster forming activity of the MTOGs from *Pseudocentrotus depressus* eggs *in vitro*, little inhibitory effect of the HP 1 and 2 antibodies was observed.

3.6. Microinjection of HP 1 from Clypeaster japonicus Eggs

To investigate the role of the 52-kD HP 1 antigen in living eggs, we injected with the HP 1 antibody into

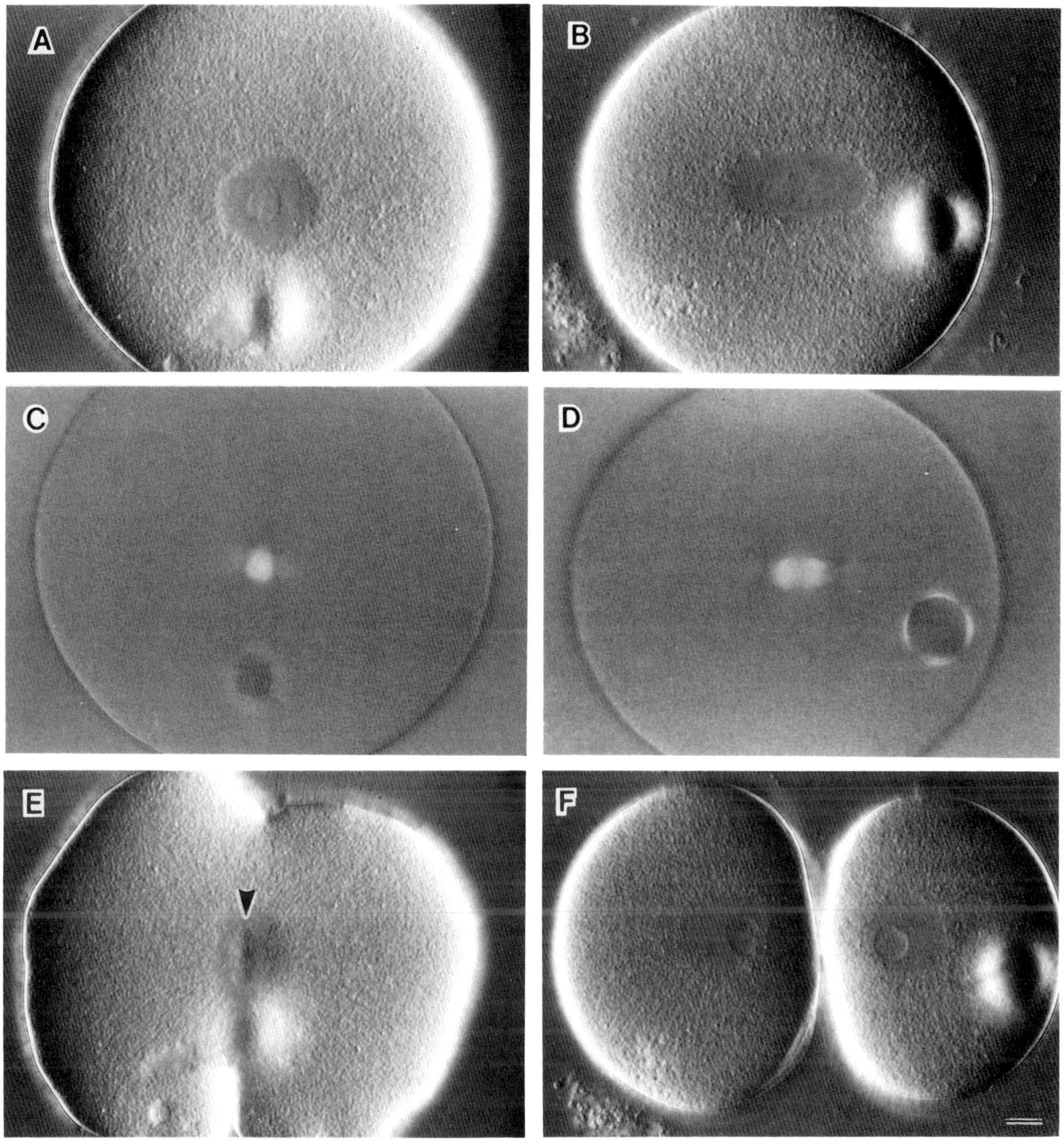

Fig. 6. Effect of the HP 1 antibody injected into *Clypeaster japonicus* egg at prometaphase (45–60 min after insemination). Eggs were injected with 25 pl of 20 mg/ml HP 1 antibody (*A*, *C*, and *E*) and 25 pl of 20 mg/ml rabbit IgG for the control (*B*, *D*, and *F*). Differential interference (*A*, *B*, *E*, and *F*) and polarized light micrographs (*C* and *D*, corresponding to *A* and *B*) were taken 9 min (*A*), 9 min 30 sec (*C*), 9 min 45 sec (*B*), 10 min 20 sec (*D*), 17 min (*E*), and 20 min (*F*) after the injection. Arrowheads indicated the undivided nucleus. Note the short spindle in the egg injected with the HP 1 antibody (*C*). Bar, 10 μm

Clypeaster japonicus eggs. When injected into the eggs before prophase, formation of the MA was not observed, thereby causing total suppression of cleavage. Figure 6 shows polarized-light (*C* and *D*) and differential interference (*A*, *B*, *E*, and *F*) micrographs of an egg injected with the HP 1 antibody (*A*, *C*, and *E*) at prometaphase and a control egg injected at the same stage with 20 mg/ml goat gamma globulin dissolved in the same buffer (*B*, *D*, and *F*). Injection of the HP 1 antibody into the prometaphase egg in which the growth of astral microtubules had already begun did not disturb the growth of the astral microtubules. However, the formation of the spindle was suppressed. Figure 6 *C* shows that the MA of the egg injected with the HP 1 antibody has asters and a very short spindle in contrast to the control eggs (Fig. 6 *D*). Furthermore, the eggs which had been injected with the HP 1 antibody could not complete cell division (Fig. 6 *E*), and the incomplete cleavage furrow regressed eventually. Fig. 6 *F* shows the control eggs divided into two blastomeres. It was difficult to assess whether the chromosome movement occurred or not in such a short spindle. The nucleus in the HP 1 antibody injected eggs stayed single and failed to divide into two (6 *E*, arrowhead).

4. Discussion

To prepare monoclonal antibodies against the 51-kD protein, we immunized mice with the PC-fraction which contained the 51-kD protein predominantly. Four clones (HP 1, 2, 5, and 6) which recognized a 51,000 molecular weight protein were obtained. The 51,000 molecular weight antigen was distributed in all fractions that contained the 51-kD protein. Furthermore, localization in the MA of the 51,000 molecular weight antigen was the same as that of the 51-kD protein. Therefore it was concluded that the 51,000 molecular weight antigen detected by HP 1, 2, 5, and 6 corresponds to the 51-kD protein.

Besides the predominance of the 51-kD protein, the PC-fraction contained minor protein components (less than 40% of the total protein in the fraction prepared from the isolated MAs). We obtained three kinds of monoclonal antibodies against these minor components. From immunoblotting experiments, their antigens were a 63,000 (HP 3 and 8), a 140,000 (HP 4), and 100,000–110,000 (HP 7) molecular weight proteins. Preliminary observations by immunofluorescence microscopy showed that the 140,000 molecular weight antigen localizes in the MAs. Other antigens were shown to exist uniformly in eggs. The PC-fraction may contain minor components which may contribute to aster forming activity through an interaction with the 51-kD protein (TORIYAMA *et al.* 1988). It might be possible that the 140,000 molecular weight protein is such a protein. Studies of the localization of this antigen during the cell cycle are now in progress.

The monoclonal antibodies against the 51-kD protein HP 1 and 2 cross-reacted with 52-kD proteins in MAs of *Anthocidaris crassispina, Temnoplurus hardwicki,* and *Clypeaster japonicus*, and a 51-kD protein in MAs of *Pseudocentrotus depressus*. These proteins were considered to be homologous to the 51-kD protein of *Hemicentrotus pulcherrimus*. Difference of reactivity between species, such as *Anthocidaris crassispina* and *Temnoplurus hardwicki*, may be due to the amino acid sequence of the HP 1 and 2 epitopes. In spite of the difference in reactivity, the HP 1 and 2 epitopes was supposed to be conserved generally in several kinds of sea urchin eggs.

We could not detect any protein immunoreactive with the HP 1 and 2 antibodies in the centrosome fraction of CHO cells. Two possibilities were considered to interpret this result. One is the absence of the 51-kD protein-like protein in CHO cells, and the other is the absence of the epitopes recognized by the HP 1 and 2 antibodies for a centrosomal protein in CHO cells. Further biochemical analysis of the protein in CHO centrosome fraction remains to be carried out.

HP 1 and 2 cross-reacted with the 52-kD protein in *Clypeaster japonicus* eggs. Localization of the 52-kD protein in *Clypeaster japonicus* eggs examined by immunofluorescence microscopy indicated that the 52-kD antigen was localized in the basal regions of astral microtubules and the half spindles. This localization generally agreed with that of the 51-kD protein in *Hemicentrotus pulcherrimus* MAs. We have discussed the functions of the 51-kD protein from its localization in *Hemicentrotus pulcherrimus* eggs through cell cycle (OHTA *et al.* in press). The 51-kD protein in *Hemicentrotus pulcherrimus* was considered to have a role in the nucleation of microtubules as well as in the formation of microtubule matrices, such as the spindle matrix. In *Clypeaster japonicus* eggs, the 52-kD antigen, which was homologous to the 51-kD protein in *Hemicentrotus pulcherrimus*, should have the same function as does the 51-kD protein. Localization of the 52-kD antigen generally supported this idea.

Microinjection of an antibody into the living cell is a means to investigate the cellular functions of the antigen (MABUCHI and OKUNO 1977). However, it often requires high concentration of antibody. The antibody concentration of the affinity purified polyclonal antibodies which we obtained was not high, so we used a gamma globulin fraction of monoclonal antibody which showed a high specificity to the 52-kD protein in *Clypeaster japonicus* eggs. The growth of astral microtubules was not inhibited by injection of the HP 1 antibody when the injection was carried out after the eggs reached prometaphase. This stage is characterized by the initiation of astral microtubules from the centrosomes to a substantial extent. However, the formation of the spindle was largely suppressed. This result leads us to an assumption that the 51-kD protein is possibly involved in the formation of spindle microtubules with a role of stabilizing the spindle architecture. In contrast, when injected into the eggs before prophase, the formation of the MA was not observed. We have already shown by electron microscopy that MTOGs begin to assemble around the centrioles to form centrosome from prophase, initiating small asters focusing on small clusters of MTOGs at both the poles. The size of the cluster of MTOGs reaches maximum at metaphase. The apparent contradiction that the antibody did not inhibit aster forming activity of the cluster of MTOGs *in vitro* might be explained by assuming that the antibody injected into the egg before prophase blocked the assembly of MTOGs into the centrosome. Further observations of the changes in the localization of the MTOGs by microinjection of the antibody remained to be performed.

Acknowledgements

We are grateful to Dr. T. Arai for his gift of myeloma pai cell line. We thank Misaki Marine Laboratory for supplying sand dollars and Kominato fisherman's association for collecting sea urchins.
This work was supported by Grants-in-Aid from Ministry of Education, Science and Culture of Japan (60065005 and 62540538), and Special Coordination Funds for Promoting Science and Technology Agency of Japan.

References

Bornens M, Paintrand M, Berges J, Marty MC, Karsenti E (1987) Structural and chemical characterization of isolated centrosomes. Cell Motil Cytoskel 8: 238–249

Bradford MM (1976) A rapid and sensitive method for the quantitation of microgram quantities of protein utilizing the principle of protein-dye binding. Anal Biochem 72: 248–254

Brinkley BR (1985) Microtubule organizing centers. Ann Rev Cell Biol 1: 145–172

Endo S (1979) The clusters of granular material around the centriole during mitosis. Cell Struct Funct 4: 71–74

— (1980) Further observations of the clusters of granular material around the centriole in the sea urchin egg: Changes in distribution during mitosis. Dev Growth Differ 22: 509–516

— Toriyama M, Sakai H (1983) The mitotic apparatus with unusually many microtubules from sea urchin eggs treated with hexyleneglycol. Dev Growth Differ 25: 307–314

— — — (1985) Microtubule organizing granules (MTOGs) in mitotic sea urchin egg. In: Ishikawa H, Hatano S, Sato H (eds) Cell motility: mechanism and regulation. Univ of Tokyo Press, pp 403–414

Gosti-Tetsu F, Marty M, Berges J, Maunoury R, Bornens M (1986) Identification of centrosomal proteins in a human lymphoblastic cell line. EMBO J: 2545–2550

Hamaguchi Y, Toriyama M, Sakai H, Hiramoto Y (1985) Distribution of fluorescently labeled tubulin injected into sand dollar eggs from fertilization through cleavage. J Cell Biol 100: 1262–1272

Hiramoto Y (1974) A method of microinjection. Exp Cell Res 87: 403–406

Köhler G, Milstein C (1975) Continuous cultures of fused cells secreting antibody of predefined specificity. Nature 256: 485–496

Laemmli UK (1970) Cleavage of structural proteins during the assembly of the head of bacteriophage T4. Nature 227: 680–685

Lowry OH, Rosebrough NJ, Farr AL, Randall RJ (1951) Protein measurement with the Folin phenol reagent. J Biol Chem 193: 265–275

Mazia D (1984) Centrosomes and the mitotic poles. Exp Cell Res 153: 1–15

Mabuchi I, Okuno M (1977) The effect of myosin antibody on the division of starfish blastomeres. J Cell Biol 74: 251–263

Mitchison T, Kirschner MW (1984) Microtubule assembly nucleated by isolated centrosomes. Nature 312: 232–237

O'Farrell PZ, Goodman HM, O'Farrell PH (1977) High resolution two-dimensional electrophoresis of basic as well as acidic proteins. Cell 12: 1133–1142

Ohta K, Toriyama M, Endo S, Sakai H (1988) Localization of mitotic-apparatus-associated 51-kD protein in unfertilized and fertilized sea urchin eggs. Cell Motil Cytoskel (in press)

Pickett-Heaps JD (1969) The evolution of the mitotic apparatus: An attempt at comparative ultrastructural cytology in dividing plant cells. Cytobios 1: 257–280

— Spurck T, Tippit D (1984) Chromosome and the spindle matrix. J Cell Biol 99: 137s–143s

Showman RM, Foerder CA (1979) Removal of the fertilization membrane of sea urchin embryos employing aminotriazole. Exp Cell Res 120: 253–255

Toriyama M, Endo S, Sakai H (1984) Aster formation *in vitro* is nucleated by granules isolated from the mitotic apparatus. Cell Struct Funct 9: 213–224

— Ohta K, Endo S, Sakai H (1988) 51-k protein, a component of microtubule-organizing granules (MTOGs) in the mitotic apparatus involving in aster formation *in vitro*. Cell Motil Cytoskel 9: 117–128

Towbin H, Staehelin T, Gordon J (1979) Electrophoretic transfer of proteins from polyacrylamide gels to nitrocellulose sheets: Procedure and some applications. Proc Natl Acad Sci USA 76: 4350–4354

Protoplasma (1988) [Suppl 1]: 122–132

Changes of the Chromosomal Spindle (Kinetochore) Fibers and the Behaviour of Bivalent Chromosomes in Grasshopper Spermatocytes After Irradiation of the Kinetochores with an Ultraviolet Microbeam

K. Izutsu*

Department of Pathology, Mie University School of Medicine, Tsu, Mie-ken

Received March 1, 1988
Accepted June 14, 1988

Dedicated to Professor Dr. Noburo Kamiya on the occasion of his 75th birthday

Summary

Selected small parts of primary spermatocytes of the grasshoppers, *Mongolotettix japonicus* and *Trilophidia annulata* were irradiated with an ultraviolet (UV) microbeam and the subsequent processes in division were followed by phase-contrast and polarization optics. When a kinetochore of a metaphase bivalent was irradiated, the birefringence of the associated kinetochore fiber was lost. Then the bivalent shifted to the pole toward which the other unirradiated kinetochore was oriented. Such a conspicuous shift was not seen when the irradiation was given to a portion of the associated kinetochore fiber alone. The kinetochore-irradiated half-bivalent underwent anaphase movement only when a thin birefringent fiber was reorganized between the irradiated kinetochore and the pole. Frequently, no anaphase movement occurred in the half-bivalent whose kinetochore was irradiated at the end of metaphase. When the kinetochore of a half-bivalent was irradiated in anaphase, the associated kinetochore fiber lost its birefringence, the chromosome continuing to move with some lag. Such a half-bivalent was shown to be pulled poleward by a reduced force. When part of an anaphase kinetochore fiber was irradiated, the motion of the half-bivalent associated with the fiber either did not change or speeded up during the irradiation, the partner half-bivalent being often accelerated.

Keywords: Chromosome movement; Grasshopper spermatocyte; Kinetochore fiber; Ultraviolet microbeam irradiation.

Abbreviation: UV ultraviolet.

1. Introduction

Kinetochores play a pivotal role in mitosis and meiosis (Schrader 1953, McIntosh 1985, Nicklas 1987). These particular regions of the chromosome are essential not only in the formation of the spindle microtubular organization, but also in the function of the completed spindle. At present, however, the molecular mechanism involving the kinetochore activity is almost unknown. Based on light microscopical observations and ultraviolet(UV)- and laser microbeam irradiation experiments, the chromosomal arrangement on the equator in metaphase has been interpreted that the two kinetochores of each mitotic chromosome (two pairs of kinetochores for each bivalent chromosome in meisosis) are pulled toward the opposite poles through the kinetochore spindle fibers attached to the kinetochore, which are bundles of kinetochore microtubules, and that the two attractive forces working on the kinetochores, equal in magnitude and opposite in direction, are in equilibrium (Östergren 1951, Izutsu 1961, McNeill and Berns 1981). As regards the role of kinetochores in anaphase, aside from the importance as the site of attachment of kinetochore microtubules to the chromosome, evidence has been shown that the kinetochore actively moves along the stationary kinetochore microtubules toward the poles in anaphase (Gorbsky *et al.* 1987). The poleward pull working on the kinetochores by the kinetochore fibers exists in metaphase and anaphase, and also in prometaphase, although the mechanism of the motive force remains to be solved. The problem seems to be the cardinal one concerning the mechanism of chromosome movement in cell division.

The present study has been undertaken to investigate

* Correspondence and Reprints: Department of Pathology, Mie University School of Medicine, Tsu, Mie-ken 514, Japan.

the function of the kinetochores and the possible mechanism of force generation by the kinetochore-kinetochore fiber complex in metaphase and anaphase, based on correlative observations of the changes in kinetochore fibers and in the behaviour of chromosomes after irradiation of the kinetochore or associated kinetochore fiber with a UV microbeam. Thus, one of the kinetochores or the associated kinetochore fibers of a bivalent chromosome in primary spermatocytes of grasshoppers was selectively irradiated. The subsequent position and movement of the chromosomes and changes in the kinetochore fibers were followed by phase-contrast and sensitive polarization microscopy, respectively. The grasshopper cells used were large and optically clear. Their spindle and chromosomes were large and the chromosomes relatively small in number. Kinetochore pairs of the bivalents in metaphase and anaphase were at the ends nearest the poles. The considerable distance between the kinetochore pairs of each bivalent permitted one to selectively irradiate one of the kinetochore pairs with a microbeam.

This study is an extension of previous studies (IZUTSU 1959, 1961), in which irradiated cells were studied by ordinary light microscopy and, therefore, the spindle fibers could not be observed. In the present study, use of phase-contrast and polarization optics, combined with cinemicrography, enabled observation of the spindle and chromosomes in the irradiated cells and quantitative analyses of the chromosome movement in relation to changes of the spindle fibers.

2. Materials and Methods

2.1. Experimental Materials

Primary spermatocytes of the Japanese grasshoppers, *Mongolotettix (Chrysochraon) japonicus* and *Trilophidia annulata* were used. The karyotype of the *Mongolotettix* consists of 8 pairs of autosomes and one univalent X chromosome. The *Trilophidia* has 11 pairs of autosomes and one univalent X.

Cell preparations. These cells were prepared into hanging drop preparations using the fluorocarbon oil, FC-47 (Minnesota Mining & Mfg. Co., St. Paul, Minnesota) by the method described previously (IZUTSU *et al.* 1977). For studies of microbeam-irradiated cells by phase-contrast microscopy a quartz coverslip of 0.35 mm thickness was used instead of a glass one, because the Zeiss Ultrafluar phase-contrast objective used both for observation and microbeam irra-

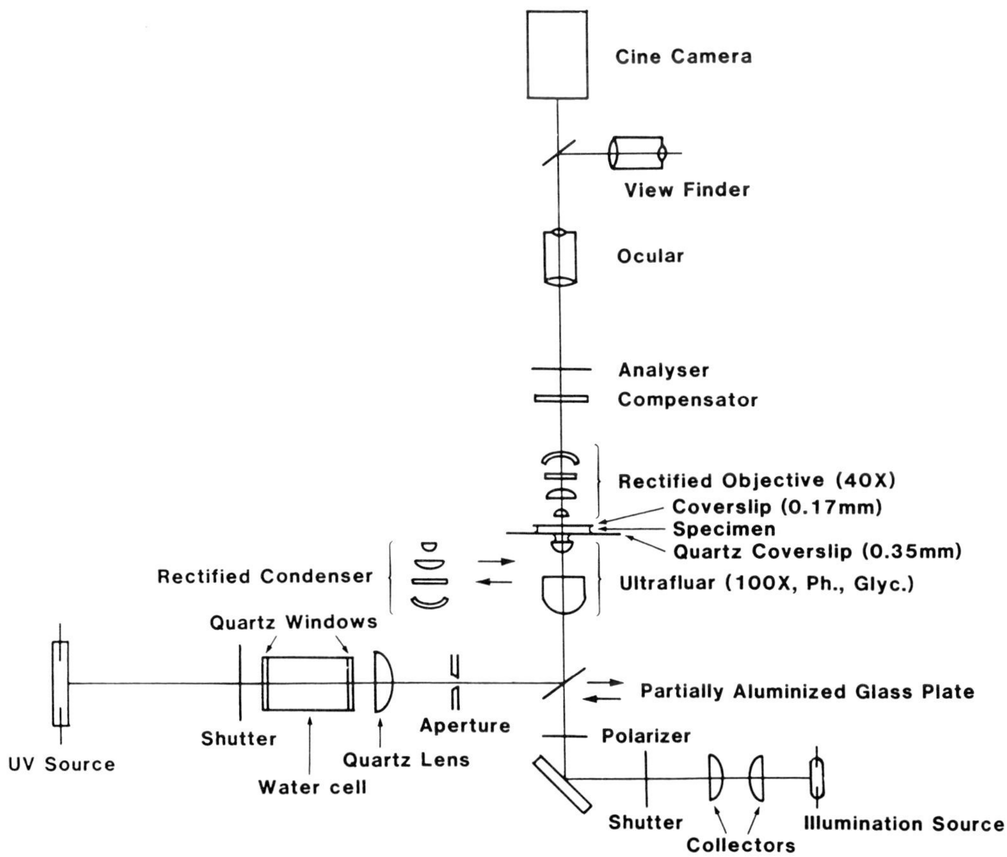

Fig. 1. Schematic diagram of the apparatus used for ultraviolet microbeam irradiation and observation by sensitive polarization microscopy. For details see text

diation was designed to use with a quartz coverslip of the above-mentioned thickness. For studies by polarizations microscopy of irradiated cells, two types of preparations were used. One was similar to that which was adopted for observation by phase-contrast microscopy but the thickness of the quartz coverslip used was 0.17 mm. The other is illustrated in the Fig. 1. In this type of preparation a monolayer of cells was sandwiched along with FC-47 between a glass coverslip and a quartz one 0.35 mm thick using as spacers Pyrex glass fibers 70 microns thick. The sandwich was put, with the quartz coverslip down, over a hole of a stainless steel slide and then fixed with paraffin wax to the slide.

2.2. *UV Microbeam Irradiation*

Both for observation by phase-contrast and microbeam irradiation, an irradiation apparatus of the Uretz type described previously by the author (Izutsu 1959) was used with the following modifications: An Ultrafluar phase-contrast objective (× 100, glycerin immersion; Zeiss, Federal Republic of Germany) was used for viewing and irradiation. This objective was achromatically corrected in the range from 2,310 to 7,000 Å, thus enabling aiming of microbeam at a visible wave length. An annular diaphragm was installed under the substage condenser. In addition, as the UV source a 450 watts high-pressure mercury arc (UM-452, Ushio, Japan) was used. A ten seconds exposure was sufficient to give a "paling reaction" of the chromosomes, which was already visible immediately after irradiation. The irradiation time adopted in the present study ranged from fifteen to thirty seconds.

For microbeam irradiation of the cells to be observed and recorded with a polarization optics one of the following two methods was adopted. (1) When the preparations in which the quartz coverslip was on the upper side were used, cells were observed and photographed before irradiation by the time-lapse technique. Meanwhile, the position of the preparation on the mechanical stage was read on the axes of the abscissa and ordinate. Just before irradiation, a small amount of distilled water was dropped onto the region of the quartz coverslip overlying the cell concerned and then, a small fragment of quartz coverslip 0.17 mm thick was put on the same part of the coverslip, the fragment being held to the coverslip by capillary action. The preparation was then transferred onto the mechanical stage of the Uretz type irradiation apparatus. Positioning of the cell to the center of the field was made by the mechanical stage based on a formula relating the coordinates of the mechanical stage of the polarization microscope to those of the irradiation apparatus. Then a small amount of glycerin (n_D = 1.454, 23 °C) was put on the fragment for immersion of the front lens of the Ultrafluar objective. After irradiation using this objective, the coverslip fragment was removed from the lower coverslip and the water on the coverslip was blotted dry with filter paper. The preparation was then transferred back to the polarization microscope for the subsequent observation and cinemicrographic recording. (2) When preparations as illustrated in Fig. 1 were used, irradiations were made with the system which is diagrammed in the same figure. This system consisted of a sensitive polarization microscope as was used by the author in our previous study (Izutsu *et al.* 1977) and an irradiation apparatus which was incorporated into the former. An image of the aperture is projected through a substage Ultrafluar objective (× 100) onto the specimen. When a circular aperture 0.4 mm in diameter was used a focal spot of 3.5 µm in diameter was produced. In some cases rectangular apertures 0.4 mm wide and of varying lengths were used. Cells to be irradiated were observed through the polarization optics using a rectified strain-free condenser (Nikon, N.A. 0.52) and a rectified × 40 objective (Nikon, N.A. 0.65) and photographed by the time-lapse technique. For microbeam irradiation the substage condenser was changed for the Ultrafluar objective. Then the objective was again changed for the rectified condenser for the subsequent observation and recording through the polarization microscope. Irradiations and observations were made at room temperatures ranging from 25 to 31 °C.

2.3. *Cinemicrographic Analysis*

Cinematographic recordings by phase-contrast optics using a × 100 Ultrafluar objective were made on Fuji Minicopy 16 mm negative film by the time-lapse technique at 8 seconds cycle, and those by polarization optics were made using a Nikon × 40 rectified objective (N.A. 0.65) on Kodak Plus-X 16 mm negative film at a time-lapse of the same rate. Positive copies of the films were analyzed by the method described previously (Izutsu *et al.* 1977).

3. Results

In the first meiotic metaphase and anaphase of the spermatocytes used in the present study each bivalent had a pair of adjacent kinetochores at its polar ends. Usually the two kinetochores composing the pair functionally behave as a single kinetochore and it was rather rare for them to become oriented to the opposite poles (amphitelic orientation). Therefore, each pair of kinetochores at the polar ends of a bivalent will be described in the following as a kinetochore of the bivalent. When any part of the cell was irradiated with a UV microbeam, changes were observed that were common to the exposure of all cellular parts. These include diminution in the pole-to-pole distance and abnormalities of cytokinesis. The former occurred constantly, but the latter was not always encountered and apparently caused by a failure of the elongation of both the spindle and cell body in late anaphase and early telophase. The decrease in spindle length and frequency of the cytokinetic abnormalities increased in proportion to the dose of UV delivered. With the dose used in the present study, cytokinetic abnormalities rarely occurred.

3.1. *Irradiation of a Kinetochore in Metaphase I*

Since selective irradiation of kinetochores inevitably involves exposure of a part of the associated kinetochore fibers, in most cases of kinetochore irradiation the kinetochores were brought to the center of the microbeam spot in order to ensure the aiming of the microbeam.

When a kinetochore of a bivalent was irradiated in metaphase in this way, the irradiated chromosomal portion became pale with phase contrast (Fig. 2) and, then, the bivalent moved to the pole to which the unir-

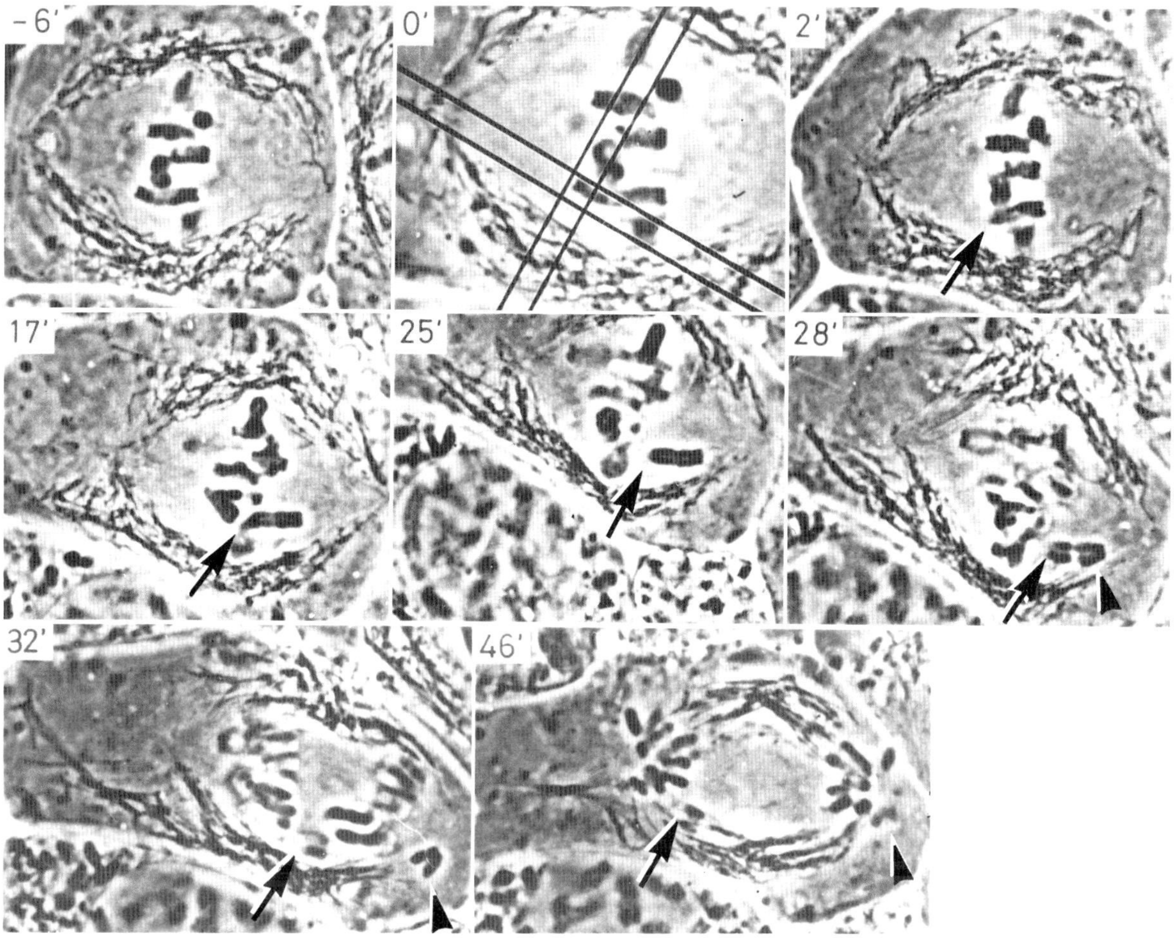

Fig. 2. Behaviour of a bivalent which had one kinetochore irradiated in metaphase as observed by phase-contrast microscopy. Time in minutes after the irradiation is indicated at the upper left corner of each photo. The irradiated portion corresponds to the circle which can be inscribed in the square formed by the aiming cross hairs in the ocular (O'). The arrow indicates the irradiated and paled kinetochore region, and the arrowhead the partner half-bivalent of the irradiated chromosome

radiated kinetochore was oriented. Usually, such a bivalent remained stationary near that spindle pole until anaphase. Occasionally, however, the unirradiated kinetochore then changed orientation to the opposite pole, and moved to this pole, or resulted in amphitelic orientation temporarily and the whole bivalent returned to the equator. In anaphase the irradiated bivalent divided into half-bivalents which moved to the opposite poles as is graphically shown in Fig. 3. Except for the temporary amphitelic orientation of the unirradiated kinetochore, a new finding in the present study, these observations completely confirm the results of previous studies (Izutsu 1959, 1961). A time-lapse film recorded using phase-contrast optics revealed the amphitelic orientation of the unirradiated kinetochore region: the two kinetochores of the unirradiated half-bivalent were seen pointing towards the opposite poles.

A total of 28 cells were irradiated at a kinetochore of a bivalent in the first meiotic metaphase. All but three cells in which irradiation was made at the end of metaphase underwent shift of the bivalent toward the pole further from the exposed site. Figure 3 shows changes in distances of the kinetochores and spindle poles from the equator in the irradiated cell shown in Fig. 2. After irradiation the exposed bivalent moved slowly to the pole toward which the unirradiated kinetochore faced. The speed of the anaphase movement of the irradiated half-bivalent was somewhat higher than that of the unirradiated half-bivalent, being at a maximum when the irradiated half-bivalent passed across the equator. In the experiment shown in Fig. 4, also, a kinetochore and an adjacent small part of the associated kinetochore fiber of a bivalent was irradiated in metaphase. The behaviour of this cell was followed by polarizing microscopy. After exposure, no kinetochore fiber was

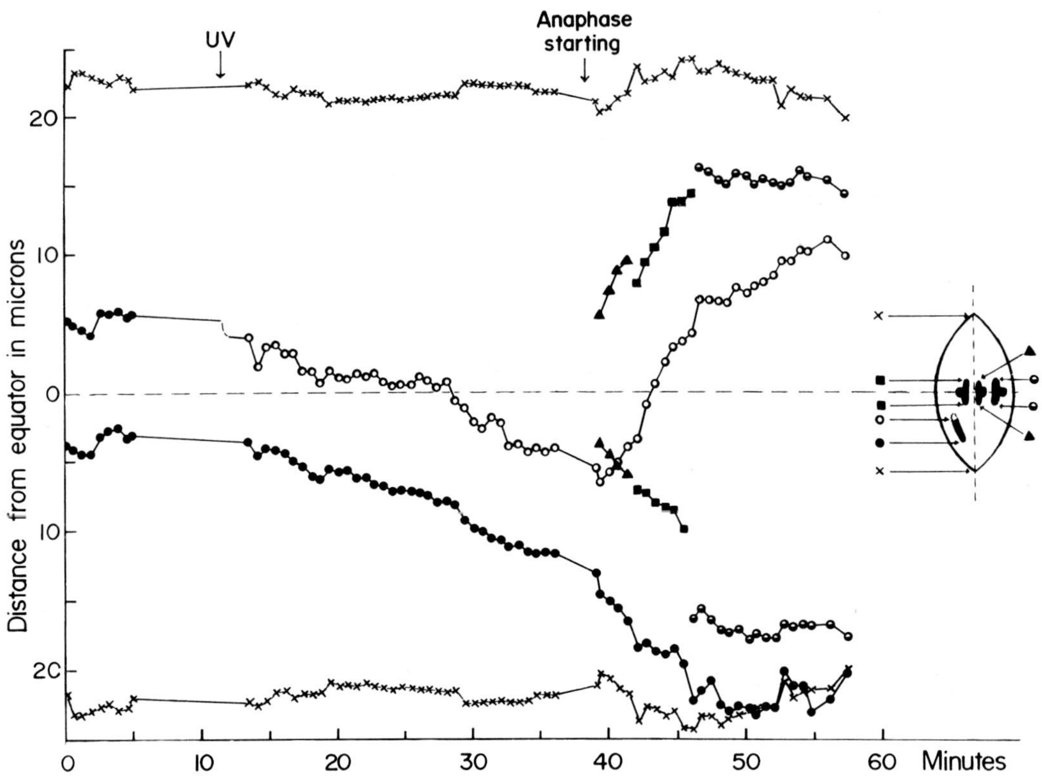

Fig. 3. Distances of kinetochores and spindle poles from the equator after microbeam irradiation of one kinetochore of a bivalent of the cell in metaphase shown in Fig. 2. A bivalent off the equator in the schematic diagram in the right had been irradiated at the upper kinetochore which was marked with an open circle. Details in text

observed to be associated with the irradiated kinetochore, even when the azimuth was changed so that maximum detectability was obtained. The bivalent shifted to the pole with which the unirradiated kinetochore was connected with its associated kinetochore fiber. In anaphase the bivalent divided into half-bivalents, each of which moved to the opposite poles at almost the same time as the unirradiated bivalents did (Figs. 4, 5, and 6). During the anaphase movement of the irradiated half-bivalent a weakly birefringent fiber was observed connected to the irradiated kinetochore. The speed of the kinetochore separation was similar between the irradiated and unirradiated bivalents. Figure 6 also indicates that the irradiated kinetochore moved to the pole at a speed almost comparable to that of the unirradiated kinetochore.

When that part of a half-bivalent which was located between its kinetochore and the joint between the half-bivalent and the partner one was irradiated in metaphase, the irradiated, paled portion was gradually drawn out to a considerable extent during the rest of metaphase. This fact combined with the shift of a kinetochore-irradiated bivalent mentioned above indicates that metaphase bivalents are being pulled at both the kinetochoric ends from the opposite poles.

When a kinetochore of a bivalent was irradiated at the end of metaphase, no shift of the bivalent was observed from the equatorial plane. The bivalent divided into half-bivalents in anaphase. Frequently, the kinetochore-irradiated half-bivalent did not move and remained on the equator. It was then squeezed by the cleavage furrow. Description was made in previous papers (Izutsu 1959, 1961) on the same behaviour of a bivalent whose kinetochore was irradiated at the end of metaphase I. In such cells no kinetochore fiber was observed at the irradiated kinetochore. In this connection, as was previously reported (Izutsu 1976) it has been consistently observed that whenever any half-bivalent whose kinetochore was irradiated in metaphase underwent anaphase movement, a birefringent kinetochore fiber had been reorganized at the kinetochore. Figure 7 shows distribution in the time of irradiation for a total of 28 cells in which one kinetochoric end was irradiated in the first meiotic metaphase. Solid bars represent those cells in which irradiation suppressed anaphase movement of the irradiated half-bivalent.

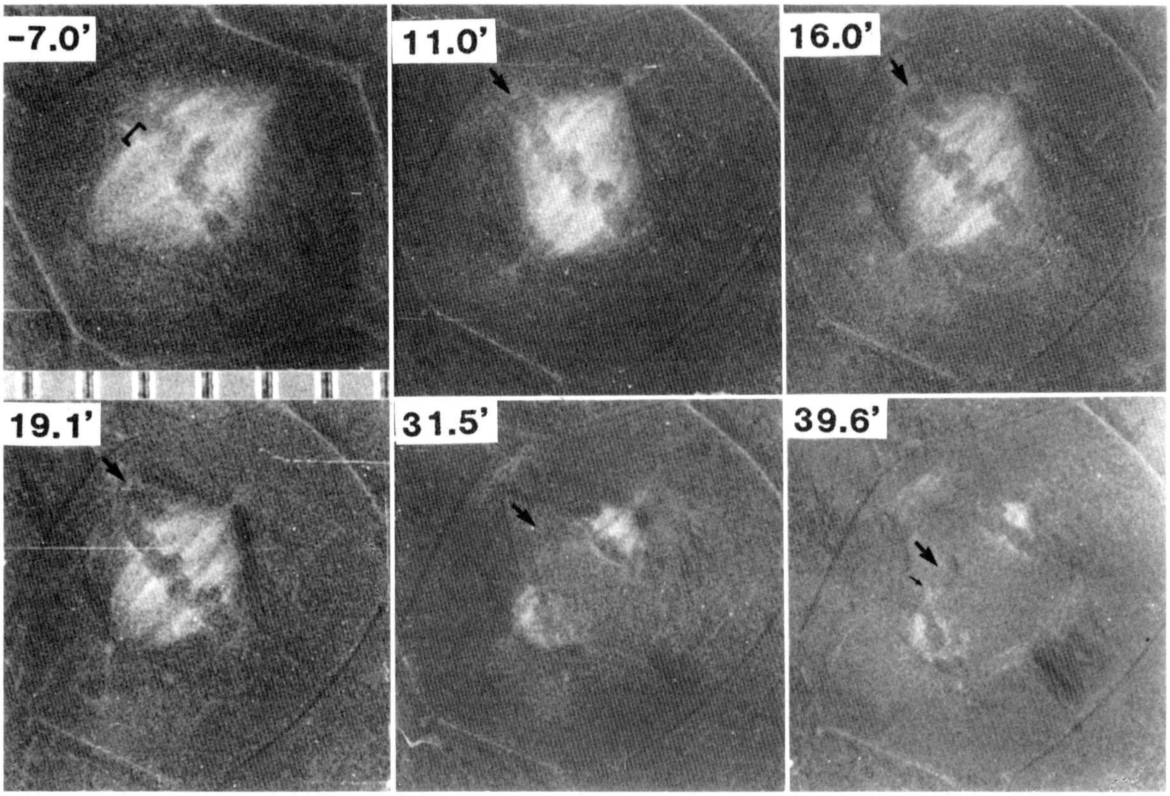

Fig. 4. Behaviour of a bivalent and its associated kinetochore fibers after selective irradiation of one, of the kinetochores in metaphase. Polarization microscopy. The large arrow indicates the kinetochore-irradiated half-bivalent and the small one, the reorganized, thin kinetochore fiber

Fig. 5. Traces of the cell shown in Fig. 4. The irradiated bivalent is drawn in black

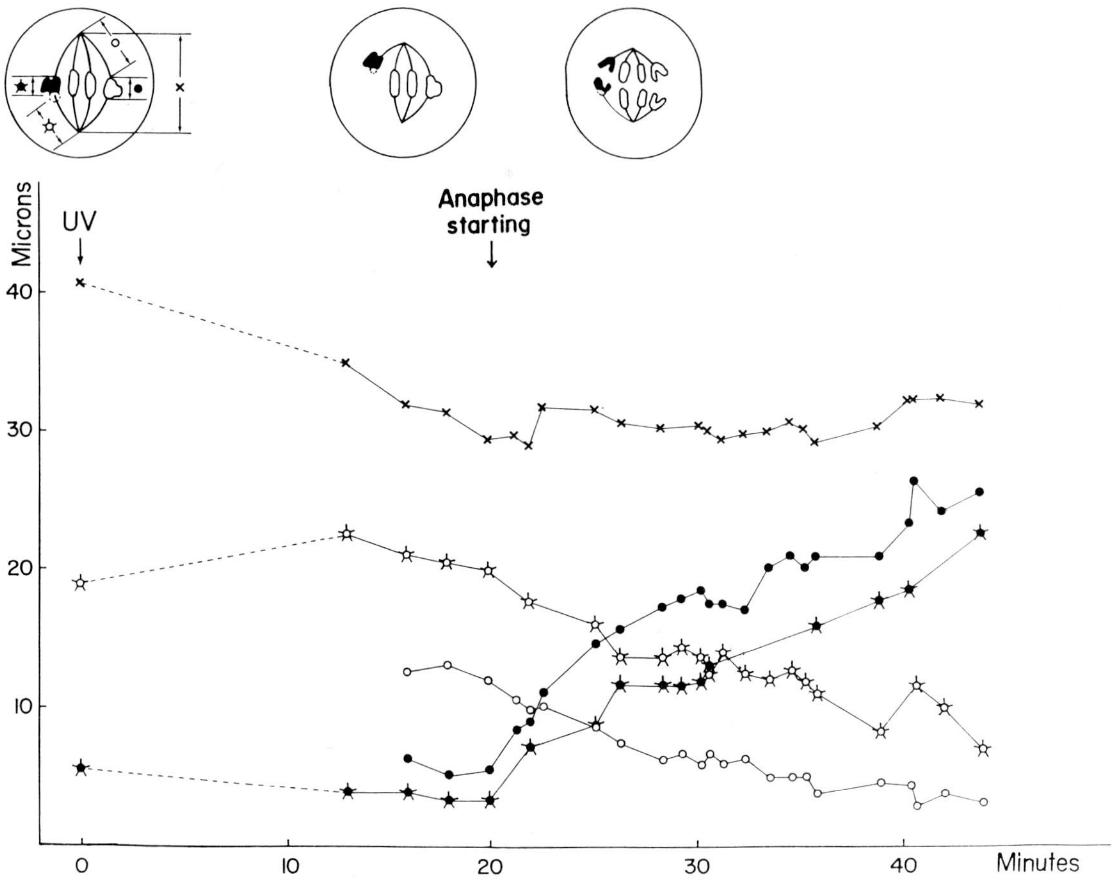

Fig. 6. Graphs showing changes in the interkinetochore distances (–●–, –★–), kinetochore fiber lengths (–○–, –☆–) and the spindle length (– × –) in the same cell as shown in Figs. 4 and 5. The irradiated bivalent is drawn in black. Symbols for various distances are shown in one of the schematics at the top

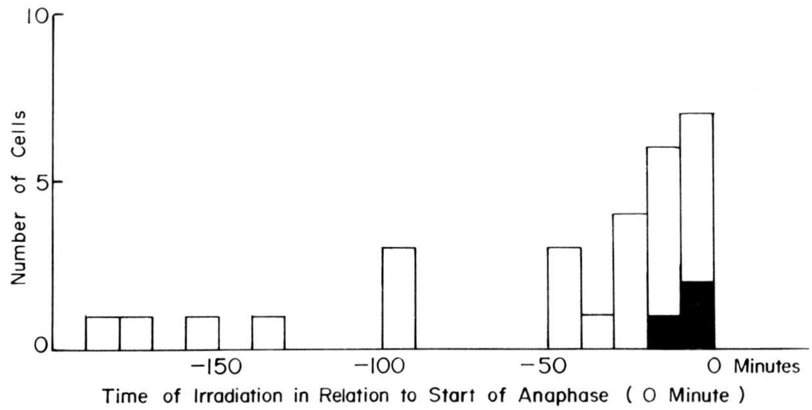

Fig. 7. Histogram showing the time of irradiation of one of the kinetochores of a metaphasic bivalent and the anaphase chromosome movement. Solid bar represents cells in which irradiation suppressed anaphase movement of the exposed half-bivalent

From this figure it can be seen that anaphase movement of the irradiated half-bivalent occurred in all cells that were irradiated more than 20 minutes ahead of the start of anaphase. In other words, anaphase movement may be halted if irradiation is made within 20 minutes of the start of anaphase.

3.2. Irradiation of a Kinetochore of a Bivalent in Anaphase I

When the kinetochore of a half-bivalent was irradiated during anaphase, birefringence of its the associated kinetochore fiber was reduced or lost. However, the

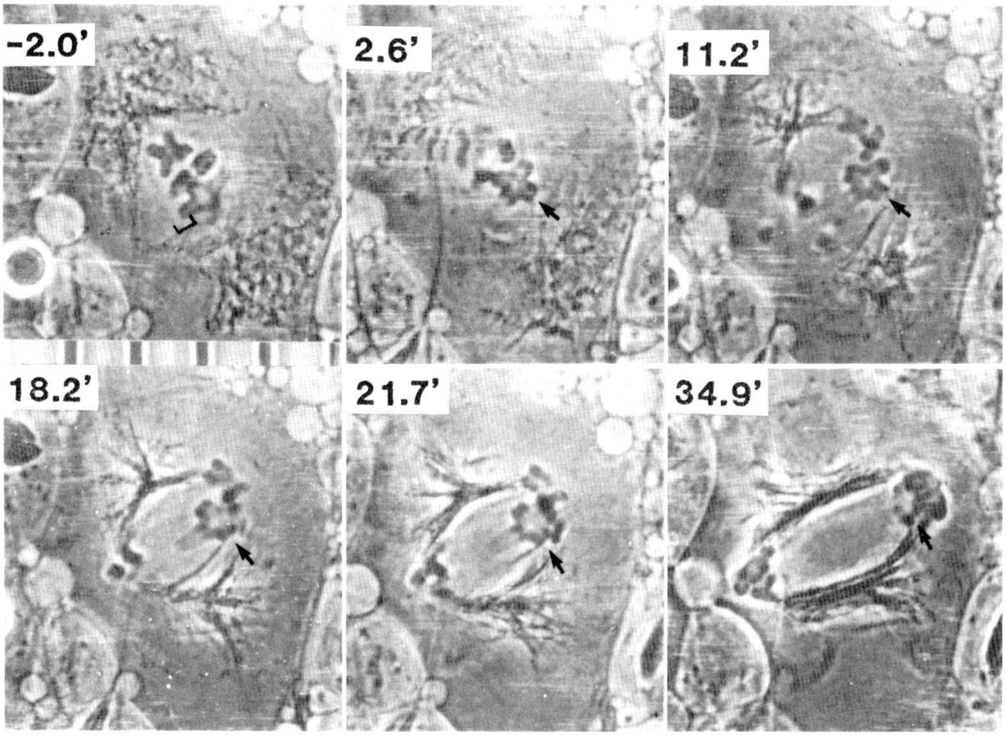

Fig. 8. Behaviour of a bivalent in anaphase, which was made stickly at the junction of the two half-bivalents in late metaphase and one kinetochore (marked with a bracket) of which was irradiated in early anaphase. For explanation see text

chromosome continued to move to the pole, although the speed was somewhat slower. The following experiment shown in Fig. 8 indicates that such a kinetochore-irradiated half-bivalent is pulled poleward by a reduced force. The slightly paled portion of this bivalent had been made sticky by a small dose of irradiation at the end of metaphase. In early anaphase, the lower kinetochore of this bivalent was irradiated. The bivalent, then, was pulled toward the upper pole and finally fused with the upper chromosome group. In the meantime, the irradiated, paled kinetochore region was observed at the equatorial region still oriented to the lower left pole. We can assume that the force acting on the upper kinetochore was stronger than that on the lower, irradiated kinetochore.

3.3. Irradiation of a Kinetochore Fiber of a Bivalent in Metaphase I

When a small part of a kinetochore fiber was irradiated with UV, it was rather rare to see a localized change at the exposed site by the polarization optics.

In the present study at least a few minutes elapsed from the completion of irradiation until the irradiated site was first observed by polarization microscopy. Any change which had occurred immediately after the exposure might had been repaired before observation of the irradiated kinetochore fiber. Even the lesion rarely observed at the irradiated site gradually became obscured within several minutes. In the grasshopper spermatocytes used in the present study, therefore, areas of reduced birefringence as were reported for crane fly spermatocytes (Forer 1965) were not produced. Nor was it possible to observe the movement of the "lesion" toward the polar end of the kinetochore fiber.

3.4. Irradiation of the Kinetochore Fiber of a Half-bivalent in Anaphase I

When part of the kinetochore fiber of a bivalent was irradiated with UV, the kinetochore fiber disappeared or the birefringence decreased along its whole length. These changes were partially reversed during anaphase. In about half of the cells subjected to this type of treatment the anaphase movement of the half-bivalent with the irradiated kinetochore fiber did not change. In the other half of the cells, the anaphase movement of the half-bivalent was accelerated during irradiation. In most of the latter cases the speed of the partner half-bivalent was also accelerated to a similar degree.

4. Discussion

In interpreting the results mentioned above the sensitivity of the polarization microscope used in the present study is of fundamental importance. Since it was usually observed that a bivalent kinetochore, whose associated kinetochore fiber became hardly detectable as the result of kinetochore irradiation, still remained to be oriented to the spindle pole, it seems probable that at least single microtubules cannot be detected by the optics used in the present study. On the other hand, comparison of the distribution of birefringent material in living grasshopper spermatocytes with that of microtubules visualized by immuno-gold staining in the same cells has revealed an excellent coincidence (IZUTSU *et al.* to be published). Therefore, the sensitivity might be comparable to that of immuno-gold staining for tubulin, especially when polarization microscopy and time-lapse cinemicrography are combined.

As was noted in the results and also reported previously (IZUTSU 1959, 1961), following irradiation of one kinetochore of a metaphase bivalent, the bivalent shifted toward the pole to which its unirradiated kinetochore was directed. Such a conspicuous shift in the position of the bivalent was not observed when radiation was delivered solely to a portion of one of the associated kinetochore fibers. Instead, the bivalent shifted slightly and for a short period to the pole with which the other associated kinetochore fiber was connected. In addition it has been shown that the position of metaphase bivalents was not influenced when they were exposed outside the kinetochore (IZUTSU 1959, 1961). Also, elongation of the irradiated chromosomal portion lying between the two kinetochores of a metaphase bivalent observed in the present study indicates that bivalents in metaphase are being pulled at both the kinetochoric ends toward the opposite poles. These facts indicate that the kinetochore is the primary element which determines the position of the metaphase bivalent on the equatorial plane. Furthermore the disappearance of the kinetochore fiber associated with the irradiated kinetochore of the metaphase bivalent point to the role of the kinetochore fiber as well as the kinetochore in maintaining the metaphase bivalents on the equatorial plane.

The results of the irradiation of bivalent kinetochores in prometaphase or metaphase is in agreement with that of laser microirradiation of the kinetochore of a single chromatid in mitotic prometaphase or metaphase in rat-kangaroo cells (MCNEILL and BERNS 1981). After irradiation, the whole chromosome moved toward the pole to which the unirradiated kinetochore was oriented and remained near that pole until anaphase. In mitosis, therefore, bipolar tension produced by two functional kinetochores is necessary for the alignment of a chromosome on the equatorial plane. However, point-irradiation of kinetochores of *Haemanthus* endosperm by BAJER and MOLÈ-BAJER (1961) did not inactivate these organelles. Thus chromosomes with kinetochores irradiated during prophase, prometaphase, or metaphase did not show noticeable abnormalities in their movements until anaphase when the chromosome tended to remain in an equatorial position.

The bivalent whose kinetochore was irradiated in prometaphase or metaphase divided into half-bivalents in anaphase and both the irradiated and unirradiated half-bivalent underwent anaphase movement, during which a thin kinetochore fiber was observed connecting the irradiated kinetochore with the spindle pole toward which the irradiated half-bivalent was moving. In the experiment shown in Figs. 4 and 5 the time needed from the disappearance of the kinetochore fiber to its reappearance was about 20 minutes. We may safely assume that with the UV dose delivered in the present study the kinetochore-kinetochore fiber complex can recover from the radiation damage after about 20 minutes or so. In addition, it has been generally found that half-bivalents whose kinetochore was irradiated in metaphase underwent anaphase movement only when a birefringent fiber was reorganized between the irradiated kinetochore and the pole. These facts might explain the finding that anaphase movement occurred in all cells that were irradiated more than 20 minutes ahead of the start of anaphase and the movement was often suppressed when irradiation was made within 20 minutes of the start of anaphase (Fig. 7). Thus, in order for the anaphase movement of the kinetochore-irradiated chromosome to take place, kinetochore activity might have to recover from the radiation damage.

When observed by polarizing microscopy, the kinetochore fiber appeared continuous from the kinetochore to one of the spindle poles. Therefore it is considered to be anchored to the pole. In the primary spermatocyte of the grasshopper studied by NICKLAS *et al.* (1982) fewer than half the kinetochore microtubules extended all the way to the pole in anaphase. The present study has given no clue to the solution of the problem as to whether the kinetochore is passively pulled toward the pole through its associated kinetochore fiber or moves poleward along stationary kinetochore microtubules which depolymerize during anaphase from their kinetochore-attached end.

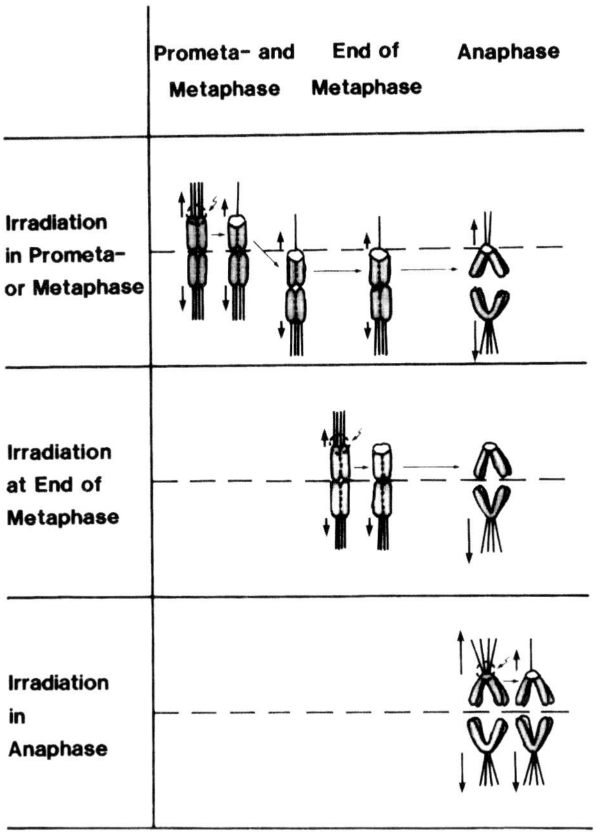

Fig. 9. Interpretation of the results of microbeam irradiation at one of the kinetochores of a bivalent. The upper row depicts the effect of irradiation in prometaphase and metaphase, the middle row, irradiation at the end of metaphase and the lower row, irradiation in anaphase. The broken line represents the spindle equator and the vertical arrows, the forces acting on the kinetochores

A tentative interpretation of the results obtained in the present study is schematically shown in Fig. 9. The following points were taken into account in making these drawings. (1) As was stated above, it has been the author's consistent finding that whenever a half-bivalent whose kinetochore was irradiated in metaphase was undergoing anaphase movement, a birefringent kinetochore fiber had been reorganized at the kinetochore. (2) In normal cells the number of kinetochore microtubules composing a kinetochore fiber seems not to increase, at least not conspicuously, in the transition from metaphase to anaphase (ITO 1980, IZUTSU *et al.* to be published). On the other hand, the maximum force the spindle can produce on the kinetochore toward the pole can increase dramatically in anaphase (NICKLAS 1983). Therefore, the force generated by each kinetochore microtubule during anaphase must be much greater than during prometaphase and metaphase. (3) It has been shown in the present study that half-bivalents whose kinetochore was irradiated during anaphase were pulled poleward with a reduced force (Fig. 8), but they continued moving toward the pole. On the other hand, the birefringence of the associated kinetochore fiber of such a kinetochore was markedly reduced. Although the author has not made electron microscope studies on the kinetochore fiber whose kinetochore was irradiated with UV, probably very few microtubules remain intact after irradiation of their kinetochores in anaphase, although they may not be detected by the optics used in the present study. (4) It was a very conspicuous fact that kinetochore irradiation at the end of metaphase frequently inhibited anaphase movement of the irradiated half-bivalent whereas irradiation in anaphase did not. Apparently, the kinetochore-kinetochore fiber system is more resistant in anaphase than at the end of metaphase. The difference in the suceptibility of the metaphase kinetochore and the anaphase one to UV irradiation remains to be explained.

Acknowledgements

The author would like to thank Dr. TOSHIMICHI YOSHIDA and Mrs. MARI HARA for their kind help with manuscript preparation.

References

BAJER A, MOLÈ-BAJER J (1961) U.V. microbeam irradiation of chromosomes during mitosis in endosperm. Exp Cell Res 25: 251–267

FORER A (1965) Local reduction of spindle fiber birefringence in living *Nephrotoma suturalis* (Loew) spermatocytes induced by ultraviolet irradiation. J Cell Biol 25: 95–117

GORBSKY GJ, SAMMAK PJ, BORISY GG (1987) Chromosomes move poleward in anaphase along stationary microtubules that coordinately disassemble from their kinetochore ends. J Cell Biol 104: 9–18

ITO T (1980) Ultrastructure of the centriole and the mitotic and meiotic spindles in grasshopper spermatogenic cells. Mie Med J 30: 95–110

IZUTSU K (1959) Irradiation of parts of single mitotic apparatus in grasshopper spermatocytes with an ultraviolet microbeam. Mie Med J 9: 15–29

— (1961) Effects of ultraviolet microbeam irradiation upon division in grasshopper spermatocytes. II. Results of irradiation during metaphase and anaphase. Mie Med J 11: 213–232

— (1976) Effects of ultraviolet microbeam irradiation of parts of dividing cells on the spindle fiber behaviour and chromosome movement. J Cell Biol 70: 67 a

— SATO H, NAKABAYASHI H, AOKI N (1977) The behaviour of spindle fibers and movement of chromosomes in dividing grasshopper spermatocytes. Cell Struct Funct 2: 119–133

— BAJER A, MOLÈ-BAJER J, YOSHIDA T, to be published

MCINTOSH JR (1985) Spindle structure and the mechanisms of chromosome movement. In: DELLARCO VL, VOYTEK PE, HOLLAENDER A (eds) Aneuploidy. Plenum Press, New York, pp 197–229

McNeill PA, Berns MW (1981) Chromosome behaviour after laser microirradiation of a single kinetochore in mitotic PtK_2 cells. J Cell Biol 88: 543–553

Nicklas RB, Kubai DF, Hays TS (1982) Spindle microtubules and thier mechanical association after micromanipulation in anaphase. J Cell Biol 95: 91–104

— (1983) Measurements of the force produced by the mitotic spindle in anaphase. J Cell Biol 97: 542–548

— (1987) Chromosomes and kinetochores do more in mitosis than previously thought. In: Gustafson JP, Appels R, Kaufman RJ (eds) Chromosome structure and function: the impact of new concepts (18th Stadler Genetics Symposium). Plenum Press, New York (in press)

Östergren G (1951) The mechanism of co-orientation in bivalents and multivalents. Hereditas 37: 85–156

Schrader F (1953) Mitosis, 2nd edn., Columbia University Press, New York

Protoplasma (1988) [Suppl 1]: 133–152

Division of Plastids by a Plastid-Dividing Ring in *Cyanidium caldarium*

T. MITA[1] and T. KUROIWA[2,*]

[1] Department of Food and Nutritional Sciences, University of Shizuoka, and [2] Department of Biology, Faculty of Science, University of Tokyo

Received March 11, 1988
Accepted April 15, 1988

Dedicated to Professor Dr. NOBURO KAMIYA on the occasion of his 75th birthday

Summary

The fine structure and function of the plastid-dividing (PD) ring and its fate during the life cycle have been studied in a synchronized population of cells of the unicellular alga *Cyanidium caldarium* RK-1. Cells were examined by using epifluorescence microscopy after staining with 4′,6-diamidino-2-phenylindole (DAPI) and rhodamin-phalloidin, and by electron microscopy after freeze-substitution fixation and immuno-gold cytochemistry. The young cell contains a single chloroplast, a cell nucleus and a mitochondrion, and these organelles divide in that order just prior to cytokinesis. As a result of two endospore divisions, the mother cell divides to form four daughter endospore cells each of which contains a cell nucleus, chloroplast, and mitochondrion. Cytochalasin B (20 µg/ml) does not inhibit the division of cell nuclei but markedly inhibit the division of plastids. By contrast, cremart (10 µg/ml) does not inhibit the division of plastids but does inhibit division of cell nuclei. Using the rhodamin-phalloidin staining technique, we have clarified that a bundle of actin-like filaments is located around the constricted isthmus of the dumbbell-shaped chloroplast. Under the electron microscope, the PD rings, of about 60 nm and 30 nm in width, are seen in the cytoplasm around the equatorial region of the dividing chloroplasts during the first and second chloroplast divisions, respectively. The PD ring consists of numerous fine filaments of about 4.5 nm in diameter which are aligned circumferentially. The fine filaments of the PD ring appear to be formed from small somewhat-electron dense granules, 40–90 nm in diameter, just before the formation of the constricted isthmus. The ring progressively generates a deep furrow with dimensions similar to those of the PD ring. Finally, the contents of chloroplast are separated into two parts, as if by the PD ring, to form two daughter chloroplasts. These results and the data obtained by immuno-gold cytochemistry and electron microscopy suggest that the PD ring consists of actin-like filaments and is an important participant in the division of the chloroplast.

Keywords: Chloroplast division; Plastid-dividing ring; Actin-like filaments; Unicellular alga; *Cyanidium caldarium*.

* Correspondence and Reprints: Department of Biology, Faculty of Science, University of Tokyo, Hongo, Bunkyo, Tokyo 113, Japan.

Abbreviations: CN cell nucleus; M mitochondrion; CP chloroplast; MN (mt-nucleus) mitochondrial nucleus (nucleoid); CPN (cp-nucleus) chloroplast nucleus (nucleoid); SEB somewhat-electron dense body; SEG somewhat-electron dense granule.

1. Introduction

It is well-known that mitochondria and plastids contain organelle-specific DNA (RABINOWITZ and SWIFT 1970) and divide into daughter organelles by fission. Such organelle-specific DNAs are not naked, but are associated with specific, basic proteins and organized into organelle nuclei (nucleoids) that resemble bacterial nuclei (KUROIWA 1974, 1976, 1982). The nuclei of mitochondria and young proplastids are located in the central area of these organelles. Since the mitochondria of *Trypanosoma acanthocephalin* (RENGER and WOLSTENHOLME 1972) and *Physarum polycephalum* (KUROIWA 1982) contain multiple copies of organelle-specific DNA, the mitochondrial nuclei (mt-nuclei) are electron dense. In such cases, the division of organelles can be unequivocally classified into two main processes, namely, the division of the organelle nucleus and organellekinesis (plastidkinesis and mitochondriokinesis) (KUROIWA 1982). These three latter terms have been adopted as appropriate analogs of cytokinesis.

It is of great interest to us to elucidate the a structural basis of the mechanism of organellekinesis, although very little information about this mechanism is available at present. When micro-plasmodia of *P. polycephalum* were treated with cytochalasin B at a concentration of 50 µg/ml, dumbbell-shaped mitochondria did not appear but, instead, spherical and ovoid mitochondria, which contained two electron-dense mt-nuclei, appeared, suggesting that mitochondriokinesis had been unsuccessful even though the mt-nucleus was able to divide (KUROIWA and KUROIWA 1980). Thus, actin-like filaments may be involved in mitochondriokinesis. However, we have been unable to discern any fine structures that are intimately related to mitochondriokinesis. Probably, only a few actin-like filaments, which cannot be visualized under the electron microscope after fixation by conventional methods, may be important for mitochondriokinesis.

Since the plastids are considerably larger than mitochondria, more bulky apparatus may be required for plastidkinesis than for mitochondriokinesis. SUZUKI and UEDA (1975) were the first report on the appearance of electron dense material, which they consider to be a septum, at the narrow bridge between daughter proplastids or daughter amyloplasts, after the middle stages of plastid division in *Pisum sativum*. Similar electron-dense deposits have been observed at the narrow neck of dividing chloroplasts and have been described as "plaques" or "bar forms" in the dividing chloroplasts of *Atriplex semibaccata* and *Sesamum indicum* var. gluca (LEECH *et al.* 1981) and *Phaseolus vulgaris* (WHATLEY 1980). Subsequently, CHALY and POSSINGHAM (1981) reported that electron-dense deposits, which are parts of an annulus, are located between the outer and inner envelopes at the narrow bridge between daughter proplastids of *Pisum sativum*. *Phaseolus vulgaris*, *Lyoperisicon esculentum*, *Lactuca sativa*, *Citrullus lanatus*, *Hordeum vulgare*, *Zea mays*, *Triticum aestivum* and *Pinus radiata*. Common features of these earlier observations can be summarized briefly as follows: electron-dense deposits appear only at the narrow bridge of the dumbbell-shaped plastid just before separation of daughter plastids and are never seen at the early stage of plastid division; and electron-dense deposits are located inside the plastids themselves.

More recently, MITA *et al.* (1986) identified a novel cellular structure, which we have called the plastid-dividing (PD) ring, in the red alga *Cyanidium caldarium*. The PD ring is not located inside the plastids but is observed in the cytoplasm close to the outer limiting envelope that is seen around the constricted isthmus of dividing chloroplasts. Similar PD rings have been observed in the moss *Funaria hygrometrica* (TEWINKEL and VOLKMANN 1987a 1987b) and the higher plant *Avena sativum* (HASHIMOTO 1986). HASHIMOTO (1986) reported the presence of an electron-dense "double" ring structure (PD ring doublet) around the narrow neck of dividing plastids. The inner and outer rings of the doublet coat the inside (stromal side) of the inner envelope and the outside (cytoplasmic side) of the outer envelope, respectively. HASHIMOTO (1986) also describes that the ring structures can be recognized by electron microscopy at the final stage of plastid division.

There appear to be discrepancies in results from different species of plant with respect to the timing of the appearance and the localization of the PD ring in dividing plastids. Furthermore, it is still unclear how the PD ring is organized in space, in time, within or outside the plastids, and in relation to other parts of the cytoplasm. Its component parts are unknown. The aim of the present experiments was to answer these problems by epifluorescence microscopy after staining with DAPI (WILLIAMSON and FENNELL 1975) or rhodamin-phalloidin, and by electron microscopy after freeze-substitution fixation (HEUSER and SALPETER 1979) and an immunoelectron microscopy using immuno-gold technique (ROTH 1986).

Preliminary reports of this study have been published elsewhere (MITA *et al.* 1986, KUROIWA and MITA 1987); this report extends scope of the previous work and includes discussion of molecular mechanisms of chloroplastkinesis (plastidkinesis).

2. Materials and Methods

2.1. Synchronous Culture

Cyanidium caldarium RK-1 cells were cultured in modified Allen's medium (ALLEN 1959), to which 35 g/l Jarmarin powder (artificial seawater; Jarmarin Laboratory, Osaka) was added, and which was adjusted to pH 3.5 with H_2SO_4. Cells were grown in 15-ml, L-type culture tubes and the cell density was adjusted to about 10^7 cell/ml in a total volume of 8 ml of medium. Culture tubes were incubated under continuous light at 45 °C and shaken at about 30 revolutions/min. Percoll density-gradient centrifugation was employed for the collection of synchronized cells. Percoll and culture medium that contained cells were mixed in a ratio 3 : 1 and the mixture was centrifuged at 10,000 × **g** for 20 min. The uppermost band of the gradient was collected. About 80% of cells in this fraction were young cells which were derived from the recent division of large mother cells. These cells were harvested by centrifugation at 3,000 × **g** for 5 min, washed in fresh medium, and transferred to tubes that contained fresh medium for the synchronous culture of cells. The number of cells and the frequencies (%) of single cells, two-endospore cells, and four-endospore cells were determined by examination of about 200 cells at each time point. Mean percentages were calculated from results of three experiments.

2.2. Epifluorescence Microscopy for Observations of DNA and Actin Filaments

For the observations of cell nuclei and chloroplast nuclei (cp-nuclei), the cells were collected at intervals of 4 h after initiation of synchronous cultures, fixed in 1% glutaraldehyde dissolved in buffer S

(0.25 M sucrose, 1 mM EDTA, 0.6 mM spermidine, 0.05% 2-mercaptoethanol, 10 mM tris-HCl, at pH 7.6) for 10 min and stained with DAPI at a final concentration of 1 µg/ml, according to the method described previously (KUROIWA and SUZUKI 1980). For observations of actin filaments, the collected cells were stained with 0.3 µM rhodamin-conjugated phalloidin (Molecular Probes, Inc., Junction City, OR, U.S.A.). The stained samples were observed under an Olympus BHS-RFK epifluorescence microscope as described previously (KUROIWA and SUZUKI 1980). Observations of fluorescence were made with a UVFL 100 objective, using a UV excitation filter (365 nm) in combination with a 420 nm suppression filter for DAPI, and a red excitation filter (530–545 nm) in combination with a 590 nm suppression filter for rhodamin-conjugated phalloidin. Photographs were taken at a magnification of × 850 on 35 mm Fuji 400 film.

2.3. Treatment with Cremart and Cytochalasin B

Twenty four hours after the initiation of synchronous cultures, cells were incubated for 30 h in medium supplemented into 10 µg/ml cremart, an inhibitor of microtubule, dissolved in methanol, or cytochlasin B, an inhibitor of actin filaments, a final concentration of 20 µg/ml, and then harvested. The cytochalasin B was dissolved in DMSO and the final concentration of DMSO or methanol added to the medium was not greater than 0.4%. The cells were fixed in 0.6% glutaraldehyde for 10 min and stained with DAPI.

2.4. Electron Microscopy After Rapid Freeze-substitution Fixation

Cells were fixed at various times after initiation of synchronous cultures by the rapid freeze-substitution fixation method of HEUSER and SALPETER (1979). Briefly, synchronously cultured cells were collected by centrifugation and placed in contact with a cooled block of copper at the temperature of liquid helium (—270 °C) using Slamer (Polaron-Meiwa, Ltd. Japan). Frozen samples were transferred to absolute acetone that contained 2% OsO_4 and incubated at —80 °C for 48 h. After this substitution, samples were transferred to an incubator at —20 °C for 2 h. They were then kept at 0 °C for 2 h, and then warmed to room temperature. After a brief rinse in absolute acetone, cells were embedded in Epon 812. Thin sections were cut and stained with 1% uranyl acetate for 30 min and then with lead citrate for 4 min. These sections were examined with a JEM-200CX electron microscope (JEOL, Tokyo, Japan). Volumes of cells and chloroplasts, as ellipsoids, were calculated from the dimensions of sections of cells fixed at various stages of life cycle.

2.5. Immunoelectron Microscopy

For immuno-electron microscope, cell fixation was carried out according to the method described by NAGELE *et al.* (1985). Briefly, cells were fixed at the temperature of liquid helium (—270 °C) and substituted in a mixture containing 3.5% glutaraldehyde in 10 mM cacodylate-buffered 90% ethanol (pH 7.2). After 48 h substitution, cells were embedded in Lowicryl K4M according to the method of ROTH (1986). Thin sections were floated on droplets of a solution of actin-specific antibodies, raised against actin from smooth muscle of chicken gizzards, and then incubated with second antibodies coupled to 5 nm gold particles (Janssen Life Sciences, Beerse, Belgium) which were used at a dilution of 1 : 20. We previously demonstrated, by immunoblotting of an actin-specific monoclonal antibody to smooth muscle of chicken gizzards, that the cells of *C. caldarium* contain an actin-like protein which reacts with antibodies raised against actin from chicken gizzards (unpublished data).

3. Results

3.1. Effects of Cremart and Cytochalasin B on the Division of Chloroplasts

The synchronous cultures were initiated using young cells. These young cells are small daughter cells with volumes of approximately 4 µm^3 that are obtained by Percoll density-gradient centrifugation of seven-days-old cultures. Figure 1 *a* shows numbers of cells at different times after the initiation of a synchronous culture. The number of cells increases stepwise 44 and 60 h after initiation of the incubation and finally reaches a value of about four times the initial number. The increase at the second step is markedly greater than that at the first step. The basic life cycle of *Cyanidium caldarium* is shown in Figs. 1 *b* and 10 and is derived from data described previously (NAGASHIMA and FUKUDA 1981, MITA and KUROIWA 1986). If we start with an examination of young cells (*I* in Fig. 1), we see that the cell contains a cell nucleus of about 0.1 µm^3 in volume and a chloroplast of about 1 µm^3 in volume. Growth takes place up to the stage of ripened cells and these are mostly quadrinucleate after the second endospore divisions (*II–IV* in Fig. 1). The final step involves the formation of fully matured cells and the production of four new daughter cells from each mother cell. The chloroplast, cell nucleus, and mitochondrion divide in order just before the first and second endospore divisions (MITA *et al.* 1986). The relative proportions of cells that contain in two endospores and of cells that contain four endospores were examined at different times after the initiation of synchronous cultures. The number of cells increased markedly 60 h after the initiation of synchronous culture (Fig. 1 *a*), and reached about four times the number of cells at the initiation of the culture. The increase in the number of cells containing two endospores (*III* in Fig. 1 *b* and *c*) occurs just before that of the cells containing four endospores (*IV* in Fig. 1 *b* and *c*). These increases occur at the same times as the decrease in number of young cells (*I* in Fig. 1 *b* and *c*) and of single cells occurs (*II* in Fig. 1 *b* and *c*), about 20 h before the increase in cell number (Fig. 1 *a*). This series of events demonstrates how the young cells grow and develop, and how the daughter cells, formed by the second endospore division, mature for about 20 h and divide to form new young cells.

The experiments involving administration of inhibitors were conducted by anticipating the timing at which furrows developed around the chloroplasts with reference to mitotic nuclear events. Figure 2 shows representative epifluorescence photomicrographs of

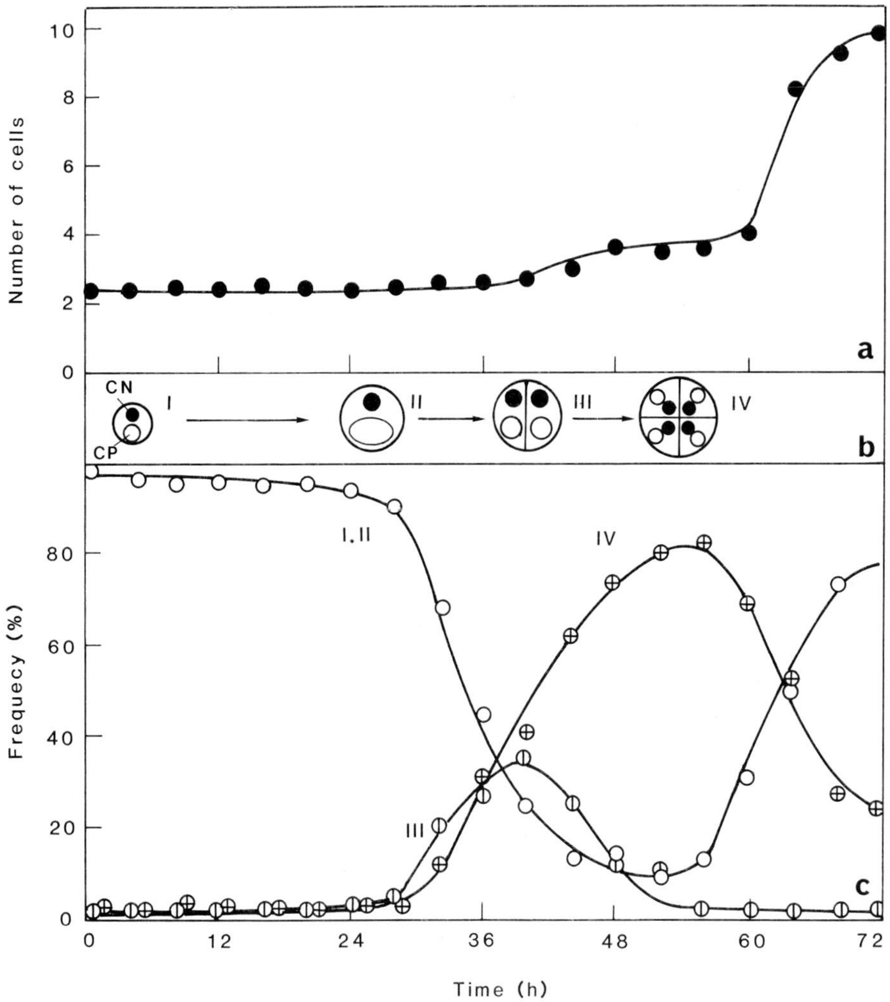

Fig. 1. Changes in cell number ($\times 10^7$ cells/ml) (*a*) and a schematic representation (*b*) and frequencies (%) (*c*) of four different types of cell: young cells (*I*) and single cells (*II*), two-endospore cells (*III*) and four-endospore cells (*IV*), at various times after the initiation of synchronous cultures of *Cyanidium caldarium* prepared by Percoll density-gradient centrifugation. One mother cell divides synchronously to form four daughter endospores after two endospore divisions. The points are means of duplicated measurements

mother cells fixed after treatment without or with the inhibitors cremart or cytochalasin B. In the control preparation, when the mother cell was excited with ultraviolet light after DAPI staining, four spherical cellular nuclei and irregular-shaped cp-nuclei, emitting blue-white fluorescence, can be seen (Fig. 2 *a*) while, when the same field of the cell was excited with green light, four chloroplasts, emitting red fluorescence, can be clearly seen in the areas where cp-nuclei are located (Fig. 2 *b*). These results indicate that the untreated mother cell is composed of four discrete endospore cells, each of which contains a cell nucleus and a chloroplast. The mt-nucleus is very small and, thus, it was only rarely observed when cells were squashed with relatively high pressure on a glass slide. When cells are treated with cremart, an inhibitor of the assembly of tubulin, throughout two sequential endospore divisions, each mother cell is seen to contain one cell nucleus (Fig. 2 *c*) and four chloroplasts (Fig. 2 *d*) as predicted. By contrast, when the cells are treated with cytochalasin B, an inhibitor of the aggregation of actin into filaments, throughout two sequential endospore divisions, each mother cell is seen to contain four cell nuclei (Fig. 2 *e*) and one large chloroplast (Fig. 2 *f*). The results suggest that microtubules are not involved in the division of the chloroplast, but are involved in division of the cell nucleus, while actin filaments are not involved in division of the cell nucleus but are involved in the division of chloroplasts.

3.2. Staining of Actin Bundles Around the Chloroplast

In order to define the localization of actin filaments, the cells were fixed, 36 h after the initiation of syn-

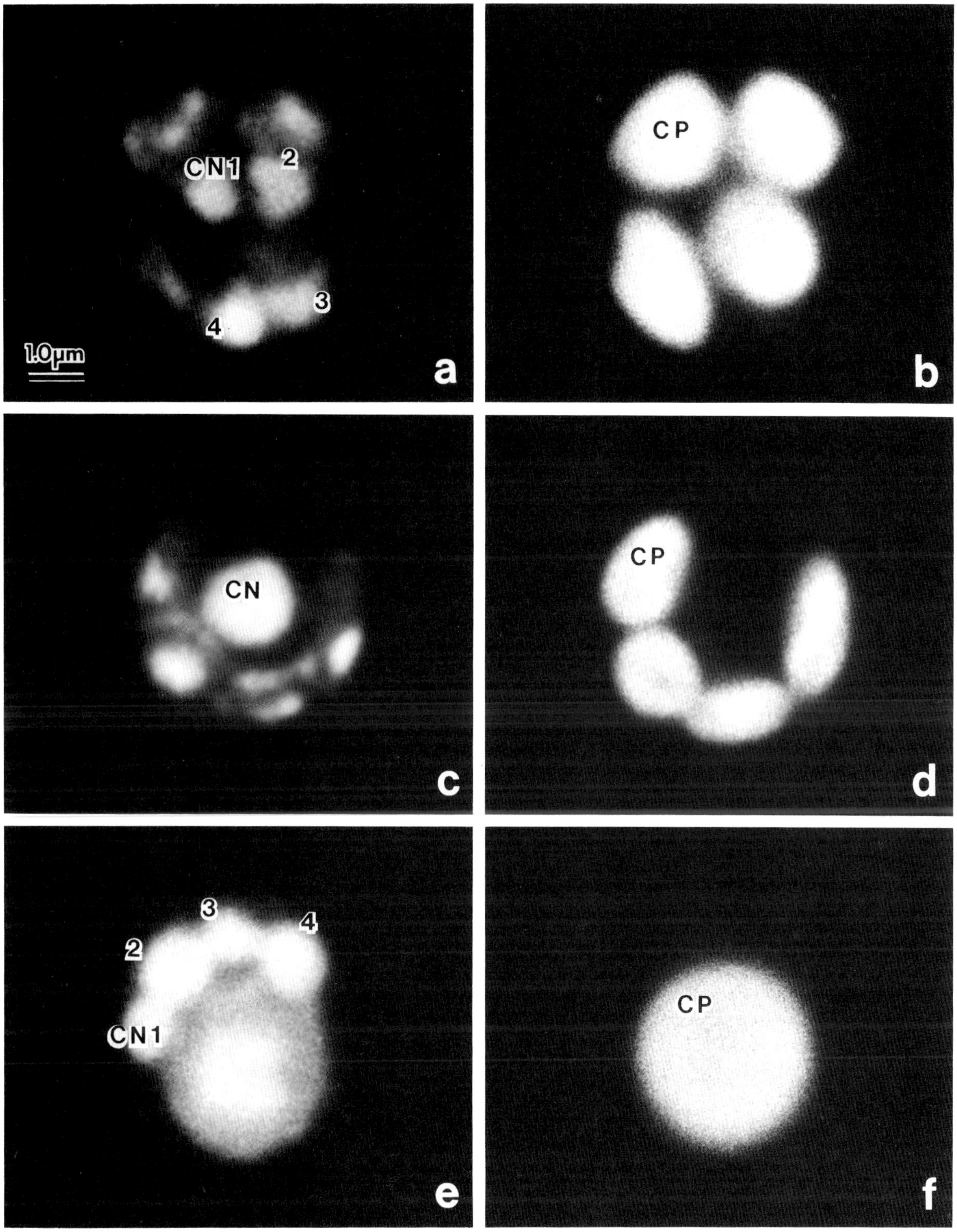

Fig. 2. Epifluorescence photomicrographs illustrating cell nuclei (*CN 1*–*CN 4*) and the chloroplast (*CP*) in mother cells fixed immediately after incubation for 30 h without inhibitors (*a, b*), with cremart at a concentration of 10 μg/ml. Fluorescence photomicrographs with UV (*a, c, e*) and green lights (*b, d, f*) were taken in the same field for each treatment. In the control (*a, b*), one mother cell is composed of four discrete endospore cells, each of which contains a cell nucleus and a chloroplast. When the cells are treated with cremart through two sequential endospore divisions, the mother cell contains one cell nucleus and four chloroplasts (*c, d*). By contrast, the cell are treated with cytochalasin B, the mother cell contains four cell nuclei and one large chloroplast (*e, f*). Bar represents 1 μm

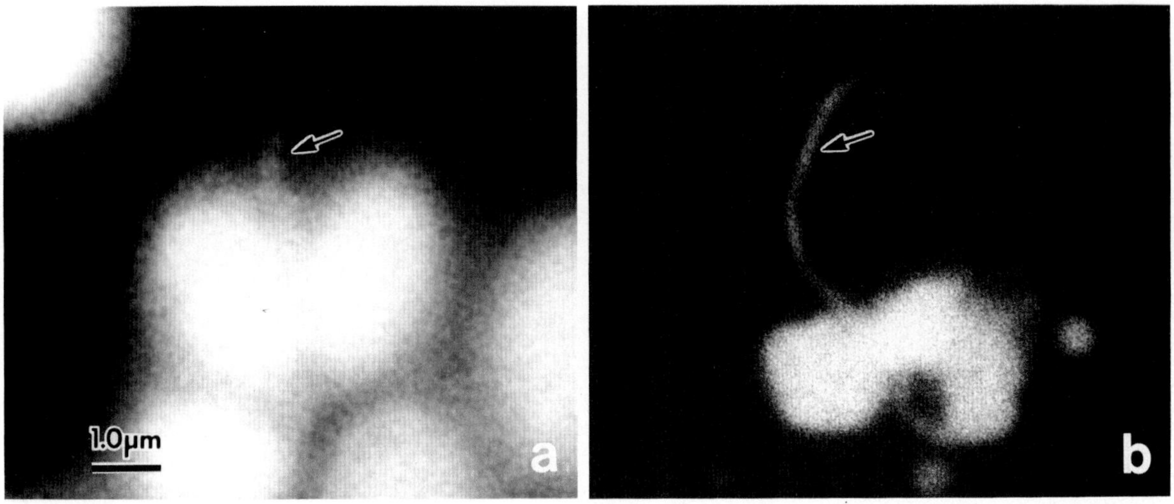

Fig. 3. Epifluorescence photomicrographs illustrating ring-like structures (arrows) around chloroplasts released from a cell squashed gently (*a*) and from a cell squashed with greater pressure (*b*), stained with rhodamine-phalloidin. When the chloroplast has been squashed under significant pressure, the ring-like structure is released from the chloroplast. Bar represents 1 µm

chronous culture, and stained with rhodamin-phalloidin (Fig. 3). A ring emitting orange colored fluorescence appeared faintly in the equatorial region of the dividing chloroplast, which emitted a red color in a gently squashed preparation (arrow in Fig. 3 *a*). In a more vigorously squashed preparation, part of ring is often released from the immediate area around the chloroplast (arrow in Fig. 3 *b*), suggesting that the ring, which consists of actin-like material, is located on the outside of the chloroplasts.

3.3. Formation of Plastid-Dividing Ring Around Chloroplast

We examined the localization and fine structure of the ring-like structure in cells fixed at different stages of the life cycle by electron-microscopic examination of cells prepared by the rapid freeze-substitution fixation method. Figure 4 *a–c* shows representative epifluorescence (Fig. 4 *a* and *b*) and electron (Fig. 4 *c*) micrographs of a young cell fixed 4 h after initiation of a synchronous culture. The cell contains one cell nucleus, one rod-shaped mt-nucleus, and one ct-nucleus (Fig. 4 *a*). The cp-nucleus is located in the central region of the spherical chloroplast (Fig. 4 *a* and *b*). Under the electron microscope, the spherical chloroplast is seen to contain concentric, circular, thylakoid membranes with phycobilisomes of about 30 nm in diameter (arrows in Fig. 4 *c*), in addition to the cp-nucleus. In the cytoplasm, a somewhat-electron dense body (SEB) and a mitochondrion are also visible. After 20 h, the cell, the cell nucleus, and the chloroplast increase in volume about three fold (to about 12 μm^3), two fold (to about 0.2 μm^3), and three and a half fold (to about 3.5 μm^3), respectively. As the shape of the chloroplast changes, and a football-like structure develops, concentric circular thylakoid membranes begin to separate into two parts (Fig. 4 *d*). A deposit with a cross section similar to that of the PD ring is faintly visible at the equatorial region of the chloroplast (arrow in Fig. 4 *d*).

With the formation of the distinct invagination of the envelope of the chloroplast, 24 h after the initiation of synchronous culture, the somewhat-electron-dense deposits appear as compressed dots, 60 nm in their longest dimension, opposite one another, on the outside of the envelope, on either side of the constricted isthmus (arrows in Fig. 5 *a*). At higher magnification, the deposits

Fig. 4. Epifluorescence photomicrographs (*a, b*) and electron micrographs of intervening sections (*c, d*) of a cell nucleus (*CN*), a chloroplast (*CP*), a chloroplast nucleus (*CPN*), a mitochondrion (*M*), a mitochondrial nucleus (*MN*), and a somewhat electron-dense body (*SEB*), in cells fixed 4 h (*a–c*) and 24 h (*d*) after the initiation of synchronous culture. *a* and *b* are images of three nuclei (*a*) and chloroplast (*b*) taken in the same field under excitation by UV (*a*) or green light (*b*) after DAPI staining. In young or single cells, the cell nucleus, the mitochondrion and the chloroplast form a line (*a–d*). The cp-nucleus is located in the central area of concentric thylakoid membranes with phycobilisomes about 30 nm in diameter (arrow in *c*). When the chloroplast increases in volume and becomes a football-like structure, the PD ring is faintly visible on a part of the outer envelope at the equatorial region of the chloroplast (arrow in *d*). Bars represent 1 µm

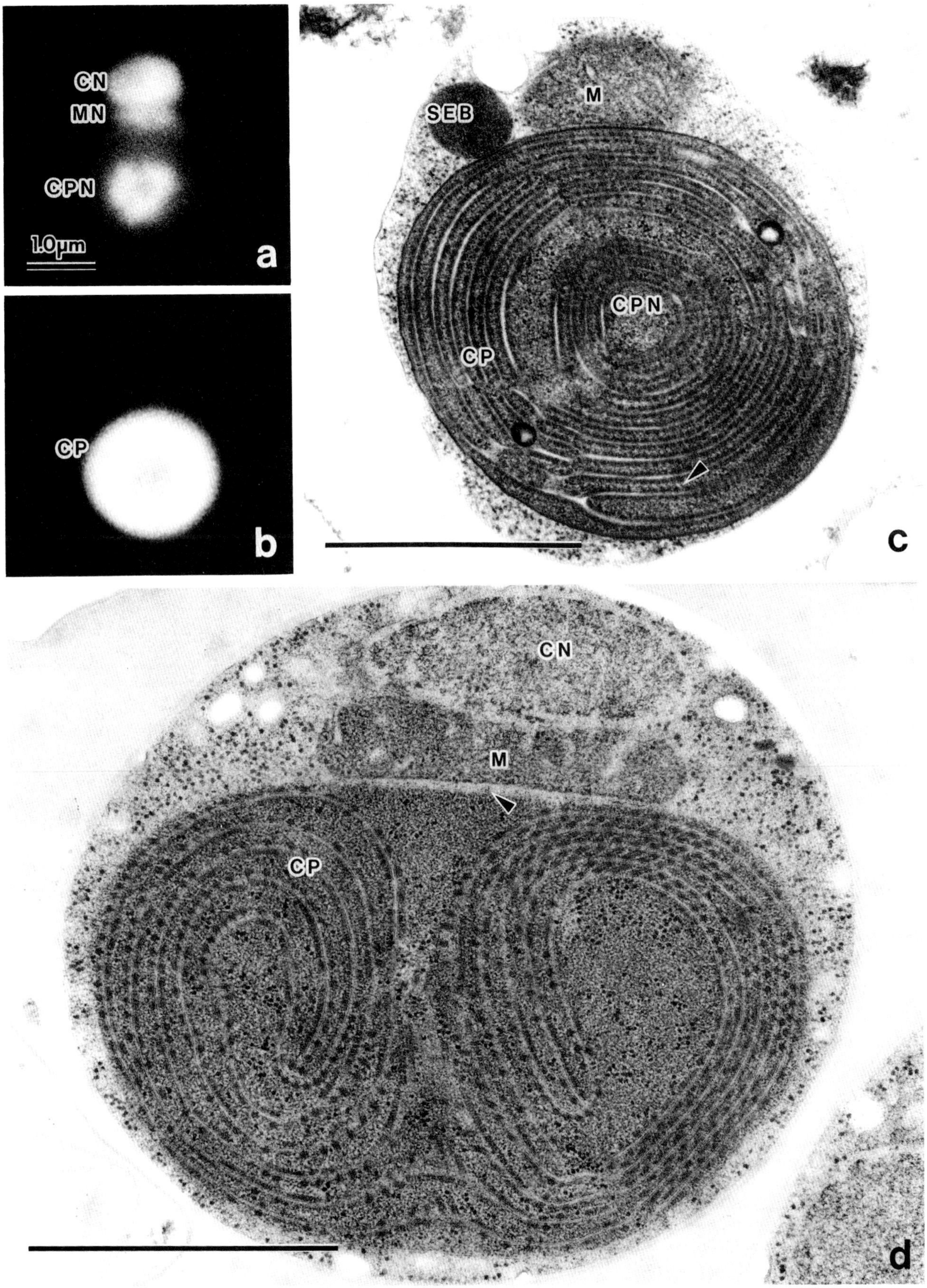

Fig. 4

of the PD ring are seen to be closely associated with the outer envelope and appear to press against the envelope at the point of constriction of the isthmus of the chloroplast (Fig. 5 *b* and *c*). The matrix of the chloroplast, beneath the inner envelope, becomes electron-dense as an inner belt (arrows in Fig. 5 *b* and *c*).

When the PD ring was serially sectioned in a direction perpendicular to the plane of division, at the edge of the same chloroplast as that shown in Fig. 5 *a–c*, a dense deposit was seen as a bar, about 60 nm in width, at the edge of the PD ring (arrows in Fig. 5 *d* and *f*). The bar consists of fine filaments, each about 4.5 nm in diameter, which are aligned parallel to the longitudinal direction of the bar (arrowheads in Fig. 5 *f*). This arrangement suggests that the PD ring is a bundle of fine filaments. Many somewhat-electron-dense granules (SEG), each 40–90 nm in diameter, are distributed along the bundle of filaments. Judging from the density of the granules, we can guess that the small granules may be generated by fragmentation of SEB. Often the end of the fine filament appears to coincide almost exactly with the edge of the SEG (arrowheads in Fig. 5 *f*), suggesting that fine filaments are made from the SEG.

3.4. Contraction of the PD Ring

By the time the PD ring starts to contract, the small SEGs have completely disappeared. Figure 6 *a–f* shows sequential thin sections through the constricted isthmus of a dividing chloroplast, 28 h after initiation of synchronous culture. It appears that the electron-dense deposits at the bridge between daughter chloroplasts are distributed as a ring or belt-like structure, 60 nm wide and 50 nm thick, along the outside of the outer envelope of the chloroplast (arrows in Fig. 6). The PD ring is filamentous but it is difficult to see any individual fine filaments. As a belt of microfilaments, the PD ring has an overall width of about 60 nm, a size which corresponds very precisely to the size of the concave depression of the constricted furrow itself. It is as if the PD ring generates the constricting force that separates the daughter chloroplasts (arrows in Fig. 6). SEBs appear in the cytoplasm between daughter chloroplasts. Figure 7 shows a thin section of the cell fixed 32 h after the initiation of synchronous culture. Even though the division of chloroplast has reached an almost final stage, nuclear division and mitochondrial division have not yet started. After succeeding contractions, the PD ring becomes thicker and more electron-dense than at the early stage of the division of the chloroplast. At the final stage, the width of the deposits that make up the constricted PD ring appears to be somewhat greater than that of the deposits at the early and middle stages of chloroplast division (arrow in Figs. 7 *a* and 11). However, the width of the PD ring does not deviate very much from a value of the 60 nm, regardless of the stage of division, the steepness of the walls of the furrow between the daughter chloroplasts, and even among chloroplasts of vastly different volume. Our results suggest that the PD ring may generate contractile force that is involved in the division of the chloroplast.

At higher magnification, the cross sections of the PD ring reveal clearly that the PD ring is located on the cytoplasmic side of the outer envelope (Fig. 7 *b*). Inside the inner envelope, some electron-dense deposits (arrowheads in Fig. 7 *b* and *c*) can also be seen but their width and thickness do not change from early stage of the chloroplast division until it is complete. The PD ring, when seen in sections cut in a direction parallel to the plane of division, clearly comprises a circular belt which appears to be composed of tightly packed, fine filaments (Fig. 7 *c*).

3.5. Conversion of the PD Ring to SEDs

Figure 8 shows the PD ring at the final stage of chloroplast division in a cell fixed 32 h after the initiation of synchronous culture. The electron density of the PD ring become markedly high and what is left of the ring remains as a conspicuous, electron-dense structure between daughter chloroplasts which are approximately 3.5 μm^3 in volume (Fig. 8 *a*). When the chloroplast has ceased dividing, a centriole-like plaque with microtubules develops outside the cell nucleus (double arrow-

Fig. 5. Electron micrographs of the intervening region (*a–c*) and the edge (*d–f*) of the PD ring sectioned serially in a direction perpendicular to the plane of division of the chloroplast in a cell fixed 24 h after initiation of synchronous culture. Somewhat-electron-dense, flat deposits appear as dots, 60 nm in width, opposite one another on either side of the constricted isthmus of the dividing chloroplast (arrows in *a*). At higher magnification, the deposits are seen to be located in the cytoplasm and to be closely associated with the outer envelope. The margin of the inner matrix beneath the inner envelope also become electron-dense when they are compressed by the PD ring (arrowheads in *b* and *c*). In the three serial sections of the edge of the PD ring, the dense deposits appear as a bar about 60 nm wide (arrows in *d–f*). Fine filaments are aligned parallel to the longitudinal direction of the bar (*d–f*). Small, somewhat-dense granules (*SEG*), 40–90 nm in diameter, appear along the bundle of fine filaments and SEGs can be seen to be associated with the fine filaments (arrowheads in *f*). Bars represent 0.1 μm

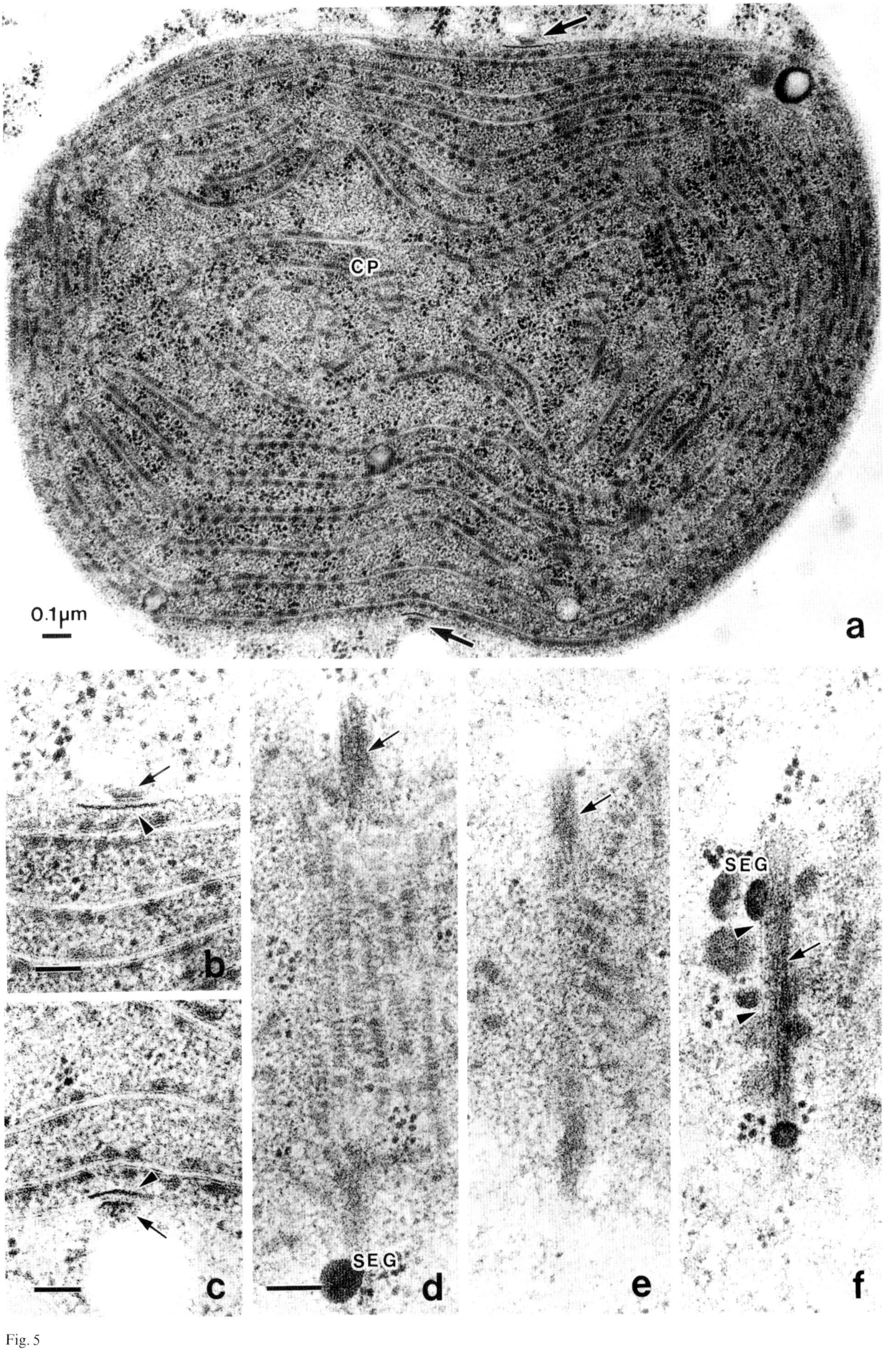

Fig. 5

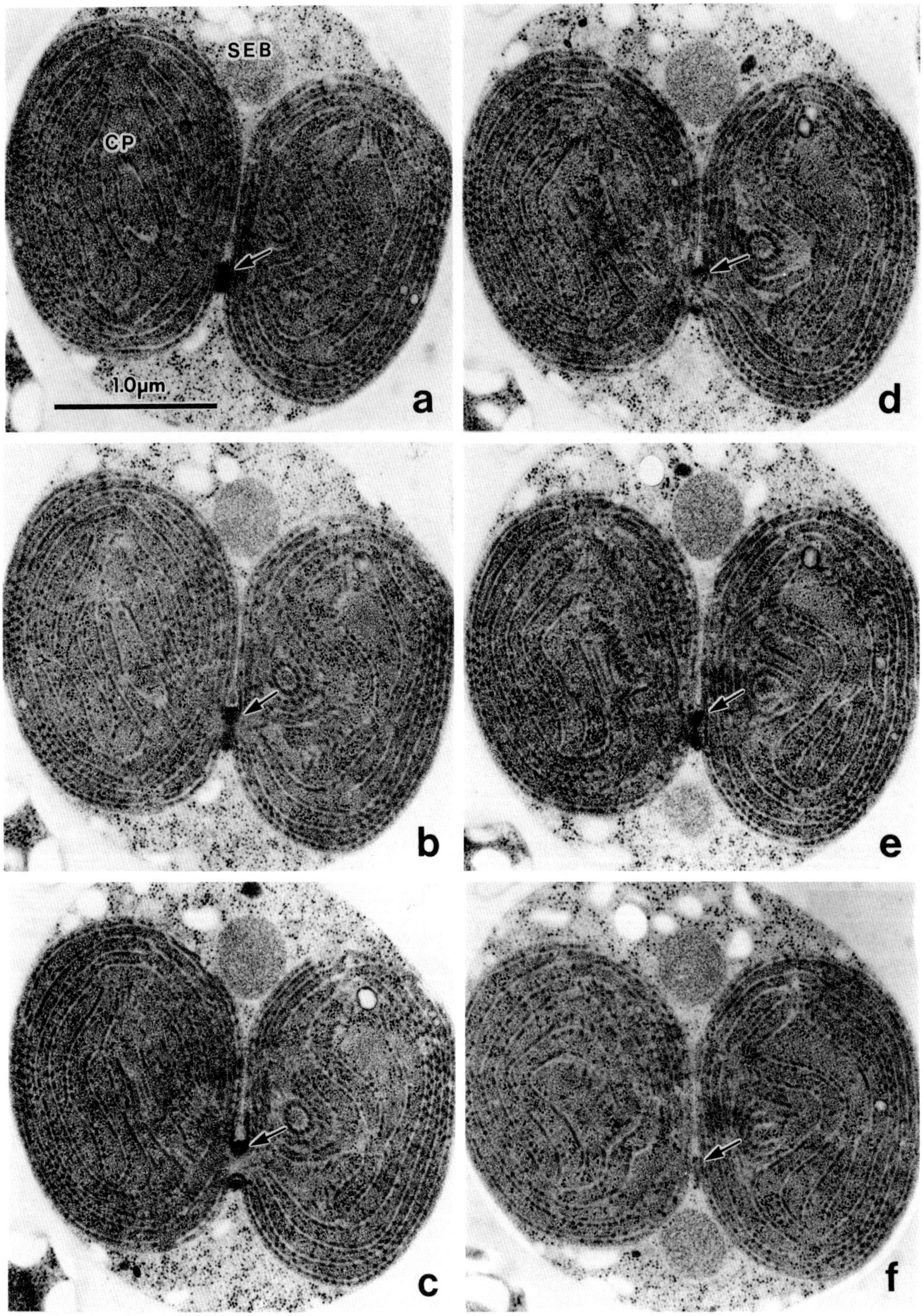

Fig. 6. Electron micrographs of six serial thin sections cut in a direction perpendicular to the plane of division of the chloroplast in a cell fixed 28 h after the initiation of synchronous culture. The PD ring is filamentous but it is difficult to distinguish individual fine filaments. As a belt of microfilaments, the PD ring has an overall width of about 60 nm which corresponds very precisely to the width of concave depression of the constricted furrow itself (arrows). Bar represents 1 µm

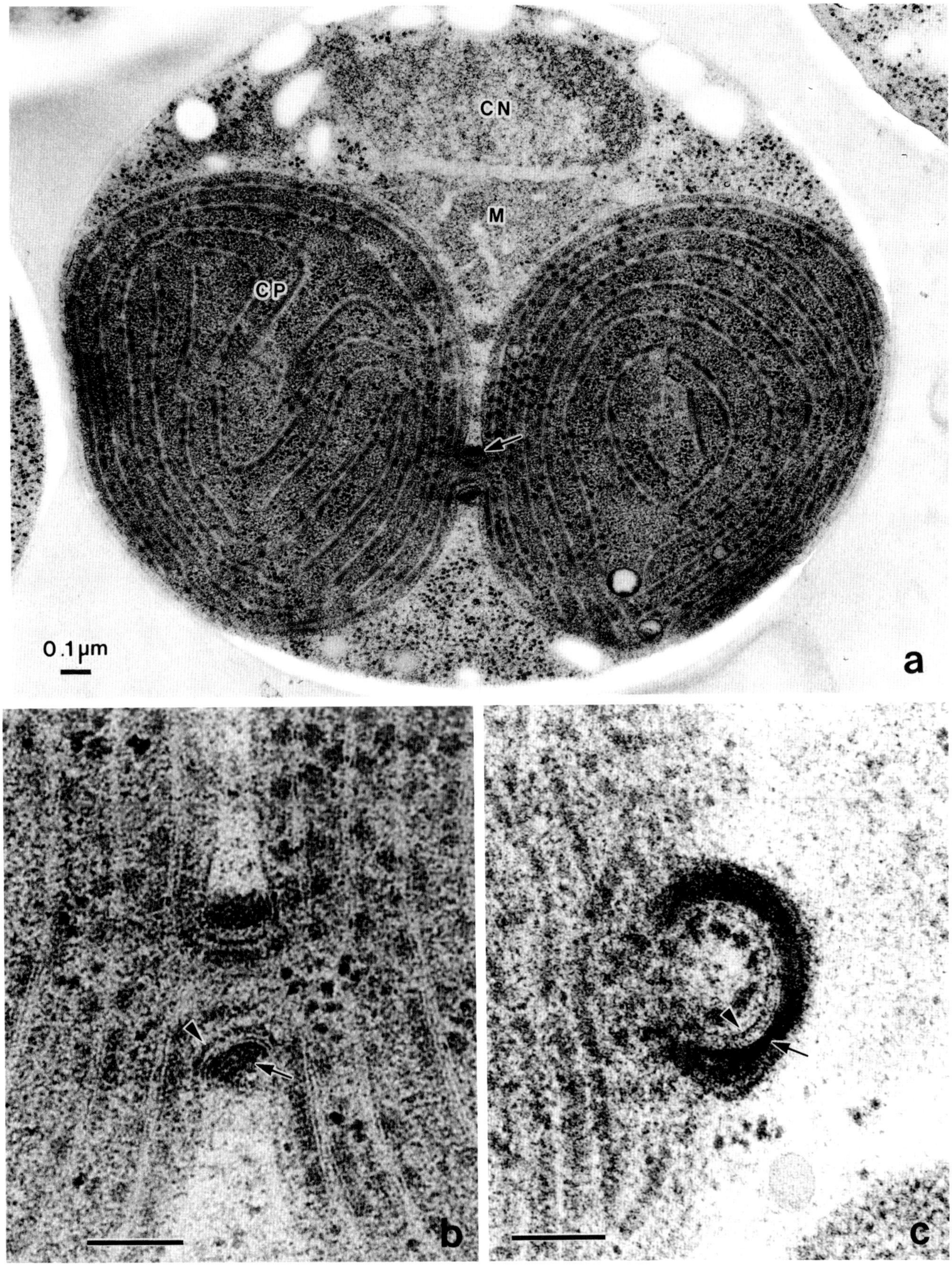

Fig. 7. Electron micrographs of sections cut in directions perpendicular to (*a, b*) and parallel to the plane of division (*c*) at the constricted isthmus of the dividing chloroplast, in cells fixed 32 h after the initiation of synchronous cultures. When the chloroplast is progressively pinched, the PD ring become more-electron dense and the width of the deposits is similar to that of the narrow furrow of the envelope (arrow in *a*). At higher magnification, the PD rings can be seen clearly to be located on the cytoplasmic side of the outer envelope and to increase in thickness (arrows in *b* and *c*), while the inner electron-dense belt (arrowheads in *b* and *c*) does not change from that observed at the early stage of the division. The bars represent 0.1 μm

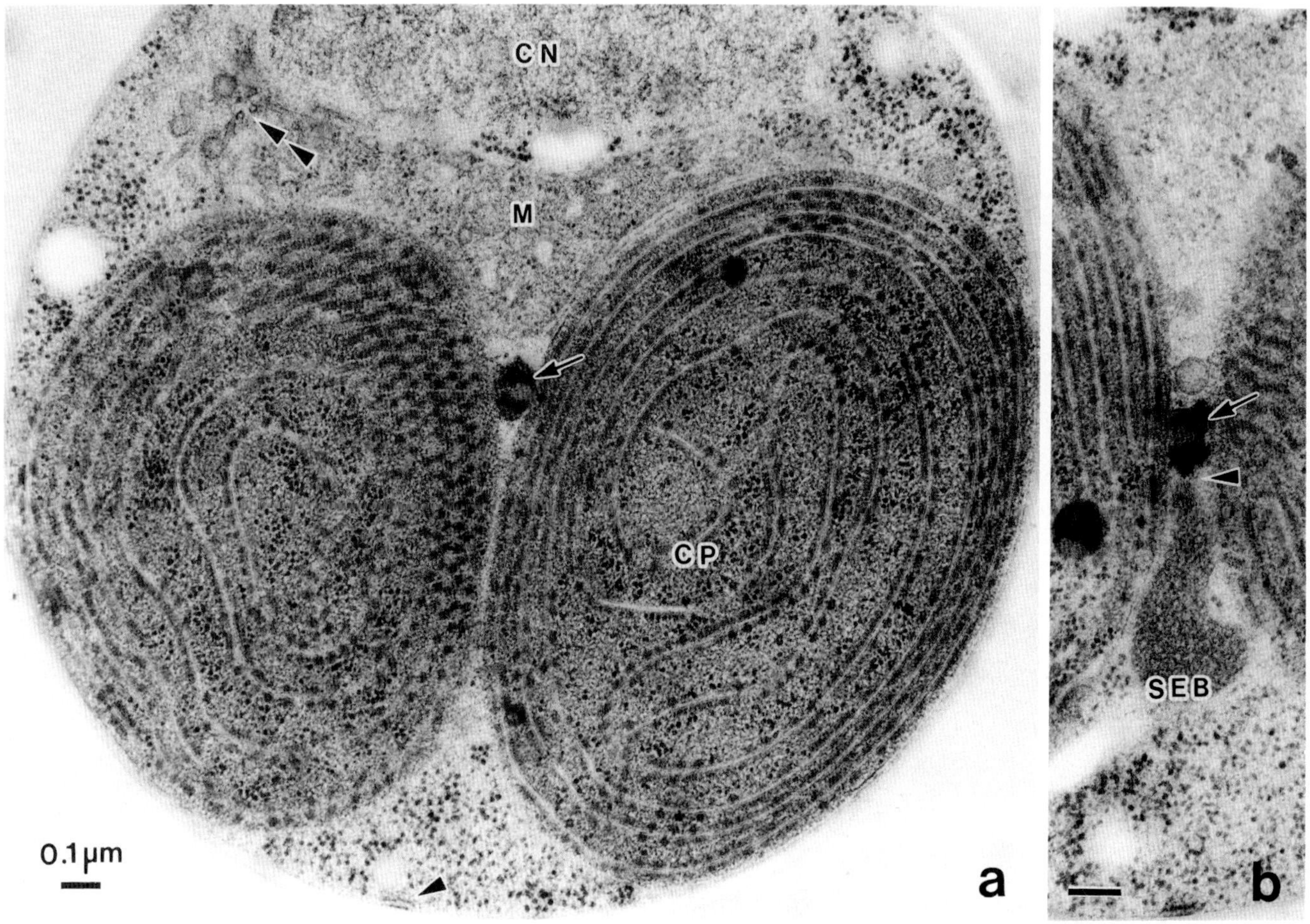

Fig. 8. Electron micrographs of the PD rings of the chloroplasts sectioned in a direction perpendicular to the plane of division of the chloroplast in cells fixed 32 h after the initiation of synchronous culture. The electron-dense PD ring appears on the tenuous bridge between daughter chloroplasts (arrows). The centriole-like plaque (double arrowheads) at the pole of the cell nucleus and the electron dense deposit (arrowhead) beneath the cell membrane, which are required for cytokinesis are now visible. Soon after this point, the chloroplast divides completely and the association between the PD ring and the balloon-like, SEB is seen (arrowhead in *b*). The bars represent 0.1 µm

heads in Fig. 8 *a*) and an electron-dense deposit beneath the cell membrane also appears (arrowhead in Fig. 8 *a*), suggesting the beginning of the division of the cell nucleus and cytokinesis required for the first endospore division. Soon after the chloroplast has divided, the remnants of the PD ring can be seen, associated with a slender part of the balloon-like SEB (arrow in Fig. 8 *b*). Since the PD ring disappears completely before initiation of the second endospore division, it is possible that it is converted into the SEB at the point of this association. At the very least, the association suggests that there is an interaction between the PD ring and SEB.

When the second endospore division of the cell begins, 40 h after the initiation of synchronous culture, the fattened deposits indicative of the cross section of the PD ring appear again on each narrowing bridge between daughter chloroplasts (arrows in Fig. 9 *a*). However, the width and thickness of the PD ring appear to be smaller than those of the PD ring present during the first chloroplast division. With the progression of the chloroplast division, the PD ring, which appears

Fig. 9. Electron micrographs of the PD rings of chloroplasts sectioned in directions perpendicular to the planes of division of the chloroplasts in two-endospore cells fixed 40 h (*a*) and 44 h (*b*) after the initiation of synchronous culture. The PD ring appears on the constricted isthmus of dividing chloroplasts during the second chloroplast divisions (arrows). The width (30 nm) and thickness (15 nm) of the PD rings are about half the size of those involved in the first chloroplast division. However, as the division of chloroplasts progresses, the PD ring increases in both electron density and thickness (arrow in *b*). Bars represent 0.1 µm

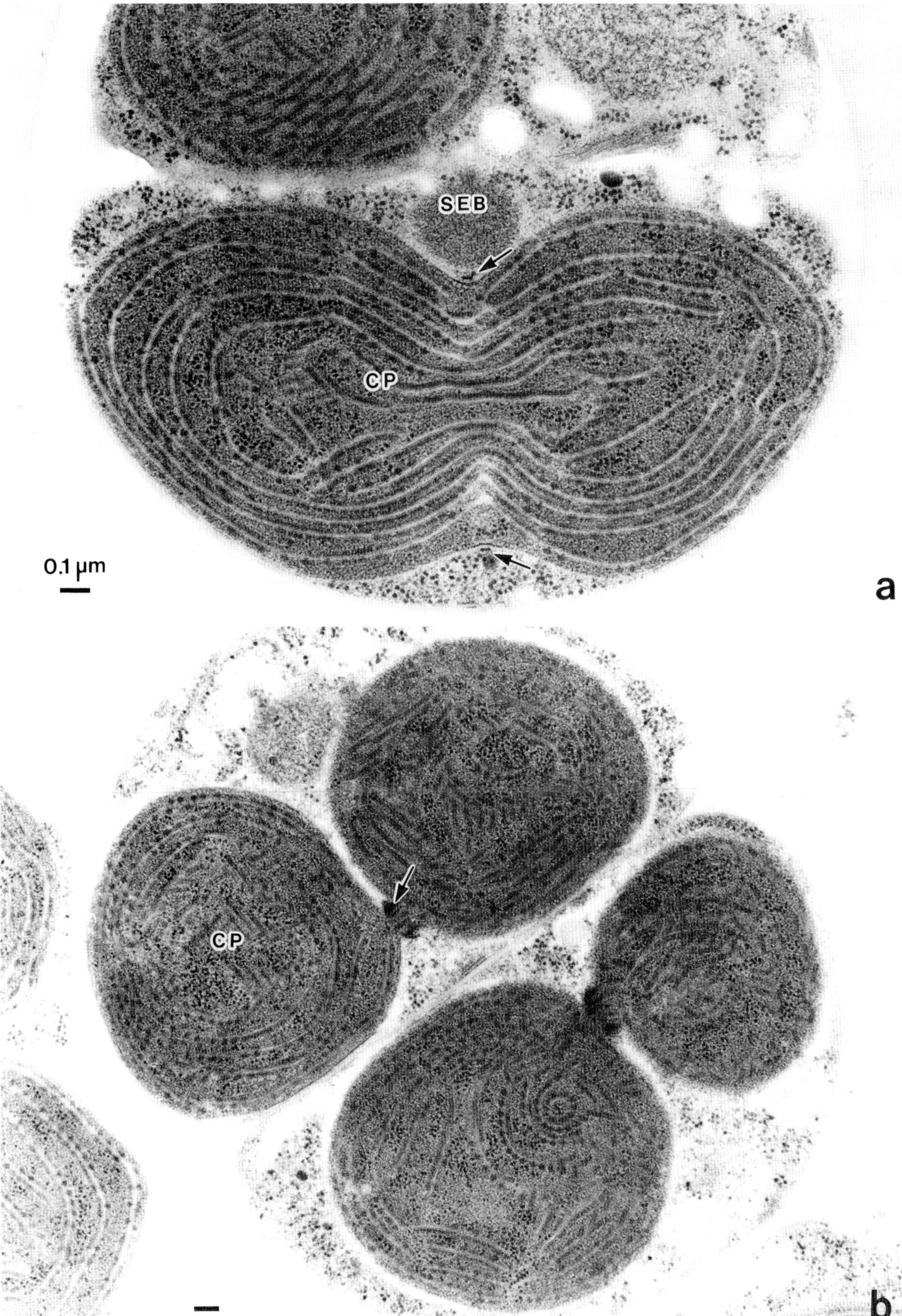

Fig. 9

at the narrow bridge of the dividing chloroplast in each of the two-endospore cells, becomes thick and its electron density increases, so that the PD ring resembles the ring involved in the first chloroplast division (Fig. 9 *b*). Figure 10 shows various sections which were cut in a direction parallel to the planes of division of the dividing chloroplasts at the early (Fig. 10 *a* and *b*), middle (Fig. 10 *c*), and late stages (Fig. 10 *d*) of the second chloroplast division. At the early stage, parts of the PD ring are absent in a few regions, suggesting that small fragments of the PD ring have been deposited in various parts of the equatorial region of the chloroplast and then fused to form a circular ring (arrows in Fig. 10 *a* and *b*). The increase in the electron density and thickness of the PD ring takes place as the contraction of the PD ring progresses (arrows in Fig. 10 *a–d*) but the thickness and electron density of the inner belts, within the inner envelopes do not change (arrowheads in Fig. 10 *a–d*). Figure 11 shows the relationship between the width and the thickness of the PD ring and the diameter of the plane of division of dividing chloroplasts at various stages of chloroplast division. The width of the PD ring does not change after the initial stage of the contraction but the thickness increases constantly during the contraction of the PD ring. These results suggest that the actual physical components of the PD ring remain constant during the contraction. The thickness (15 nm) and width (30 nm) of the PD ring during the second division of the chloroplast are about half as large as those of the first division. Since the dividing chloroplasts are about 1.7 μm^3 in volume, or about half the size of those involved in the first division, the size or volume of the PD ring appears to be intimately related to the volume of the dividing chloroplast. Judging from these results, we can assume that the mode of the second chloroplast division is basically similar to that of the first one.

3.6. *Immunoelectron Microscopy for Identification of Actin*

Figure 12 shows typical electron micrographs of cells after treatment with chicken gizzard actin-specific antibody. Gold particles are seen over the SEB (Fig. 12 *a*) and the balloon-like, SEB but do not appear to be present over the PD ring (Fig. 12 *b*). The results strongly support hypothesis that the SEBs contain actin. However, the results do not exclude the possibility that the PD ring contains actin because the number of gold particles that appeared over the sections was smaller than the predicted number. If the application of the technique can be improved, gold particles may be found over the PD ring.

3.7. *Behaviour of the PD Ring During the Cell Cycle*

Figure 13 shows a schematic representation of the division of chloroplasts in *C. caldarium*, based on the results described above. The first chloroplast division involves at least four different stages in term of its morphological characteristics. The initial event of the division is characterized by an increase in volume of spherical chloroplasts and by the conversion of spherical chloroplasts to football-like structures (chloroplast growth stage). Then small SEGs appear around the equatorial region of the football-like chloroplasts. These SEGs become associated with the fine filaments which make up the PD ring (PD ring formation stage). The PD ring contracts around the plane of division of the chloroplasts (PD ring contraction stage). After contraction of the PD ring and formation of two daughter chloroplasts, the PD ring appears to be converted into SEBs and finally disappears (PD ring conversion stage). Similar events seem to take place during the second cycle of chloroplast division, except that the volume of the dividing chloroplasts and the width and thickness of the PD rings are reduced by about 50%. This correlation suggests that the amount of material in the fine filaments of the PD ring depends on the volume of the dividing chloroplast.

4. Discussion

4.1. *Identification of the Components of the PD Ring*

We propose that at least one component of the PD ring is made up of actin filaments, on the basis of the following pieces of evidence. (1) Cytochalasin B, an inhibitor of the formation of actin filaments, inhibited the division of chloroplasts without inhibiting division of the cell nucleus (Fig. 2). (2) The ring-like structure around the chloroplasts can be stained with the rhodamin-conjugated phalloidin (Fig. 3). (3) Fine filaments, which are observed in the PD ring, are very similar in diameter to actin filaments (Fig. 5). However, our proposal is not supported by the results of our immuno-gold staining experiment: a few gold particles appeared on the SEBs and none of the PD ring (Fig. 11). However, the interpretation of the data obtained by the immuno-gold electron microscopy is very difficult. The absence of immuno-gold particles on the PD ring may be the results of one of the following possibilities. (1) The PD ring may not contain actin. (2) The amount

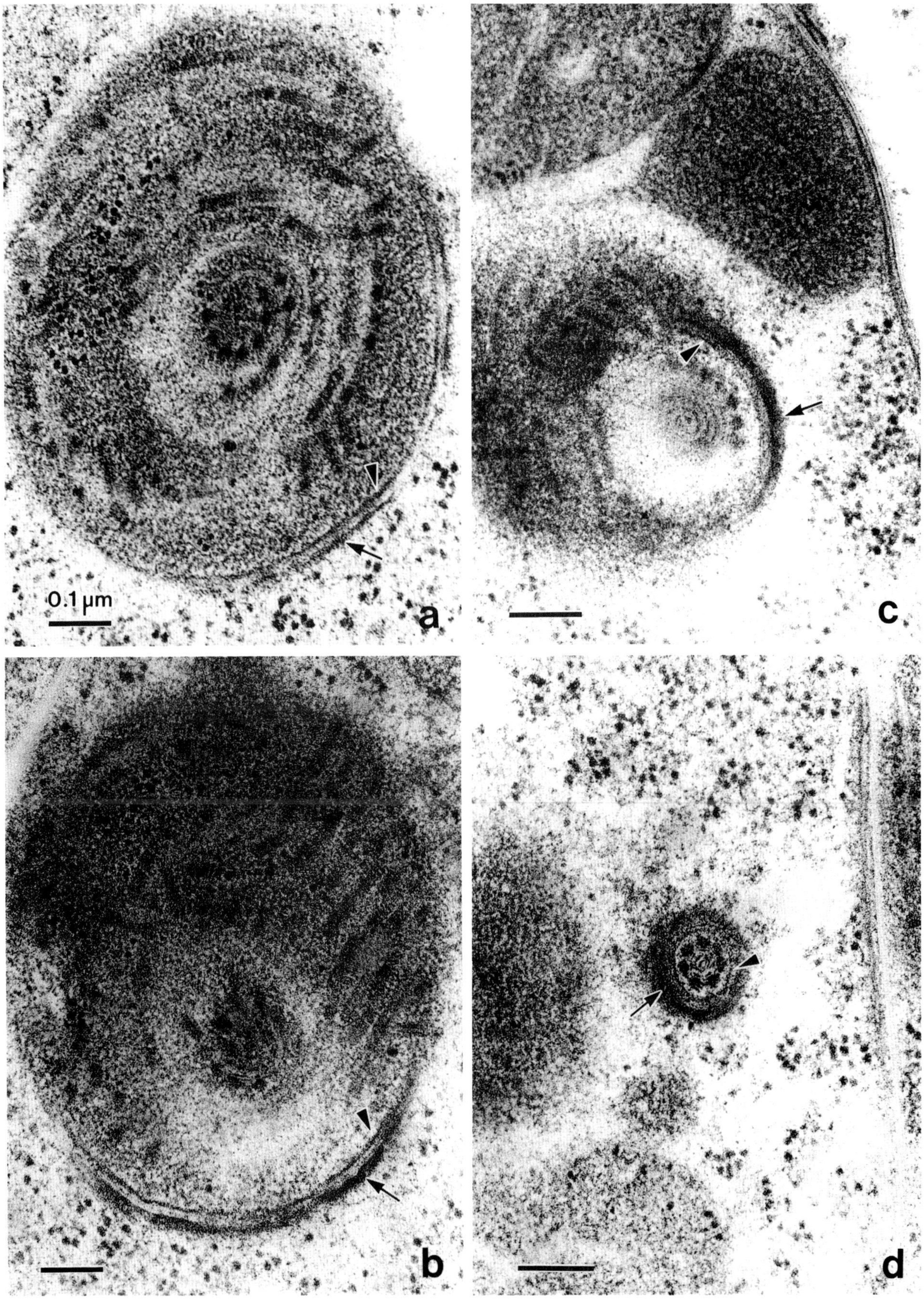

Fig. 10. Electron micrographs illustrating the circular PD rings sectioned in directions parallel to the planes of division at the constricted isthmus of the dividing chloroplasts, in two-endospore cells fixed 40 h (*a, b*) and 44 h (*c, d*) after the initiation of synchronous culture. In the second chloroplast division, fragments of the PD ring fuse to form a circular ring which then increases in thickness with the advance of the chloroplast division (arrows), while the inner matrix belts do not change in thickness (arrowheads). Bars represent 0.1 µm

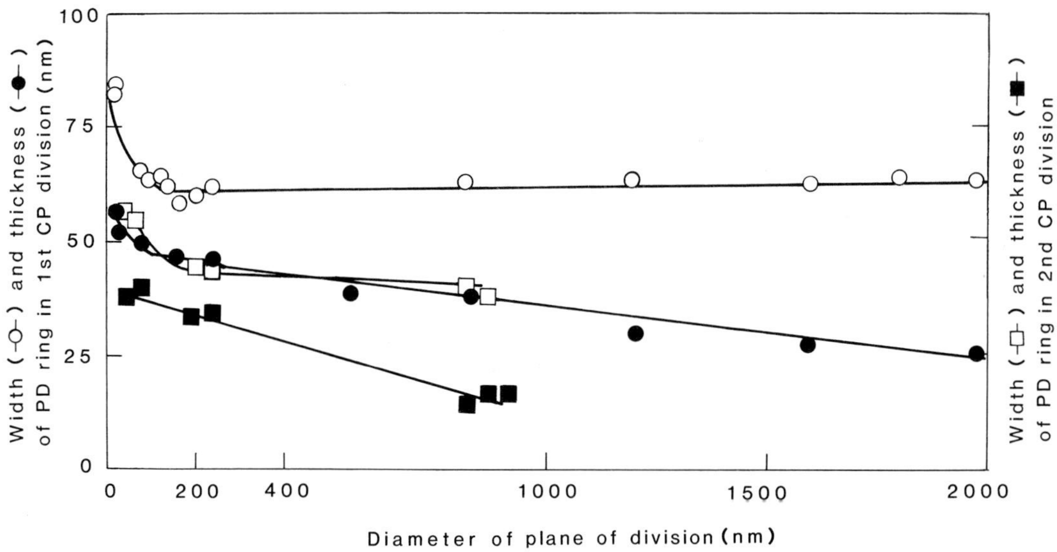

Fig. 11. Changes in the width and the thickness of the PD ring during the first and second chloroplast divisions. The width of the PD ring does not deviate far from a value of approximately 60 nm through the first chloroplastkinesis while the thickness of the ring increases constantly. However, at the final step of the division, both width and thickness increase rapidly. Similar events occur during the second chloroplastkinesis except that the width and the thickness are about half those of the ring involved in the first chloroplastkinesis

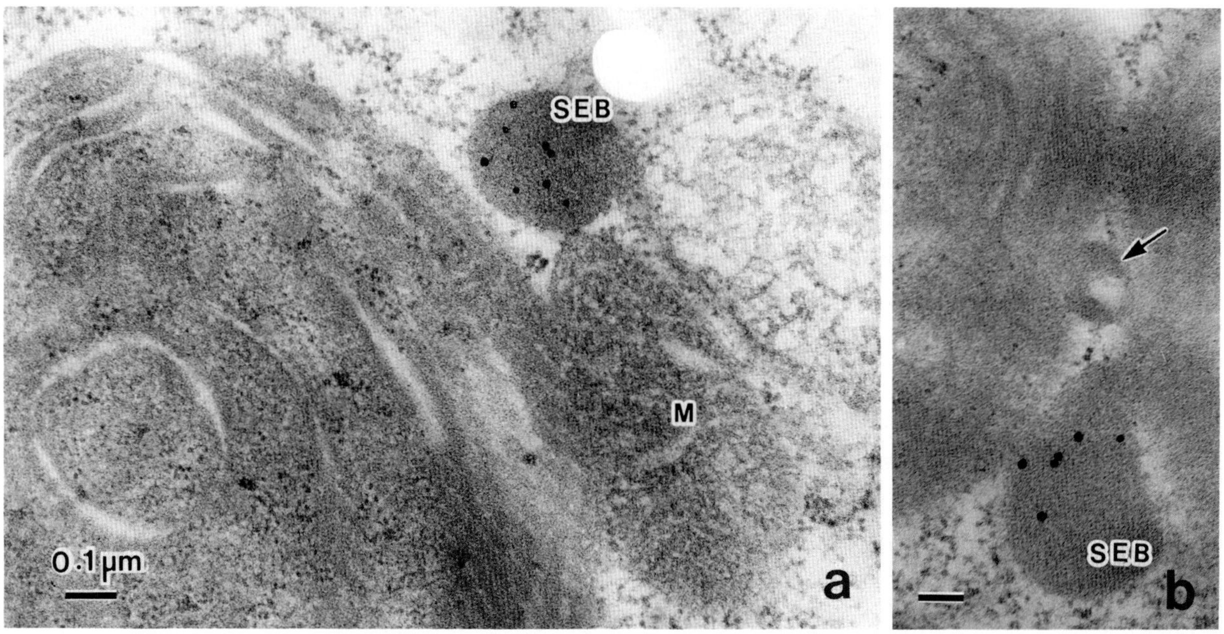

Fig. 12. Immuno-gold electron micrographs of SEB and the PD ring of a chloroplast sectioned longitudinally in the cell fixed at 24 h (*a*) and 38 h (*b*) after initiation of synchronous culture. Gold particles do not appear over the PD ring (arrow) but are seen over the SEBs. Bar represent 0.1 μm

of actin contained in the PD ring may be too small to be identified quantitatively by the present application of the immuno-gold technique. (3) The actins in the hypothetical actin filaments in the PD ring may not react qualitatively with the chicken gizzard actin-specific antibody used in the present experiments. As the number of gold particles that appeared on thin sections was very low in the present experiments, the possibilities numbered (2) and (3) cannot be excluded. Furthermore, it remains to be clarified whether the filaments of the PD ring contain a molecular species which reversibly binds heavy meromyosin, and thereby pro-

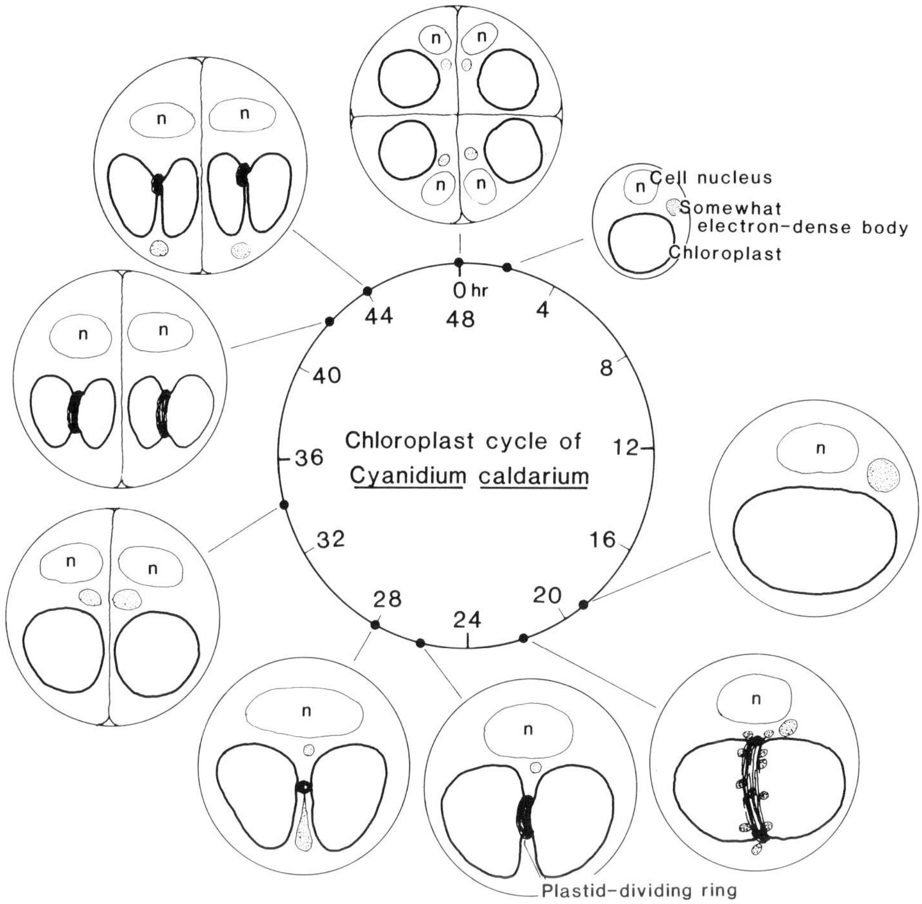

Fig. 13. Diagram of the life cycle of *Cyanidium caldarium* depicting a possible sequence of events by which the PD ring is formed, becomes aligned, causes contraction, and is eventually disassembled. The process of chloroplast division can be classified into the following four stages on the basis of morphological and temporal differences. Chloroplast growth stage: the small, spherical chloroplast increases in volume and becomes a football-like structure. PD ring formation stage: the SEB is fragmented into many SEGs, which are aligned along the equatorial region of the chloroplasts and fine filaments are formed from the SEGs in the equatorial region. The fine filaments of the PD ring align themselves according to the longest axis of their overall domain, *i.e.*, circumferentially. PD ring contraction stage: a bundle of fine filaments begins to contract and generates a deep furrow. PD ring conversionstage: after chloroplast division, the remnants of the PD ring are converted into SEBs. Similar events occur during the second cycle of chloroplast division

duces a complex that contains periodic, polymerized arrowhead patterns. Therefore, from these results, it may be more appropriate to assume that the PD ring does not contain actin but rather contains some actin-like proteins. Assuming that the PD ring consists, at least in parts, of predominantly actin-like proteins, we can consider the behaviour of the PD ring during the life cycle of *C. caldarium* as follows.

4.2. Classification of Stages of Chloroplast Division Cycle

The endospore division cycle in the red alga *C. caldarium* is basically similar to that of *Chlorella ellipsoidea* (TAMIYA 1963).

From the results of the present experiments, it has been established that chloroplasts are pinched off into a pair of daughter chloroplasts by contraction of the PD ring. The PD ring appears in the cytoplasm attached to the outer envelope of the constricted isthmus of the chloroplast and consists of actin-like filaments (Figs. 5–10). The process of chloroplast divisions can be classified morphologically into the following four stages: (1) the chloroplast growth stage, (2) the PD ring formation stage, (3) the PD ring contraction stage, and (4) the PD ring conversion stage.

4.3. Chloroplast Growth

Young cells increase markedly in volume during the first 30 h after the initiation of synchronous culture. During this period, the cell nucleus and chloroplast also synthesize DNA and their volumes double and quadrupole, respectively (MITA and KUROIWA 1986). When the spherical chloroplast changes to a football-like structure after chloroplast growth, the precursors of the PD ring become faintly visible by electron microscopy around the equatorial region of the chloroplast and a SEB is often observed near the chloroplast (Fig. 4). Ultrathin sections of Lowicryl-embedded biological material were floated on droplets of a solution of antibody, and then incubated with a second, gold-coupled antibody. Since the colloidal gold particles seen in Fig. 12 *a* illustrate the specific binding of chicken gizzard actin-specific antibodies to the SEB, but rhodamin-conjugated phalloidin does not stain the SEB (Fig. 3), it appears that the SEB contains some kind of proteinaceous particle that contains actin. Therefore, SEB shall be called "actin body".

4.4. Formation of the PD Ring

At about 24 h after the initiation of synchronous culture, the first endospore division begins. First the chloroplast divides and then the cell nucleus and mitochondrion divide sequentially and rather rapidly. Just before furrowing actually begins, an advance accumulation of fine filaments of about 4.5 nm in diameter occurs around the envelope at the cytoplasmic side of the constricted isthmus of the chloroplast and forms the PD ring. The fine filaments are very similar in diameter to actin filaments (ALBERT *et al.* 1983). Preceding the formation of the PD ring, many SEGs, which are very similar in electron density to the SEBs, are distributed around the filaments of the PD ring. Often fine filaments appear to be protruding from the SEGs. By the time the circular PD ring is completely formed, the small SEGs have disappeared (Fig. 5). Although the origin of SEGs is unknown, one possible explanation for the formation of the fine filaments of the PD ring is that small SEGs have been formed by fragmentation of the SEB, have moved to the equatorial region of the chloroplast and have there been transformed into the fine filaments. It is probable that, during formation of fine filaments, the globular, actin-like proteins contained in the SEGs are polymerized into filamentous actin-like proteins, which are the main components of the PD ring. Since the newly formed PD ring is about 60 nm wide and 20–30 nm thick, it is more like a band or belt than a complete ring. However, in the contractile ring involved in cytokinesis in animal cells, organelles, or cytoplasmic particles are neither seen near the analogous filamentous structures nor within the ring, suggesting that there is no structural precursor to the contractile ring, whose appearance is synchronous with onset of cleavage (SCHROEDER 1975). In addition, the original source of the microfilamentous subunits that form the contractile ring in such animal cells is not known (SCHROEDER 1975).

4.5. Contraction of the PD Ring

As the constriction of the PD ring progresses, the PD ring, which is composed of fine filaments, becomes tightly compacted thickness increases in electron density, and generates a deep furrow around the chloroplast with width similar to that of the PD ring (Figs. 4–7). The results suggest that the fine filaments, which have been made only at the early stage of chloroplast division, do not change in quantity during contraction of the PD ring, while the actin-like filaments, which have been formed at the early stage of the formation of the PD ring, contract continuously throughout chloroplastkinesis. The molecular mechanisms for the establishment and construction of the PD ring are not yet understood in detail. The analogy between cleavage of cells and division of plastids (plastidkinesis) gains additional credence from the discovery that actin-like, fine filaments participate in the contraction of chloroplast furrow. The circumferential orientation of fine filaments in the PD ring fits the model of a cleavage-like contractile device in the cleavage of cells (SCHROEDER 1975). The PD rings observed during the middle and late periods of chloroplastkinesis are remarkably higher in electron density than those of the contractile ring in animal cells and other actin bundles in *Physarum polycephalum* (NAGAI *et al.* 1978). This difference suggests that the PD ring may exert the strong mechanical tension necessary for deformation of the chloroplast envelope.

4.6. Conversion of the PD Ring to the SEB and Second Endospore Division

When chloroplastkinesis has been completed, a connection between the condensed PD ring and the balloon-like SEB is observed between daughter chloroplasts. Probably the materials that comprise the PD ring are converted into the SEB at the site of this association. In other words, filamentous actin-like proteins are depolymerized into globular, actin-like proteins, which may be used for the formation of the PD rings during the second endospore division. The events during the division of chloroplasts in the second endospore division are very similar to those in the first endospore division except that the chloroplasts are smaller in volume than the one involved in the first endospore division and, thus, the PD rings also are smaller than that of first endospore division.

4.7. General Occurrence of the PD Ring in Plant Kingdom

SUZUKI and UEDA (1975) observed electron-dense deposits in the matrix of proplastids and amyloplasts at the equatorial region of dumbbell-shaped plastids in *Pisum sativum* at the latest stages of plastid division and they have suggested that the electron-dense deposits constitute a septum made in the matrix of the plastid. Similar deposits, forming a ring in the interspace between the outer and inner membranes of the envelope at the furrow of dividing plastids, have been observed in higher plants by CHALEY and POSSINGHAM (1981). These authors suggested that a septum or a ring appears only during the latest stage of plastid division and is localized in dividing plastids. CHALY *et al.* (1980) also concluded, from studies with scanning electron microscope and freeze-etching, that the dumbbell-configuration of dividing chloroplasts in spinach leaves is generated from within the plastid and not from the cytoplasm. By contrast, MITA *et al.* (1986) and KUROIWA and MITA (1987) have presented in preliminary reports that the electron-dense deposits, which form a ring, are not localized in plastids but are on the cytoplasmic side of the outer envelope of dividing chloroplasts in *C. caldarium*. We have called the ring structure, which is observed in the cytoplasm, the PD ring. A similar cytoplasmic PD ring has been found in moss (TEWINKEL and VOLKMAN 1987a, 1987b) although the fine filaments in the PD ring in moss are different in terms of number and diameter from those of the PD ring in *C. caldarium*. HASHIMOTO (1986) has also carefully studied the ring structure in the higher plant *Avena sativum* and found that the ring is not localized in the interspace between the outer and inner envelopes but is in cytoplasm, like the PD ring in *C. caldarium*. In addition, he noted that there is another ring in matrix of the plastid, such that the ring in higher plants is actually a doublet (PD ring doublet). HASHIMOTO's observation conflicts markedly with those of CHALEY and coworkers. We too have observed actually HASHIMOTO's PD ring doublet in proplastids of cultured cells of *Nicotiana tabacum* (unpublished data). The structure of the inner ring in higher plants differs from the inner belt of the PD ring of *C. caldarium*. The inner belt in *C. caldarium* looks like a structure which is formed by pressure of the matrix beneath the inner envelope as a result of the contraction of the PD ring because the thickness of the inner belt does not change during the division of chloroplast.

Two major questions remain. Why is the PD ring in higher plants and moss observed only after the middle stage of the chloroplast division, and what is the mechanism for generating the contractile force required for depression of the envelope of the chloroplast, at the site of constriction of the isthmus of the dividing chloroplast? One possible answer is that, in higher plants, the substances which make up the PD ring are present at such low levels that we cannot detect them by a conventional fixation technique until when the contraction of the PD ring has begun and the more compacted structure becomes apparent. The observation of the PD ring in *C. caldarium* supports this possibility because the electron density of the PD ring is too low to allow visualization of the PD ring during the early stage of chloroplast division. When the contraction of the PD ring has advanced, it is easy to observe the ring because its thickness increases.

Acknowledgements

This work was supported in part by grants (No. 59440002, 6344-0003) from the Ministry of Education, Science and Culture of Japan (KUROIWA).

References

ALLEN MB (1959) Studies with *Cyanidium caldarium*, an anomalously pigmented chlorophyte. Arch Mikrobiol 32: 270–277

ALBERT B, BRAY D, LEWIS J, RAFF M, ROBERTS K, WATSON JD (1983) In: Molecular biology of the cell. Garland New York London

CHALY N, POSSINGHAM JV (1981) Structure of constricted proplastids in meristematic plant tissues. Biol Cell 41: 203–210

— — THOMSON WW (1980) Chloroplast division in spinach leaves examined by scanning electron microscopy and freeze-etching. J Cell Sci 46: 87–96

HASHIMOTO H (1986) Double ring structure around the constricting neck of dividing plastids of *Avena sativa*. Protoplasma 135: 166–172

HEUZER JE, SALPETER S (1979) Organization of acetylcholine receptors in quick-frozen, deep-etched, and rotary-replicated Torpedo postsynaptic membrane. J Cell Biol 82: 150–173

KUROIWA T (1974) Studies on mitochondrial structure and function in *Physarum polycephalum*.J Cell Biol 63: 299–306

— KAWANO S, HIZUME M (1976) A method of isolation of mitochondrial nucleoid of *Physarum polycephalum* and evidence of a basic protein. Exp Cell Res 97: 435–445

— KUROIWA H (1980) Inhibition of *Physarum* mitochondrial division by cytochalasin B. Experientia 36: 193–194

— SUZUKI T (1980) An improved method for the demonstration of the *in situ* chloroplast nuclei in higher plants. Cell Struct Funct 5: 195–197

— (1982) Mitochondrial nuclei. Int Rev Cytol 75: 1–59

— MITA T (1987) Mechanisms of chloroplast division. 13th Int Botanical Congr, Berlin, p 199 (Abs)

LEECH RM, THOMSON WW, PLATT-ALOIKA KA (1981) Observations on the mechanism of chloroplast division in higher plants. New Phytol 87: 1–9

MITA T, KANBE T, TANAKA K, KUROIWA T (1986) A ring structure around the dividing plane of the *Cyanidium caldarium* chloroplast. Protoplasma 130: 211–213

NAGAI R, YOSHIMOTO Y, KAMIYA N (1978) Cyclic production of tension force in the plasmodial strand of *Physarum polycephalum* and its relation to microfilament morphology. J Cell Sci 33: 205–225

NAGELE RG, KOSCIUK MC, WANG SM, SPERO DA, LEE H (1985) A method for preparing quick-frozen, freeze-substituted cells for transmission electron microscopy and immunocytochemistry. J Microsc 139: 291–301

NAGASHIMA H, FUKUDA I (1981) Morphological properties of *Cyanidium caldarium* and related algae in Japan. Jpn J Phycol 29: 237–242

RABINOWITZ M, SWIFT H (1970) Mitochondrial nucleic acids and their relation to the biogenesis of mitochondria. Physiol Rev 50: 376–427

RENGER HC, WOLSTENHOLME DR (1972) The form and structure of kinetoplast DNA of *Crithidia*. J Cell Biol 54: 346–364

ROTH J (1986) Post-embedding cytochemistry with gold-labelled reagents: a review. J Microsc 143: 125–137

SCHROEDAR TE (1975) Dynamics of the contractile ring. In: INOUE S, STEPHENS RE (eds) Molecules and cell movement. Raven Press, New York

SUZUKI K, UEDA R (1975) Electron microscope observations on plastid division in root meristematic cells of *Pisum sativum*. L Bot Mag Tokyo 88: 319–321

TAMIYA H (1963) Cell differentiation in *Chlorella*. Symp Soc Exp Biol 17: 188–214

TEWINKEL M, VOLKMANN D (1987 a) Arrangement of cytoskeletal elements during the dividing process of the plastids in the protonema of the moss *Funaria hygrometrica* Sibth. 13th Int Botanical Congr, Berlin, p 204 (Abs)

— — (1987 b) Observations on dividing plastids in the protonema of the moss *Funaria hygrometrica* Sibth. Planta 172: 309–320

WHATLEY JM (1980) Plastid growth and divison in *Phaseolus vulgaris*. New Phytol 86: 1–16

WILLIAMSON DH, FENNELL DJ (1975) The use of fluorescent DNA-binding agent for detecting and separating yeast mitochondrial DNA. In: PRESCOTT DM (ed) Methods in cell biology, vol 12. Academic Press, New York, pp 335–351

Phototaxis of Cell and Cell Organelle

Protoplasma (1988) [Suppl 1]: 155–161

Multiple Photoreceptors in Phototaxis of *Dictyostelium discoideum* Amoebae

D.-P. Häder[1, *], M. Watanabe[2], and M. Furuya[2, 3]

[1] Institute for Botany, University of Erlangen, [2] National Institute for Basic Biology, Okazaki National Research Institutes, and [3] Biology Department, Faculty of Science, University of Tokyo

Received August 6, 1987
Accepted November 21, 1987

Dedicated to Professor Dr. Noburo Kamiya on the occasion of his 75th birthday

Summary

Action spectra for amoebal positive and negative phototaxis of the cellular slime mold, *Dictyostelium discoideum* have been determined between 350 nm and 650 nm for the wild type strain NC 4, the axenic mutant strain AX 2, the parental mutant strain X 22 as well as the cobalt resistant mutant strains HU 120, HU 409, and HU 411. All of the strains are capable of both types of phototaxis and superficially the action spectra are similar. However, certain bands in prominent regions of the spectrum are either present or missing in some strains, which can be explained only by assuming that multiple photoreceptor pigments are responsible for phototactic orientation.

Keywords: Action spectra; *Dictyostelium discoideum*; Photomovement; Phototaxis; Mutants; Slime molds.

1. Introduction

In its multicellular life stage (pseudoplasmodium) the cellular slime mold *Dictyostelium discoideum* is capable of exclusively positive phototactic orientation (Bonner *et al.* 1950, Francis 1964, Poff and Loomis 1973). The zero threshold for white light was found at about 10^{-5} lx ($= 4.2 \times 10^{-8}$ W m^{-2}). The spectral sensitivity of this response extends almost throughout the visible wavelengths band (Poff and Butler 1974, Poff *et al.* 1974) and a detailed action spectrum has been calculated from fluence rate-response curves with main maxima near 420 and 440 nm and secondary peaks at 560 and 610 nm (Poff and Häder 1984). The photoreceptor pigment has been proposed to be a high spin heme protein (Poff *et al.* 1973, 1974).

The mechanism of phototactic orientation is based on a lens effect (Francis 1964, Häder and Burkart 1983 a): The internal refractive index of the cells is 1.369 and thus considerably higher than that of the external medium (air with n = 1.0). Due to this fact the cylindrical pseudoplasmodium focusses parallel lateral irradiation onto its distal surface. A turn of the slug toward the light source is thus based on a turn away from the brightest flank of the organism. The lens effect is defeated in the ultraviolet range below 300 nm where the cells show a high internal absorption. Consequently, the organism turns away from the light source in this wavelengths range (Häder 1985). It has been suggested that the cells on the brighter (distal) flank of the pseudoplasmodium produce a slug turning factor in a higher concentration than those on the darker (near) flank and that the pseudoplasmodium responds negative chemotactically to this signal (Fisher and Williams 1981, Fisher *et al.* 1981).

In its unicellular life stage (amoeba) the organism has also been found to show phototactic orientation which however differs in several respects from that of the pseudoplasmodia: At low fluence rates the amoebae move toward the light source (Häder and Poff 1979 a) while at high fluence rate the cells show negative phototaxis (Häder and Poff 1979 b). The action spectrum of these two photoresponses resemble each other but are different from the action spectrum for pseudoplasmodium phototaxis (Häder and Poff 1979 a, b, Hong *et al.* 1981). The main peak was found at about 405 nm and secondary maxima at about 580 and 640 nm.

The strategy of orientation in amoebae is also different

* Correspondence and Reprints: Institut für Botanik, Staudtstrasse 5, D-8520 Erlangen, Federal Republic of Germany.

from that in pseudoplasmodia: Experiments with narrow light beams located on different parts of the periphery indicated that the cells detect internal light gradients, which could be due to absorption and/or scattering, and orient the polarity of their movement accordingly (Häder *et al.* 1983 a).

Recently, mutant strains have been isolated and examined in their phototactic orientation in both the pseudoplasmodium and amoebal stage. The pseudoplasmodia were altered in their thermotactic behavior and also in their phototactic orientation, which resulted in a bimodal distribution since the organisms moved at a specific angle left and right from the light source (Fisher and Williams 1981). This behavior can be explained by a simple optical model (Poff *et al.* 1986) and can be described by a computer model (Häder and Burkart 1983 b, 1984). Amoebae of the same strains were also affected in their phototactic orientation: Depending on the strain and the fluence rate of lateral white light the amoebae moved in a multimodal fashion orienting in up to 12 clearly defined independent directions (Häder *et al.* 1983 b, Fisher *et al.* 1985).

Some of these mutant strains plus the wild type and an axenic strain were selected to study the wavelength dependence of both positive and negative phototactic orientation in order to determine whether the mutants are affected in their photoreceptor or in the sensory transduction chain, a question which can be answered by determination of action spectra and fluence rate-response curves.

2. Materials and Methods

2.1. Organisms and Culture

All experiments were carried out with the cellular slime mold *Dictyostelium discoideum*. The axenic mutant strain AX 2 was grown at 23 °C in darkness in rotary shake culture in 100 ml Erlenmeyer flasks containing 40 ml of a medium as described recently (Häder and Poff 1979 a). The non-axenic strains were cultivated on sterile 1.5% (w/v) water agar plates (100 mm in diameter) covered with 2 ml aliquots of an exponential culture of *Klebsiella aerogenes* growing in a peptone/yeast extract/glucose medium (Sussman 1966). Spores of about 20 sorocarps were transferred onto the plates when most of the liquid had evaporated. The spores germinated and the resulting amoebae multiplied feeding on the growing bacteria for 48 h in darkness. The parental strain X 22 and the mutant strains HU 120, HU 409, and HU 411 were kindly provided by Prof. Dr. K. Williams, Sydney (Australia).

2.2. Experimental Set-up

The axenically growing organisms were collected by centrifugation (2000 × **g**, 2 min) when in the exponential growth phase not exceeding 6×10^6 cells per ml, washed twice in a mineral salts solution (Sussman 1966) and centrifuged again to obtain a concentrated cell suspension. Amoebae of the non-axenic strains were washed off the plates and harvested as described above for the axenic amoebae.

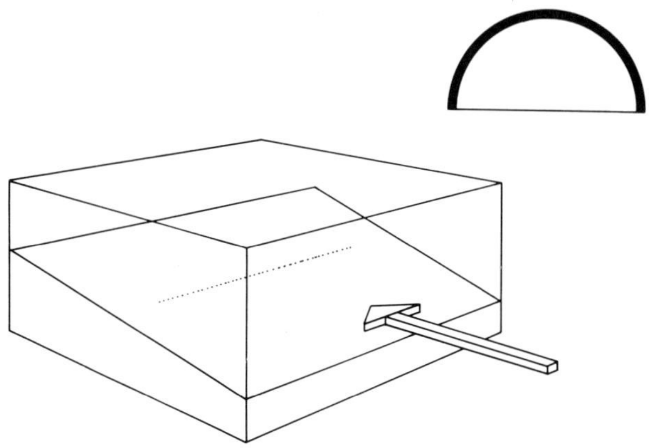

Fig. 1. Sample cuvette filled with a slanting 1% water agar onto which the cells were transferred using a hair suspended between the two ends of a horseshoe shaped metal bow (inset). Monochromatic radiation was impinging from the left (arrow)

The cells were exposed on a slightly slanting 1.5% (w/v) water agar block poured into rectangular plexiglass containers (28 × 23 × 16 mm inner dimensions) with separate lids (Fig. 1). The organisms were aligned on the agar by transferring them with a hair suspended between the two ends of a horseshoe-shaped metal bow (Fig. 1, insert).

2.3. Irradiation Procedures

The containers with the samples were placed on the experimental bench of the Okazaki Large Spectrograph (Watanabe *et al.* 1982) covering the range from 350 to 650 nm at 7.5 nm intervals and exposed for 8 h. Equal quantum flux densities were adjusted using absorptive or reflective neutral density filters. Fluence rate-response curves were obtained by stacking 6 containers on top of each other in order to expose all organisms to the same wavelength and by reducing the fluence rates at the individual samples by appropriate combinations of neutral density filters. The irradiance of the monochromatic irradiation was measured with a photon density meter HK-1 custom made by the Institute for Physical and Chemical Research, Wako-shi, Saitama, Japan (Hashimoto *et al.* 1982).

2.4. Quantitative Determination of the Results

The samples were photographed using an Olympus BO 61 photomicroscope on Kodak TX 5063 b/w film. The distance the organisms had moved toward or away from the light source was used to quantify the directedness of movement: The photographic negatives were projected by means of an optical enlarger onto a bitpad digitizer (Summagraphics Bit Pad One) and the position of each cell was entered into a microcomputer relative to the origin (inoculation line). Subsequent computer programs calculated the distance and direction travelled as well as the statistical parameters of standard deviation and standard error of the mean. For statistical treatment of directional data see Batschelet (1981) and Mardia (1972). For each data point an average of over 200 tracks was analyzed so that a total of close to 50,000 cells was calculated for this project.

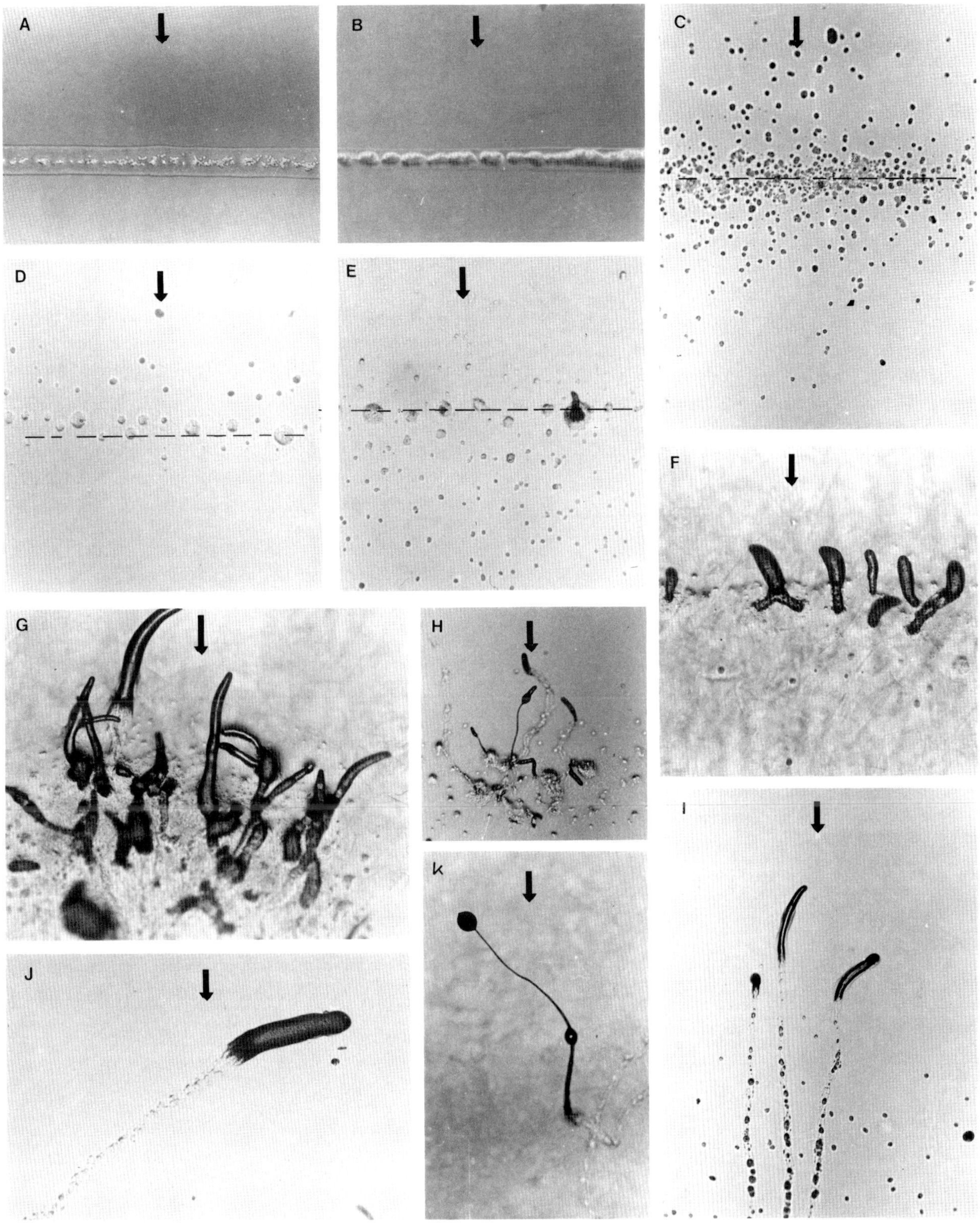

Fig. 2. Microphotographs of *Dictyostelium* amoebae immediately (*A*) after transfer onto the agar and after the first few minutes (*B*). Movement at intermediate (*C*), low (*D*) and high (*E*) fluence rates. Development involving aggregates (*E*) and pseudoplasmodia (*F*) which moved toward the light source (*G*, *H*). While the wild type strain NC 4 moved almost parallel to the light beam (*I*) the mutant strain (HU 411) deviated by about 55° from the light direction (*J*). Both had formed mature sorocarps about 24 h after inoculation (*K*). Arrows indicate direction of light source, dotted lines inoculation lines

3. Results

3.1. Behavior After Transfer of the Amoebae

After transfer onto the agar (Fig. 2 *A*) the cells expanded during the first few minutes (Fig. 2 *B*) and started moving out of the area aiming toward and away from the light source (arrow) at intermediate fluence rates (Fig. 2 *C*), mainly toward the light source at low fluence rates (Fig. 2 *D*) and mainly away from the light source at high fluence rates (Fig. 2 *E*). After about 8 h the cells formed patches which started to differentiate into aggregates (Fig. 2 *E*) which increased in size and transformed into typical pseudoplasmodia (Fig. 2 *F*). Starting from the inoculation line, the pseudoplasmodia moved toward the light source (Fig. 2 *G, H*). While the wild type strain NC 4 and the axenic strain AX 2 moved almost parallel to the light beam (Fig. 2 *I*) the mutant strains deviated by about 55° left or right from the light direction (Fig. 2 *J*). About 24 h after inoculation the organisms had formed mature sorocarps (Fig. 2 *K*).

3.2. Fluence Rate-Response Curves

Fluence rate-response curves were determined for each strain and each wavelength for both positive and negative phototaxis. Figure 3 shows a few typical examples indicating that the fluence rate-response curves started with an offset due to the random movement of cells in darkness. This offset was eliminated for the calculation of the action spectra. At higher fluence rates most of the curves showed a saturation with the exception of very ineffective wavelengths and insensitive strains.

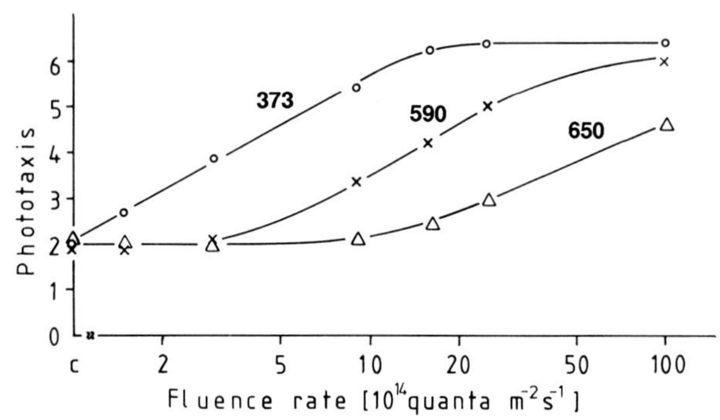

Fig. 3. Fluence rate-response curves for positive phototaxis of the wild type strain NC4 at 373 nm (○), 590 nm (×) and 650 nm (△)

3.3. Wavelength Dependence of Phototaxis in the Wild Type Strain NC 4

The wild type strain NC 4 has an action spectrum for positive phototaxis which extends almost throughout the visible spectrum up to 650 nm and also shows activity in the UV range (Fig. 4 *a*). The major peak is found near 400 nm and secondary maxima and shoulders at 373, 450, 515, 560, 590, and 635 nm. Similar peaks were found in the action spectrum of negative phototaxis (Fig. 4 *b*).

3.4. Wavelength Dependence of Phototaxis in the Axenic Strain AX 2

The action spectrum for positive phototaxis resembles that of the wild type strain in most respects. However, it lacks the peak or shoulder at 373 nm while the maximum near 400 nm is more pronounced (Fig. 5 *a*). The negative phototaxis action spectrum again is similar, but the main peak is slightly reduced in height (Fig. 5 *b*).

3.5. Wavelength Dependence of Phototaxis in the Parental Strain X 22 and the HU Mutant Strains

The strain X 22 is the parental strain for the mutants HU 120, HU 409, and HU 411. As compared to the action spectrum of positive phototaxis in the wild type strain NC 4 that of X 22 has the same main maximum

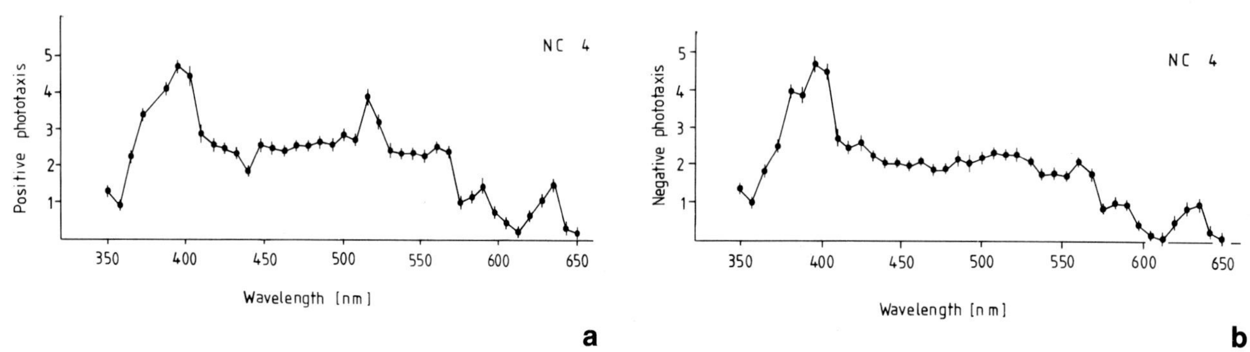

Fig. 4. Spectral response curves for positive (*a*) and negative (*b*) phototaxis of the wild type strain NC 4 amoebae

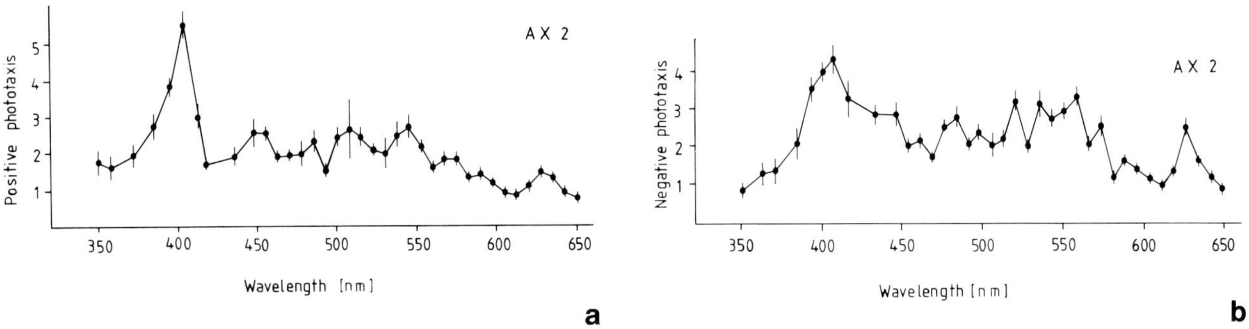

Fig. 5. Spectral response curves for positive (*a*) and negative (*b*) phototaxis of the axenic strain AX 2 amoebae

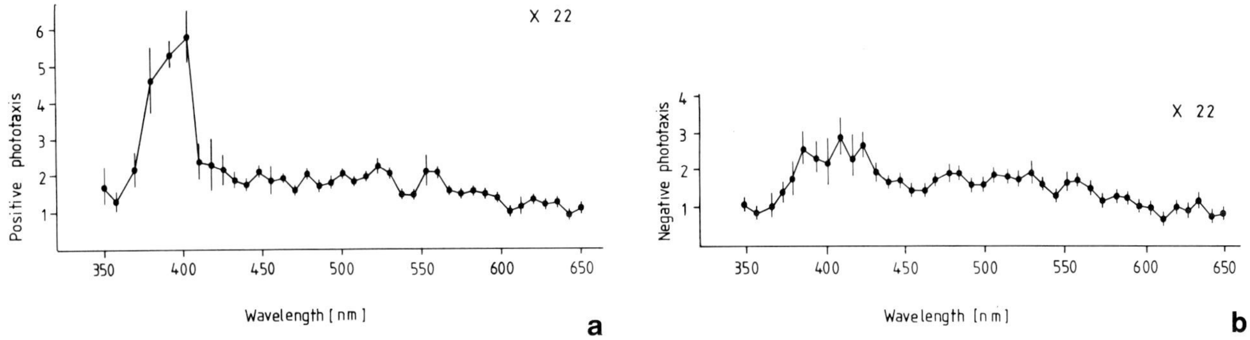

Fig. 6. Spectral response curves for positive (*a*) and negative (*b*) phototaxis of the parental strain X 22 amoebae

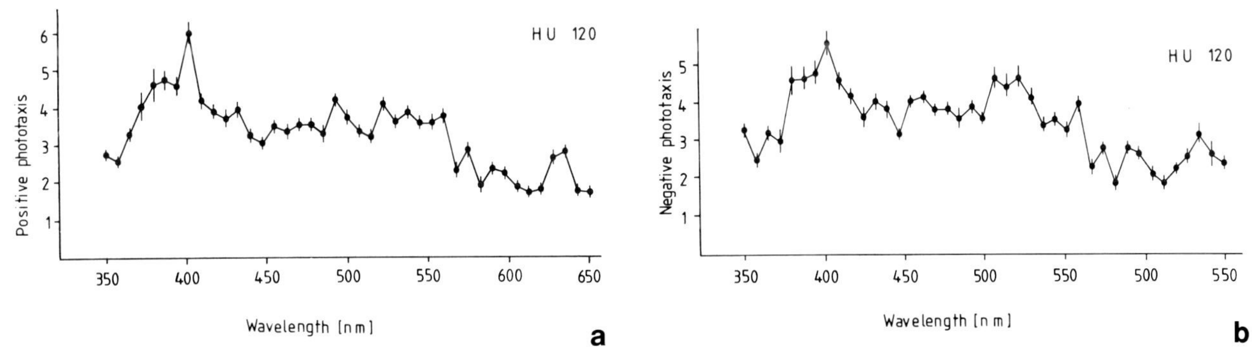

Fig. 7. Spectral response curves for positive (*a*) and negative (*b*) phototaxis of the mutant strain HU 120 amoebae

at about 400 nm and a pronounced shoulder in the near UV region but the remainder of the spectrum is drastically decreased and shows less fine structure than that of the wild type (Fig. 6 *a*). Specifically, the peaks at 450, 515, and 635 nm are far less pronounced than in the wild type. The same features are found in the action spectrum for negative phototaxis in X 22 (Fig. 6 *b*) and in addition the main maximum at about 400 nm almost disappeared. Some of the changes found may be attributed to the larger than normal statistical error in this range.

The mutants derived from the parental strain, HU 120 (Fig. 7 *a*), HU 409 (Fig. 8 *a*), and HU 411 (Fig. 9 *a*), all retain the near UV peak in positive phototaxis. While the main peak at 400 nm is very obvious in HU 120 and HU 409 it is less pronounced in HU 411. HU 409

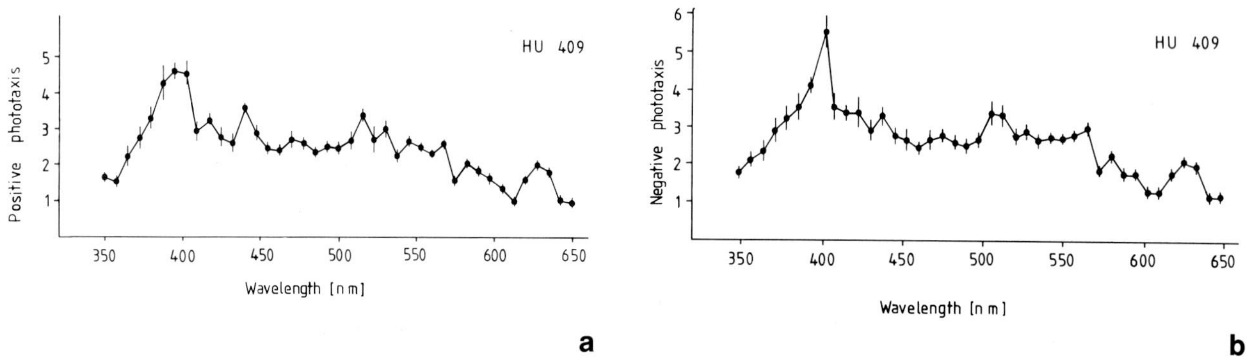

Fig. 8. Spectral response curves for positive (*a*) and negative (*b*) phototaxis of the mutant strain HU 409 amoebae

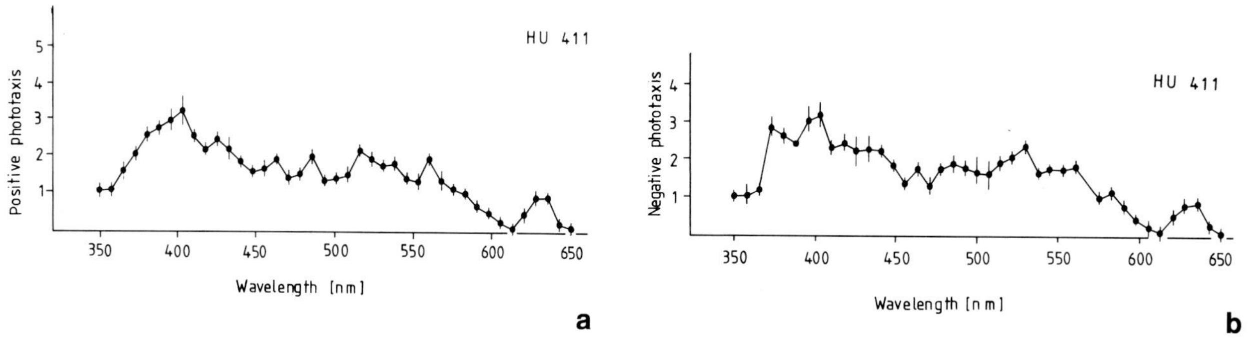

Fig. 9. Spectral response curves for positive (*a*) and negative (*b*) phototaxis of the mutant strain HU 411 amoebae

and HU 411 also lack the maximum at 590 nm, while all three mutant strains have the 635 nm peak. Similar features as in the positive phototaxis action spectra are present in the negative phototaxis action spectra of the three mutant strains HU 120 (Fig. 7 *b*), HU 409 (Fig. 8 *b*), and HU 411 (Fig. 9 *b*) though also some differences can be observed especially in the range between 350 and 400 nm.

4. Discussion

The basic assumption up to now was that positive and negative amoebal phototaxis in the cellular slime mold *Dictyostelium discoideum* is mediated by a single photoreceptor. Comparing the action spectra of the wild type, axenic and cobalt resistant mutant strains, however, shows clearly that not phototaxis as a whole is suppressed or decreased by a mutation but that individual bands are affected while the general shape of the curves is similar. Judging from the error bars in the action spectra and the high number of analyzed cells the strain-specific deviations are statistically significant and truely reflect the lack or absence of individual pigments involved in photoperception.

Considering only major differences, the results strongly indicate that in contrast to the current view positive and negative amoebal phototaxis are mediated by multiple photoreceptors which can be affected on an individual basis. These photoreceptors may be either independent from each other or operate as accessory pigments which funnel the absorbed energy to a common reaction center. If the latter explanation holds true the reaction center would be expected to absorb at longer wavelengths. The component with the longest wavelength absorbs at 635 nm; thus a P 635 could be the main reaction center for phototactic orientation in amoebae.

Comparison of the respective action spectra for positive and negative phototaxis indicates that basically the same sets of pigments seem to be responsible for both orientations. The switch between the two responses still needs to be identified. In a more detailed analysis of the individual strains the possible involvement and function of screening pigments needs to be determined. Pseudoplasmodia phototaxis should be reanalyzed on the basis of these results: It is possible that also in this case multiple photoreceptors are active. It also needs to be examined if some of the amoebal photoreceptors overlap with those responsible for pseudoplasmodia phototaxis.

Acknowledgements

This work was supported by the Deutsche Forschungsgemeinschaft (Ha 985/5-5) and the Bundesministerium für Forschung und Technologie (KBF 57). The experiments in Okazaki were carried out under the NIBB Cooperative Research Program for the Okazaki Large Spectrograph (project No. 85-519). We would like to thank M. Kubota, U. Neis, and S. Völk for skillful technical assistance.

References

Batschelet E (1981) Circular statistics in biology. Academic Press, London

Bonner JT, Clark WW Jr, Neely MCL Jr, Slifkin MK (1950) The orientation to light and the extremely sensitive orientation to temperature gradients in the slime mold *Dictyostelium discoideum*. J Cell Comp Physiol 32: 149–158

Fisher PR, Häder D-P, Williams KL (1985) Multidirectional phototaxis by *Dictyostelium discoideum* amoebae. FEMS Microbiol Lett 29: 43–47

— Smith E, Williams KL (1981) An extracellular chemical signal controlling phototactic behavior by *D. discoideum* slugs. Cell 23: 799–807

— Williams KL (1981) Bidirectional phototaxis by *Dictyostelium discoideum* slugs. FEMS Microbiol Lett 12: 87–89

Francis DW (1964) Some studies on phototaxis of *Dictyostelium*. Cell Comp Physiol 64: 131–138

Häder D-P (1985) Negative phototaxis of *Dictyostelium discoideum* pseudoplasmodia in UV radiation. Photochem Photobiol 41: 225–228

— Poff KL (1979 a) Light-induced accumulations of *Dictyostelium discoideum* amoebae. Photochem Photobiol 29: 1157–1162

— — (1979 b) Photodispersal from light traps by amoebas of *Dictyostelium discoideum*. Exp Mycol 3: 121–131

— Burkart U (1983) Optical properties of *Dictyostelium discoideum* pseudoplasmodia responsible for phototactic orientation. Exp Mycol 7: 1–8

— — (1983 b) Mathematical simulation of *Dictyostelium discoideum* pseudoplasmodia movements. J Math Biol 67: 41–57

— — (1984) Movement of *Dictyostelium discoideum* pseudoplasmodia *in vivo* and in a mathematical simulation. Plant Cell Physiol 25: 705–714

— Claviez M, Merkl R, Gerisch G (1983 a) Responses of *Dictyostelium discoideum* amoebae to local stimulation by light. Cell Biol Int Rep 7: 611–616

— Williams KL, Fisher PR (1983 b) Phototactic orientation by amoebae of *Dictyostelium discoideum* slug phototaxis mutants. J Gen Microbiol 129: 1617–1621

Hashimoto T, Yatsuhashi H, Kato H (1982) Abstracts of the Ann Meeting Jap Soc Plant Physiol, p 38

Hong CB, Häder MA, Häder D-P, Poff KL (1981) Phototaxis in *Dictyostelium discoideum* amoebae. Photochem Photobiol 33: 373–377

Mardia KV (1972) Statistics of directional data. Academic Press, London

Poff KL, Butler WL (1974) Spectral characteristics of the photoreceptor pigment of phototaxis in *Dictyostelium*. Photochem Photobiol 20: 241–244

— — Loomis WF Jr (1973) Light-induced absorbance changes associated with phototaxis in *Dictyostelium*. Proc Natl Acad Sci USA 70: 813–816

— Fontana DR, Häder D-P, Schneider MJ (1986) An optical model for phototactic orientation in *Dictyostelium discoideum* slugs. Plant Cell Physiol 27: 533–539

— Häder D-P (1984) An action spectrum for phototaxis by pseudoplasmodia of *Dictyostelium discoideum*. Photochem Photobiol 39: 433–436

— Loomis WF Jr (1973) Control of phototactic migration in *Dictyostelium*. Exp Cell Res 82: 236–240

— — Butler WL (1974) Isolation and purification of the photoreceptor pigment associated with phototaxis in *Dictyostelium*. J Biol Chem 249: 2164–2167

Sussman M (1966) Biochemical and genetic methods in the study of cellular slime mold development. In: Prescott DM (ed) Methods in cell physiology, vol 2. Academic Press, New York, pp 397–410

Watanabe M, Furuya M, Miyoshi Y, Inoue Y, Iwahashi I, Matsumoto K (1982) Design and performance of the Okazaki Large Spectrograph for photobiological research. Photochem Photobiol 36: 491–498

Protoplasma (1988) [Suppl 1]: 162–171

Reorganization of Microtubule Bundles in *Dichotomosiphon*: Its Implications in the Light-induced Translocation of Cytoplasm

T. Maekawa and R. Nagai*

Department of Biology, College of General Education, Osaka University

Received January 5, 1988
Accepted April 15, 1988

Dedicated to Professor Dr. Noburo Kamiya on the occasion of his 75th birthday

Summary

The cytoplasm in the freshwater coenocytic green alga *Dichotomosiphon tuberosus* always streams in a particular manner; multistriated and bi-directional, along the longitudinal axis of the algal body. Moreover, in response to light irradiation, the cytoplasm migrates toward the apical region and accumulates there, while it migrates toward the basal region in the dark. These movements were inhibited in the presence of colchicine (Maekawa *et al.* 1986). By immunofluorescence and electron microscopy, it was revealed that the reorganization of microtubule bundles in the endoplasm along the length of the alga is directly related to the translocation of the cytoplasm. In the dark, both small (composed of 1 to 15 microtubules) and medium bundles (16 to 30 microtubules) were evenly distributed along the whole length of the specimens (the dark type). Under light irradiation, a few small bundles were found in the apical region while large bundles (over 31 microtubules) as well as medium and small bundles, were found in the basal region (the light type). Since the number of endoplasmic microtubules increased in the basal region, microtubule polymerization may be involved in the large bundle formation. We found that one type was converted to the other type in response to light and dark treatments.

Local accumulation of the cytoplasm was induced when the alga was locally irradiated with blue light. Decomposition of bundles both at the irradiated and the apical regions was concomitantly observed.

The results suggest that reversible formation and decomposition of the endoplasmic microtubule bundles play a crucial role in the light-induced translocation of the cytoplasm in this alga.

Keywords: Cytoplasmic streaming; *Dichotomosiphon*; Light-induced translocation of cytoplasm; Microtubule bundles (light-induced reorganization).

Abbreviations: BSA bovine serum albumin; FITC fluorescein isothiocyanate; IgG immunoglobulin G; λ_{max} peak wavelength of incident light; MAPs microtubule associated proteins; PBS phosphate buffered saline; PIPES piperazine-N,N'-bis(2-ethanesulfonic acid).

* Correspondence and Reprints: Department of Biology, College of General Education, Osaka University, Machikaneyama 1-1, Toyonaka, Osaka 560, Japan.

1. Introduction

Cell organelles are capable of moving in a random or, in most instances, in an oriented fashion in response to internal and/or external stimuli. Their movements are as important for the function and the development of plant cells as they are animal cells. Light-induced translocation of the cytoplasm in the coenocytic freshwater green alga *Dichotomosiphon* shows such biological processes clearly.

Dichotomosiphon is a tip-growing, tubular alga which branches dichotomously after every 5–8 millimeters of growth. Its diameter is approximately 60 µm, but it contains a large central vacuole, which restricts cytoplasm to a thin layer. The cytoplasm of the alga always streams in a multistriated and bidirectional manner (Kamiya 1962) along the longitudinal axis of the algal body. In response to light irradiation, the cytoplasmic matrix together with the nuclei chloroplasts, and cytoplasmic granules migrates more toward the apical region and gradually accumulates there. As a consequence, the vacuole is pushed basally, and the apical region appears more deeper green than the other regions of the alga. In the dark, the cytoplasm migrates more toward the basal region than the tip and becomes concentrated there. The apical region becomes a lighter green as a result of the chloroplast evacuation. Local irradiation of the algal body causes reversible accumulation of the cytoplasm at the irradiated region.

Multistriated, bi-directional streaming continues through the whole sequence of cytoplasm movement in response to light conditions. We showed in our previous paper (MAEKAWA *et al.* 1986) that the microtubules generate the motive force for both the bi-directional cytoplasmic streaming and the light-induced translocation of the cytoplasm.

The light-induced translocation of the cytoplasm is thought to result from successive processes of intracellular systems, which are assumed to be comprised of light signal sensing and signal transduction mechanisms capable of regulating the movement and the motile apparatus. In this paper, we will present the *Dichotomosiphon* microtubule system, its changes in arrangement in response to light stimuli, and discuss factors regulating microtubule reorganization.

2. Material and Methods

2.1. Plants

Dichotomosiphon tuberosus samples were collected from a paddy field in Okinawa, Japan. These have been unialgally cultured in our laboratory. Detailed methods for culture are described in our previous paper (MAEKAWA *et al.* 1986). The culture was maintained in a light-dark cycle of 12 hour-light (2,000 lux from white fluorescent lamps, FL20S.PG, Matsushita Electrics. Kadoma) and 12 hour-dark at 21 ± 1 °C. Figure 1 shows an algal body seven days after transplantation from a mother culture. The apical region looks darker than the basal region because of cytoplasmic translocation under irradiated conditions.

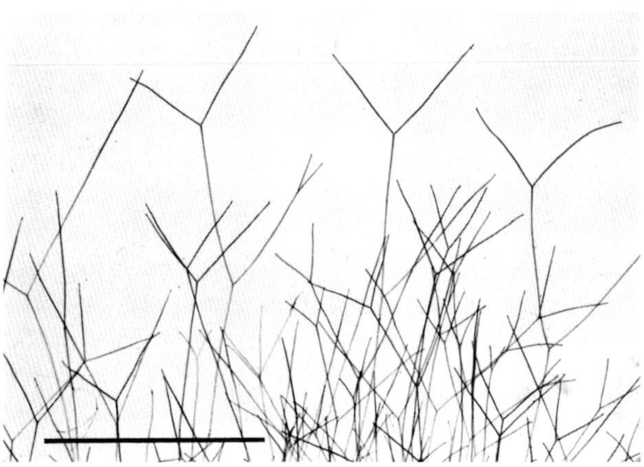

Fig. 1. The algal body of *Dichotomosiphon* under light irradiation. The apical region looks darker than the basal region because of the acropetal translocation of the cytoplasm. Bar: 1 cm

2.2. Specimen Preparation

A 1–2 cm long segment, which included two to three dichotomous branches was cut from the mother culture at the end of the light or dark period. The segment was mounted on a glass slide and a cover slip with a small amount of vaseline at each corner was placed over it. The space between the glass slide and the cover slip was filled with fresh culture medium. Each specimen was then placed in a Petri dish containing a piece of moist filter paper and kept under light or in dark at 23 ± 1 °C. Light was supplied by a bank of fluorescent lamps at 2,000 lux. The occurrence of cytoplasmic streaming and cytoplasmic translocation was confirmed with a light microscope (OPTIPHOT, Nikon Optics, Tokyo).

A microbeam for local illumination was obtained by making a small hole in a black plastic plate placed between filters (an interference and a cut-off filter described after) and the microscope field lens. The image of the hole was focused on the central plane of the cell with the condenser lens.

2.3. Monochromatic Light

Blue light (λ_{max} = 450 nm, 3×10^{18} photon/$m^2 \cdot s$), which was the most effective for accumulation of the cytoplasm, was obtained by combining an interference filter (KL-45, Toshiba Electronics, Kawasaki) and a cut-off filter (Y-44, Toshiba Electronics, Kawasaki). Green light (λ_{max} = 550 nm, 2.2×10^{17} photon/$m^2 \cdot s$), used as a safe light for the observations, was similarly obtained by combining an interference filter (KL-55) and a cut-off (0-54) filter. These filters were placed on the microscope field diaphragm. A source of light was a halogen lamp (Philips 7027, 12 V 50 W). Monochromatic light intensity was measured with a silicon photodiode (S 1337-1010BQ, Hamamatsu Photonics, Hamamatsu) positioned in the focal plane of the light microscope. Output from the photodiode was recorded for calculation of the quantum flux density.

2.4. Immunocytochemistry

Indirect immunofluorescence microscopy was done essentially as described by KITANISHI-YUMURA *et al.* (1985) and by MENZEL and SCHLIWA (1986 a) with slight modifications. After confirming that cytoplasmic translocation had taken place, segments were fixed for 2 hours at room temperature in 20 mM PIPES buffer with 3.5% paraformaldehyde and 250 mM NaCl (pH 7.0). Fixed specimens were washed for 5 min in a 250 mM NaCl and 20 mM PIPES solution (pH 7.0). Washed specimens were transferred into distilled water and cut into several pieces. Cylindrical fragments of the fixed cytoplasm spontaneously separated from the cell wall and slid out into the water. When necessary, a fine needle was used to ease the cytoplasm from the cell wall. Cytoplasm thus freed from the cell wall was transferred to a polyethyleneimine-coated coverslip, then immersed in absolute methanol at —18 °C for 5 min. After being washed with PBS (phosphate buffered saline; 137 mM NaCl, 2.7 mM KCl, 1.5 mM KH_2PO_4, 8.0 mM Na_2HPO_4, pH 7.3) for 5 min and blocked with 1% BSA/PBS for 15–30 min, specimens were treated with mouse ascites fluid (1 : 500, diluted with PBS) containing monoclonal anti-α tubulin IgG for 30 min at 37 °C. After washing in 0.03% Tween-20 in PBS for 15 min, fluorescein (FITC)-labeled sheep anti-mouse Ig (1 : 10, diluted with PBS) was applied, and the specimens were incubated again for 30 min at 37 °C. Specimens were washed again in Tween-20/PBS for 15 min and mounted in 1 mg/ml *p*-phenylenediamine-glycerol (pH 8.0) (JOHNSON and NOGUERA-ARANJO 1981). They were then monitored with an epifluorescent microscope (BHS-RFK, Olympus Optics, Tokyo) equipped with objective lenses of 20 X and 40 X (oil immersion). Micrographs were taken with Kodak Tri-X film.

2.5. Electron Microscopy

Algal segments were prefixed in 35 mM cacodylate buffer with 2% glutaraldehyde, 1% tannic acid (pH 7.0) for 2 hours at 20 °C under

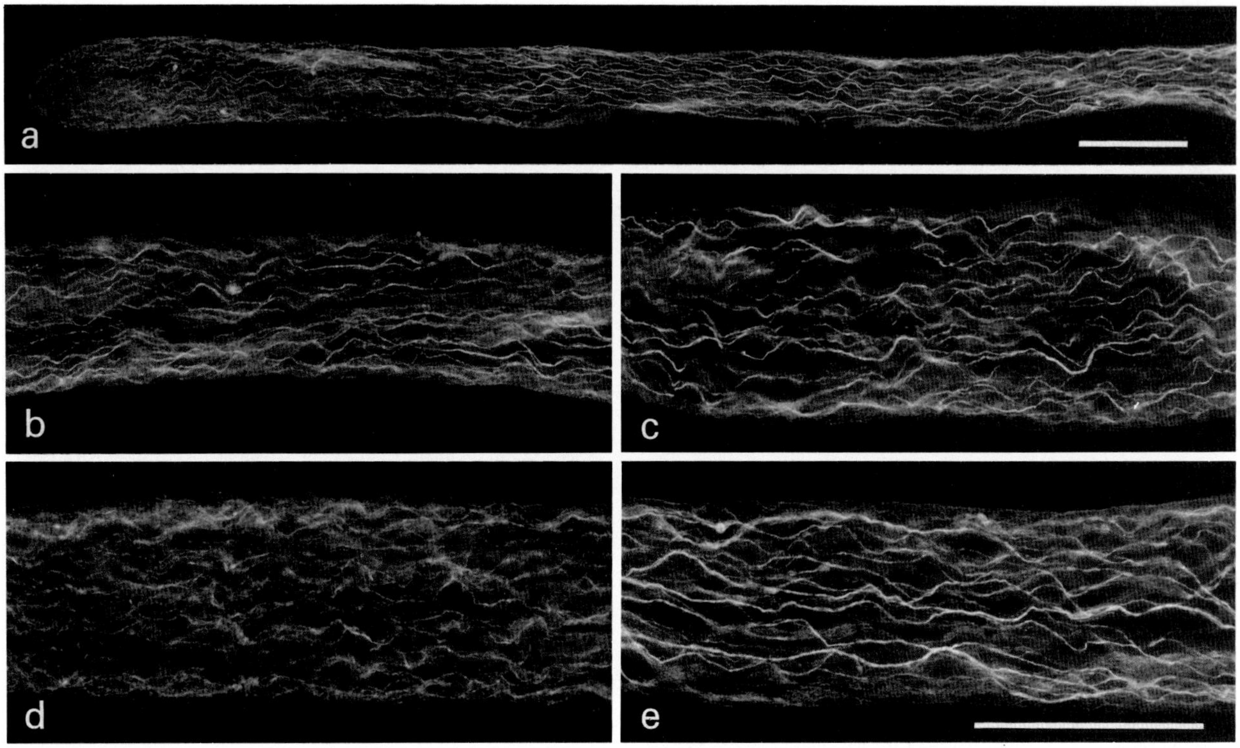

Fig. 2. Immunofluorescent staining of endoplasmic microtubule arrays. *a* The apical and subapical regions of a specimen which had been kept in the dark for 18 hours. Bar: 100 µm. *b* Higher magnification of the apical region of a specimen in the dark. *c* The basal region of the same specimen as in *b*. *d* The apical region of a specimen under light irradiation for 26 hours. *e* The basal region of the same specimen as in *d*. Bar: 100 µm

the same lighting conditions used in their preparation. 15–20 min after immersion, smaller segments were cut from the mother segments. After 2 washes in 35 mM cacodylate buffer, the segments were post-fixed for 1 hour with 1% OsO_4 in the same buffer. Specimens were dehydrated through a graded series of ethanol, transferred to SPURR's resin (1969), then cut into further smaller pieces 2–3 mm long to facilitate resin penetration. Thin sections were made on a LKB 2088 Ultrotome V. The distance from tip to section site was carefully recorded. Thin sections were stained with uranyl acetate and lead citrate and observed with either a JEM 100-C or with a JEM 100-S electron microscope at 80 kV.

2.6. Calculation of Relative Cytoplasm Area

Duplicate photographs were taken for each cross section of the cell and divided into two sets. In both sets, the cell cross sections were cut out along the cell membrane. The central vacuole was carefully removed from the photographs in one set. Comparative weighings of corresponding photographs from the two sets were taken to calculate the ratio of cytoplasmic area to total area. This ratio was regarded as the "Relative Cytoplasm Area".

2.7. Chemicals and Antibodies

Paraformaldehyde was purchased from TAAB Laboratories Equipment, Ltd. (England). Monoclonal anti-α tubulin IgG and FITC-conjugated Ig were purchased from Amersham International plc (Buckingamshire, England). The other chemicals were purchased from Wako Pure Chemical Industries (Osaka, Japan).

3. Results

Indirect immunofluorescence microscopy was done first to survey the microtubule arrays in the cytoplasm. The endoplasm and the ectoplasm each show a distinct pattern of microtubules. Endoplasmic microtubule patterns changed remarkably in response to light. Figure 2 *a* is immunofluorescence micrograph showing the microtubule arrays of segments kept in the dark for 18 hours. The fluorescent lines are wavy, variable in thickness, and appear to be continuous from the tip to the end of the segment. The thicker lines represent bundled microtubules. Figure 2 *b* and *c* shows the apical and basal microtubule arrays respectively of another specimen kept in the dark for 18 hours. The distribution of microtubule bundles stays fairly even along the length of the segment. Figure 2 *d* and *e* shows the apical and basal regions of a segment after 26 hours of constant light. Bundles in the apical region (Fig. 2 *d*) are fewer and thinner than those in Fig. 2 *b*, while the bundles in the basal region (Fig. 2 *e*) are more numerous and in many cases thicker than those in Fig. 2 *c*.

At the peripheral region of the cytoplasm, *i.e.*, in the ectoplasm, there are a large number of microtubules

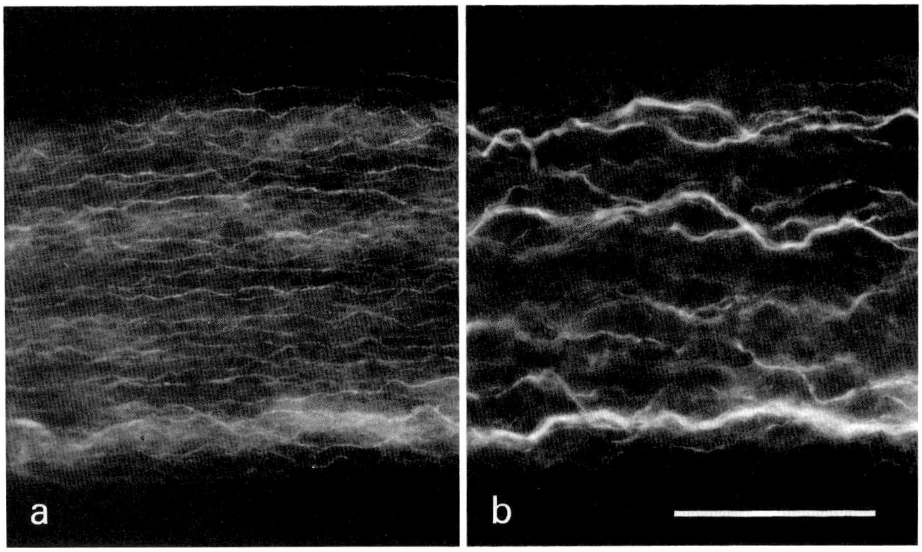

Fig. 3. Immunofluorescent staining of microtubules. *a* Ectoplasmic microtubules. *b* Endoplasmic microtubules. Bar: 50 µm

(Fig. 3 *a*), which also run in parallel with the longitudinal axis of the alga. These lines, which were revealed by electron microscopy to be mostly single microtubules, are thin, constant in width, and relatively straight compared with those in the endoplasm (Fig. 3 *b*). The array of the peripheral microtubules remained the same irrespective of light conditions (data not shown).

We further examined the microtubules by electron microscopy. Fortunately, as the diameter of the cell is about 60 µm, all of the microtubules in the cytoplasm can be surveyed in a single cross section. Figure 4 *c* is an example of a cross section 1170 µm from the tip of an alga that was irradiated for 29 hours. Both single and bundled microtubules are scattered around the endoplasmic layer. Figure 4 *a*, an enlargement of the portion marked in Fig. 4 *c* shows three different sized microtubule bundles. Most bundles were found near the tonoplast. Sometimes bridge-like structures linking neighboring microtubules were seen (Fig. 4 *b*). The ectoplasmic microtubules (arrowheads in Fig. 4 *d*) tended to be mostly single.

Bundles vary in their number of microtubules, ranging from 2 to 50, with occasional bundles made up of more than 80 microtubules. For convenience, we classified bundles in the endoplasm into three groups, by their number of microtubules: bundles composed of 1 to 15 microtubules (the thin arrow in Fig. 4 *a*), those composed of 16 to 30 microtubules (medium arrow), and those composed of over 31 microtubules (thick arrow). These groups will hereafter be called small, medium and large bundles, respectively. Figure 5 shows the frequency of the various bundle sizes 1620 µm from the tip of the same specimen as in Fig. 4.

As described before, the fluorescence microscopy showed that the patterns of microtubule array in the endoplasm changed in response to light and dark treatments. Since we expected the change have a close relationship to the translocation of the cytoplasm, we made serial cross sections from the apical region to the basal region in several specimens kept either under light irradiation or in the dark. The microtubules and microtubule bundles in each section of endoplasm were counted and bundle distribution was plotted as a function of the distance from the tip of the specimens.

Figure 6 *a* shows the distribution of endoplasmic microtubule bundles in the specimen which had been kept in light for 29 hours. Before fixation, the cytoplasm accumulated in the apical region of the specimen, and cytoplasmic movement toward the apical region continued vigorously. The arrows in the figure indicate the positions which were thin sectioned. There is a characteristic pattern in the bundle distribution. While both the bundle size and the total number of microtubules increase gradually from the tip toward the basal region, small bundles are distributed evenly throughout the whole length. Large bundles are especially conspicuous in the basal region. Accumulation of the cytoplasm in the apical region is easily seen.

In specimens kept in the dark for a longer duration (16 hours), a smaller volume of the cytoplasm streamed bidirectionally compared to those exposed to light. The distribution pattern of the bundles in such specimens

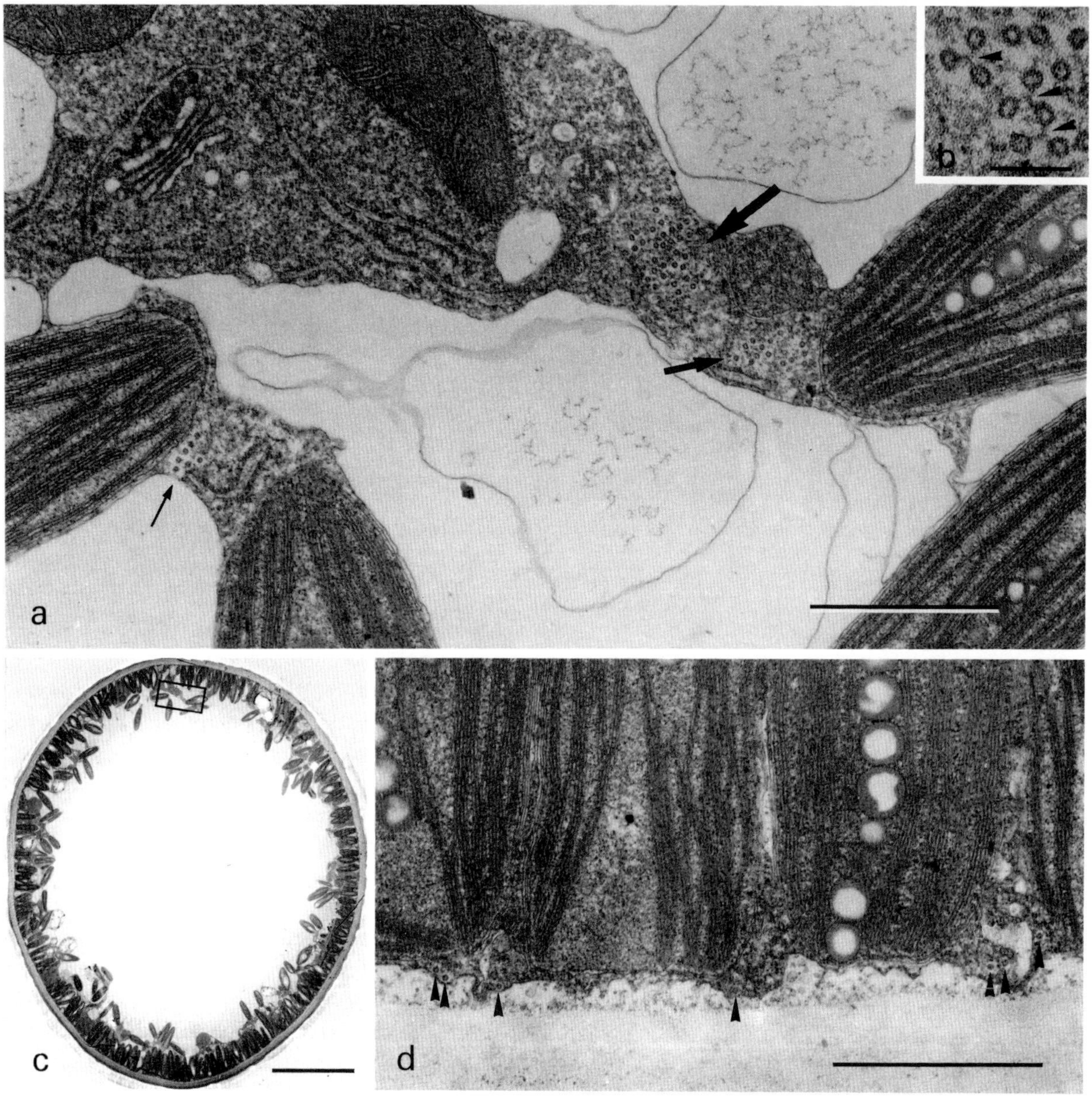

Fig. 4. Electron micrographs of a cross section of the algal body. A thin section 1170 µm behind the tip of a specimen under light irradiation for 29 hours was examined. *a* Three microtubule bundles in the portion marked in *c* at higher magnification. A thick, medium-sized and thin arrow represents a large, medium and small bundle, respectively. Bar: 1 µm. *b* Bridge-like structure (arrowheads) linking neighboring microtubules in a bundle. Bar: 0.1 µm. *c* A cross section of the algal body at lower magnification. Bar: 20 µm. *d* The peripheral microtubules (arrowheads) in the vicinity of the cell membrane. Bar: 1 µm

differed remarkably from that seen in specimens irradiated with light. An example is shown in Fig. 6 *b*. Small and medium bundles are evenly distributed from the tip toward the basal region, while large bundles do not appear anywhere we observed. We call the distribution pattern of the bundles in Fig. 6 *a* the "light type", and pattern shown in Fig. 6 *b* the "dark type".

It is reasonable to assume that the formation of large bundles at the basal region is necessary to generate the motive force for cytoplasmic translocation toward the apical region. To test this possibility, we examined specimens irradiated for 4 hours after having been kept in the dark for 24 hours. In these specimens, vigorous movement of the cytoplasm toward the tip started within 30 min of the light irradiation and continued until they were fixed for electron microscopy. Figure 7 *a* shows the distribution pattern of the endoplasmic microtubules for a specimen under these conditions. Me-

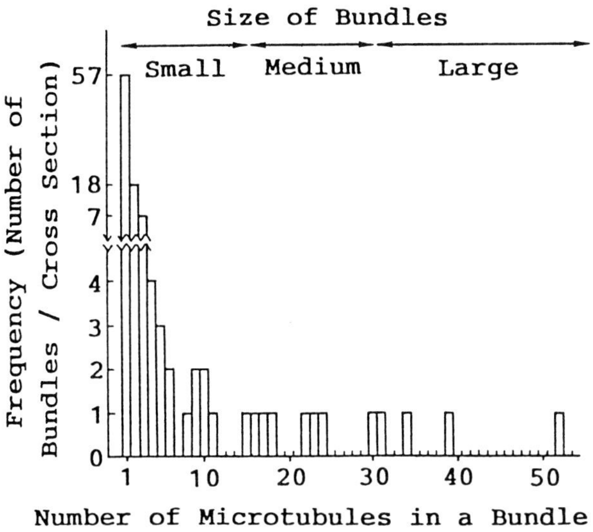

Fig. 5. The frequency of microtubule bundles. The numbers of microtubule bundles and of microtubules in each bundle were examined in a cross section at the position 1620 μm behind the tip of the same specimen as in Fig. 4. The abscissa is the number of microtubules in a bundle. The ordinate represents the number of each bundle composed of the corresponding number of microtubules shown in the abscissa

dium bundles and the total number of microtubules in the apical region decreased slightly in comparison with those in the basal region (*cf.*, Fig. 6 *b*), while cytoplasm accumulated slightly in the apical region. In addition, a large bundle of microtubules (marked by an asterisk) was formed in the basal region. Therefore, this distribution pattern of the microtubules can be understood as an intermediate closer to the light type than to the dark type.

By contrast, specimens which were transferred from light to dark storage and kept there for 4 hours had a distribution pattern closer to the dark type. As shown in Fig. 7 *b*, large bundles which should have existed at the basal region in the light disappeared (*cf.*, Fig. 6 *a*). Medium bundles remained throughout the whole length of the specimen except the tip region. The cytoplasm at the apical region is slightly evacuated. The microtubule distribution pattern in specimens which had been kept in the dark for 30 min was almost the same as the light type (data not shown).

When a small area of a specimen which had been kept in the dark was irradiated, the cytoplasm migrated from the basal region toward the irradiated region and gradually accumulated there. In order to determine the organization of the microtubule arrays, specimens were irradiated with a microbeam (diameter = 60 μm) of blue light (λ_{max} = 450 nm, 3 × 10^{18} photon/$m^2 \cdot s$) for 4 hours, and processed for immunofluorescence microscopy (Fig. 8 *a*). The apical side of the specimen is shown on the left of the figure, and the region marked by a white bracket was irradiated. In the irradiated region, a weak fluorescent background comes to the fore due to accumulation of cytoplasm, but microtubules are still visible. In the basal side of the specimen (right side of the figure) a large number of thick mi-

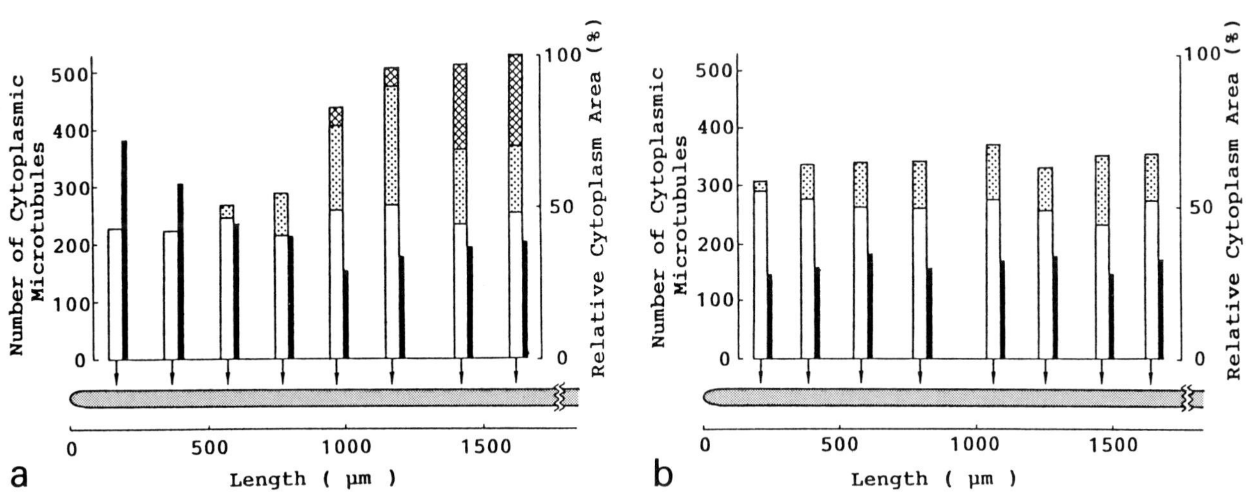

Fig. 6. Typical distribution patterns of the endoplasmic microtubules. Serial cross sections were obtained from the apical region to the basal region of a specimen (schematized in the abscissa). Each arrow corresponds to the position of a cross section examined. The ordinate in the left is the total number of endoplasmic microtubules. Open, dotted and meshed columns represent the number of microtubules making small, medium and large bundles described in Fig. 5, respectively. The ratio of the area occupied by the cytoplasm to the whole area occupied by the cytoplasm and vacuoles (closed column) is expressed as the percentage in the ordinate in the right. *a* Results from a specimen under light irradiation for 29 hours. *b* Results from a specimen kept in the dark for 16 hours

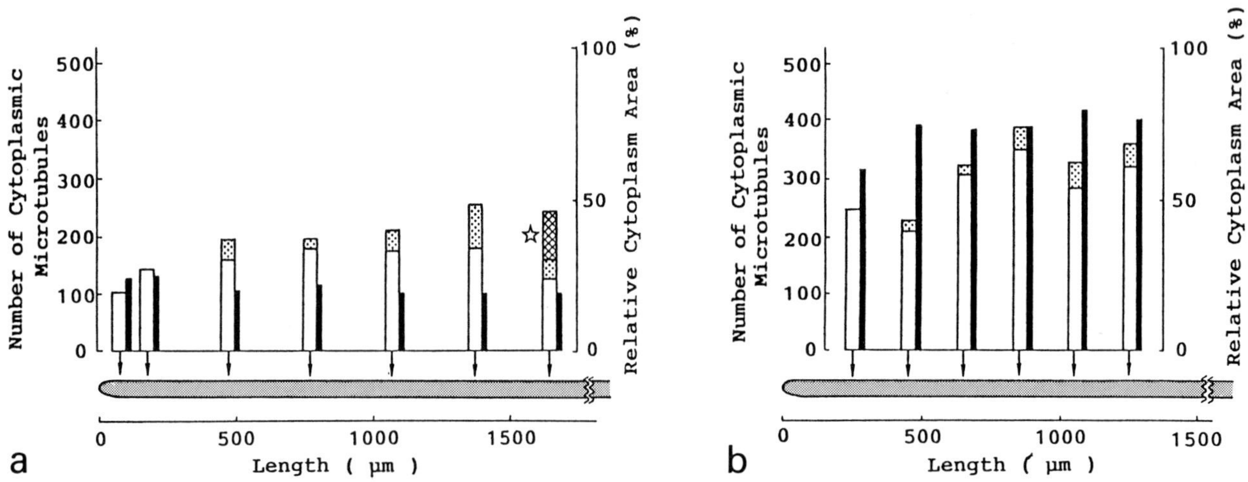

Fig. 7. Intermediate distribution patterns of the endoplasmic microtubules. Serial cross sections were examined as in the case of Fig. 6. *a* The results from a specimen irradiated for 4 hours after 24 hours dark treatment. An asterisk indicates the formation of a large bundle. *b* The results from a specimen kept in the dark for 4 hours after 22 hours light irradiation

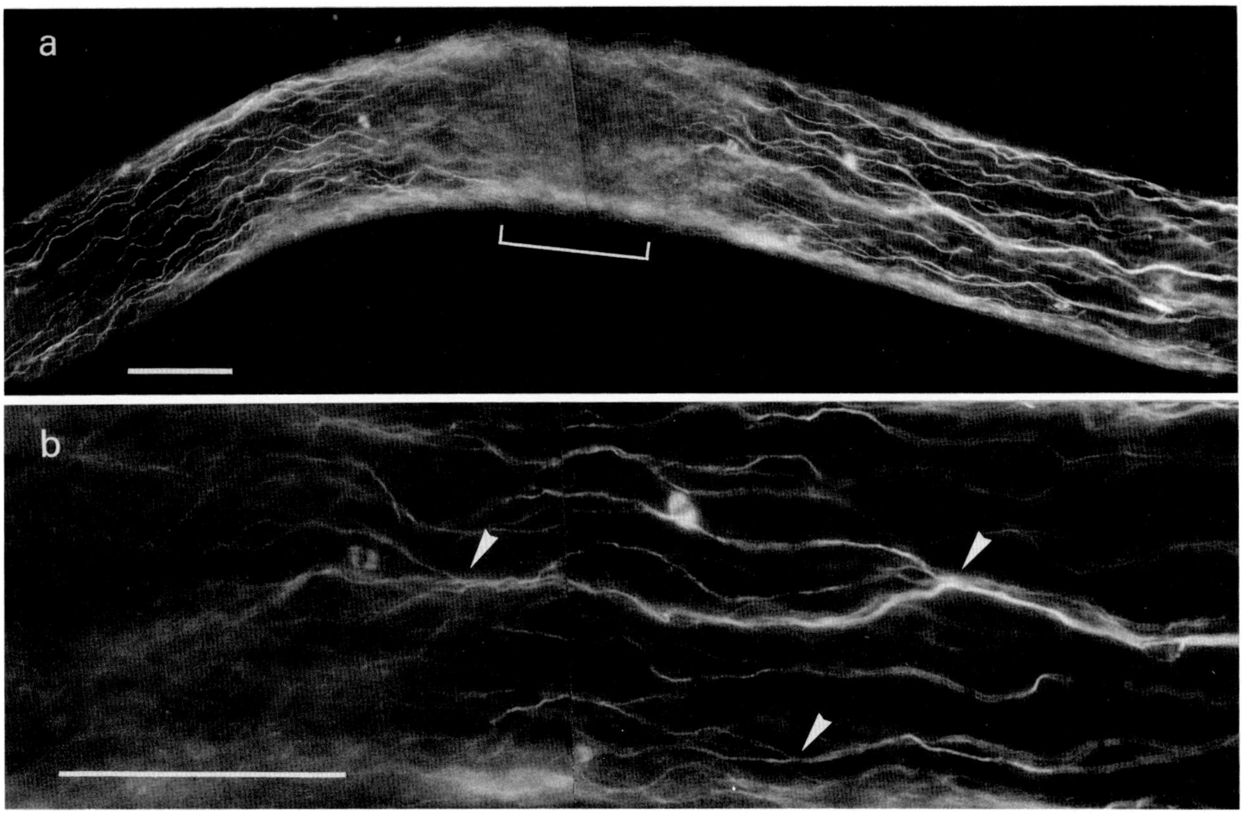

Fig. 8. Immunofluorescent staining of the endoplasmic microtubules following microbeam irradiation. The accumulation of cytoplasm was induced by the microbeam irradiation with blue light for 4 hours. The apical side is to the left. *a* Lower magnification of the irradiated region (white bracket). Bar: 50 µm. *b* Higher magnification of the basal side of the irradiated region. Arrowheads indicate the points of thick bundle ramifications. Bar: 50 µm

crotubule bundles can be observed. The bundles in the apical side appear thinner. It is worth noting that some of the thick microtubule bundles extending from the basal region appear to be disrupted in the vicinity of the irradiated region. Figure 8 *b* shows a magnified image of the basal side of the irradiated region. Here, thick bundles divide into gradually thinner bundles (arrowheads), and finally became almost undetectable at

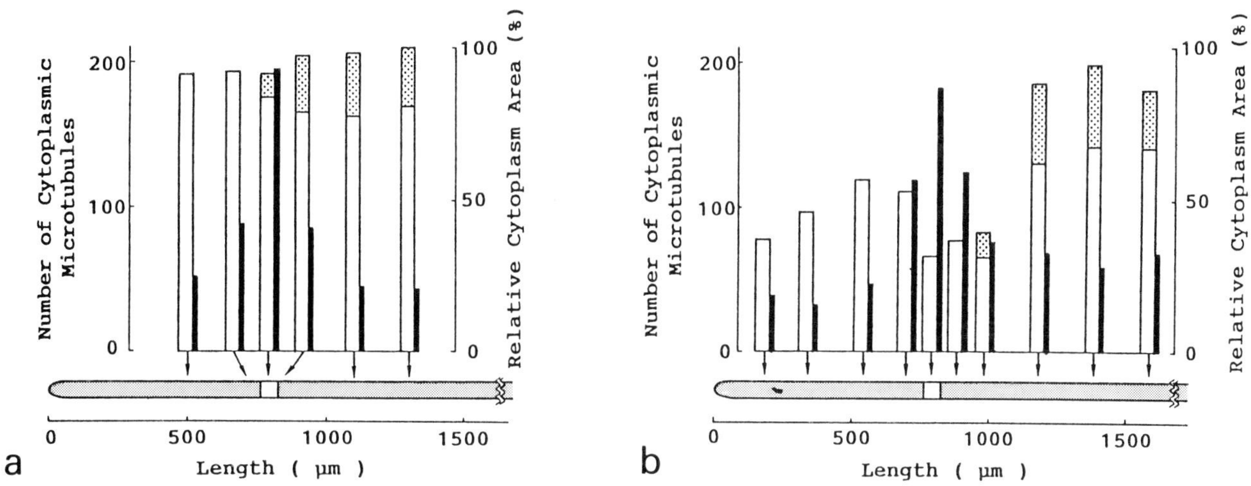

Fig. 9. Distribution patterns of endoplasmic microtubules in microbeam irradiation. Serial cross sections were examined as in the case of Fig. 6. The white square in the abscissa represents the irradiated region. *a* The results from a specimen irradiated for 1 hour. *b* The results from a specimen irradiated for 4 hours

the irradiated region. Figure 9 *a* shows the distribution pattern of the microtubules obtained by serial thin sectioning of a specimen irradiated with the microbeam for 1 hour. Of the two medium bundles on the basal side, one disappeared in the irradiated region, while the other bundle branched soon after, leaving only small bundles on the apical side. Disappearance of the medium bundles on the apical side and the irradiated region continued when the irradiation time was prolonged to 4 hours (Fig. 9 *b*). Accumulation of the cytoplasm also proceeded. It is reasonable to assume that decomposition of the medium bundles of microtubules and depolymerization of the microtubules occurred in the apical side.

4. Discussion

The translocation of cytoplasm in *Dichotomosiphon* is strictly light dependent. The cytoplasmic matrix, together with the cell organelles such as nuclei, chloroplasts, mitochondria and other small granules, migrates toward the apical region following light irradiation and toward the basal region in the dark. The light dependency of these movements is also supported by experiments using microbeam irradiation. Cytoplasm accumulation is observed only in the irradiated region, and the accumulated cytoplasm begins to disperse immediately after irradiation is terminated. Since colchicine completely inhibits this movement, endoplasmic microtubules are assumed to function in the translocation of the cytoplasm (MAEKAWA *et al.* 1986). In this study, by fluorescence and electron microscopy, we have further found that microtubule bundles are reorganized along the axis of the translocation of the cytoplasm in response to light and dark treatments. This is the first report of light-induced reorganization of microtubule bundles functioning as a motile apparatus in a cell.

Observation (Fig. 6 *a*) shows that irradiated specimens have many more endoplasmic microtubules in the basal regions than in the apical regions. Furthermore, in almost all cases, the total number of microtubules in the basal region is conspicuously larger in the light than in the dark. This strongly suggests that light irradiation stimulates microtubule formation in the basal region. These newly polymerized microtubules seem to form the large bundles. As for the apical region, changes in the total number of microtubules following light and dark conditions cannot be described because the average number of microtubules in each condition is uncertain at present. However, after specimens were irradiated with a microbeam, only small bundles and less number of microtubules remained in both the irradiated and apical regions (Fig. 9 *b*). These microbeam observations support the possibility that microtubule depolymerization in the apical region proceeds concurrently with apical decomposition of bundles when the whole algal body is irradiated.

While the driving force during light conditions for apical cytoplasmic translocation may be accelerated by the newly formed basal large bundles, cytoplasm accumulation at the apical region is related to the decomposition of bundles there. The occurrence of bundle decomposition and cytoplasmic accumulation in the irradiated area when the algal body were irradiated with a microbeam also support the above hypothesis (Fig. 9 *b*). When an algal specimen exposed to constant light is transferred into continuous dark, the light-type microtubule arrangement gradually converts into the dark-type (Fig. 7 *b*). Both the decomposition of the basal large bundles, and the formation of apical medium bundles must have proceeded in the dark. These newly formed apical bundles may drive the accumulated cytoplasm towards the basal region.

Figure 9 *b* also indicates the microtubule bundles decomposition in the unirradiated apical region, and a large amount of cytoplasm migrates acropetally in the same way as in the case of whole body irradiation. Some signal must be sent from the irradiated region, the site of light perception, to the unirradiated apical and basal regions to bring about translocation of the cytoplasm.

Cytoskeletal changes in response to light have been studied in other plant species. Microbeam irradiation with blue light induced cortical fiber reticulation at the irradiated region in the siphonaceous alga *Vaucheria* (BLATT and BRIGGS 1980, BLATT *et al.* 1980). The streaming cytoplasm consequently started to accumulate there. From the results using Cytochalasin B and phalloidin, BLATT *et al.* concluded that the cytoplasm accumulation was caused by decomposition of actin filament bundles. They further found that microbeam irradiation induced an outward electrical current from the irradiated region, and suggested that the local efflux of cations across the cell membrane mediated a process between the blue light perception and the cortical fiber reticulation (BLATT *et al.* 1981).

Tip growing organisms generate an endogenous electrical current such that positive charge flows into the tip and exits from the trunk (NUCCITELLI 1983). This is true of the case of *Dichotomosiphon* (data not shown). A tip-to-base calcium gradient was reported in tip growing plant cells using chlortetracycline, a membrane-bound Ca^{2+} indicator (REISS and HERTH 1979). JAFFE *et al.* (1975) revealed that Ca^{2+} influx at the tip, which was a component of the electrical current, was responsible for the maintenance of the calcium gradient in the case of *Lilium* pollen tube. In addition, HAUSSER *et al.* (1984) demonstrated that the calcium-binding protein, calmodulin showed a similar distribution pattern to the calcium gradient in tip growing plant cells using fluorescent calmodulin inhibitors. Ca^{2+} and/or calmodulin are possible candidates regulating the microtubule bundle reorganization in *Dichotomosiphon*.

Bridge-like structures are often seen linking individual microtubules within bundles (Fig. 4 *b*). These structures are tentatively supposed to be composed of microtubule-associated proteins (MAPs). MAPs have been purified and characterized in several kinds of animal cells (BLOOM *et al.* 1984, BULINSKI and BORISY 1980, HOLLEN ECK and CHAPMAN 1986, MUROFUSHI *et al.* 1986). They are suspected to play an important role in the formation and decomposition of microtubule bundles in this alga. The presence of these proteins in the thick microtubule bundles must be studied further.

In the developing thallus of the green alga *Caulerpa*, the chloroplasts migrate acropetally in light and basipetally in the dark (DAWES and BARILOTTI 1969). Cytoplasmic streaming has been found to require cytoplasmic microtubules (KURODA and MANABE 1983, MANABE and KURODA 1983). Although MANABE and KURODA observed a bridge-like structure in the microtubule bundles as well as in linkages between the mitochondria and microtubules, the relationship between the translocation of the cytoplasm and the motile apparatus organization in this alga is still obscure.

In *Dichotomosiphon*, addition of Cytochalasin B (100 µg/ml) reduces cytoplasmic streaming velocity for 60 minutes, but the original streaming velocity returns soon thereafter. This observation coincides with the report on *Bryopsis* with Cytochalasin D (10 µg/ml) by MENZEL and SCHLIWA (1986 b). Involvement of actin filaments and their interaction with microtubules should be explored.

Preliminary observations show that, in *Dichotomosiphon*, organelle accumulation in the locally irradiated area is most sensitive to blue light and far less sensitive to green or red light. It is possible that the same pigments as those acting in irradiation-induced chloroplast accumulation in *Bryopsis* thallus (MIZUKAMI and WADA 1981) may also act in the signal-sensing mechanism of *Dichotomosiphon*. The time lag for apical accumulation of cytoplasm was several minutes when a microbeam was placed over the tip area. When the microbeam was moved further down the longitudinal axis of the algal body, the time lag increased (data not shown). This may suggest that the photoreceptor has a spatial gradient on its localization.

Interaction of microtubules with one or more cytoplasmic translocators such as dynein (HISANAGA and PRATT 1984), kinesin (PORTER *et al.* 1987, VALE *et al.* 1985) and MAP 1 c (PASCHAL *et al.* 1987, PASCHAL and VALLEE 1987) may be involved in the motive force generation in this alga. Investigation of these proteins will provide important information for a full understanding of the light-induced translocation of the cytoplasm.

Acknowledgements

This work was supported in part by grants-in-aid from the Ministry of Education, Science and Culture of Japan.

References

BLATT MR, BRIGGS WR (1980) Blue-light-induced cortical fiber reticulation concomitant with chloroplast aggregation in the alga *Vaucheria sessilis*. Planta 147: 355–362

— WESSELLS NK, BRIGGS WR (1980) Actin and cortical fiber reticulation in the siphonaceous alga *Vaucheria sessilis*. Planta 147: 363–375

— WEISENSEEL MH, HAUPT W (1981) A light-dependent current associated with chloroplast aggregation in the alga *Vaucheria sessilis*. Planta 152: 513–526

BLOOM GS, SCHOENFELD TA, VALLEE RB (1984) Widespread distribution of the major polypeptide component of MAP 1 (microtubule-associated protein 1) in the nervous system. J Cell Biol 98: 320–330

BULINSKI JC, BORISY GG (1980) Microtubule-associated proteins from cultured HeLa cells. J Biol Chem 255: 11570–11576

DAWES GJ, BARILOTTI DG (1969) Cytoplasmic organization and rhythmic streaming in growing blades of *Caulerpa prolifera*. Am J Bot 56: 8–15

HAUSSER I, HERTH W, REISS H-D (1984) Calmodulin in tip-growing plant cells, visualized by fluorescing calmodulin-binding phenothiazines. Planta 162: 33–39

HISANAGA S, PRATT MM (1984) Calmodulin interaction with cytoplasmic and flagellar dynein: calcium-dependent binding and stimulation of adenosine triphosphatase activity. Biochemistry 23: 3032–3037

HOLLENBECK PJ, CHAPMAN K (1986) A novel microtubule-associated protein from mammalian nerve shows ATP-sensitive binding to microtubules. J Cell Biol 103: 1539–1545

JAFFE LA, WEISENSEEL MH, JAFFE LF (1975) Calcium accumulations within the growing tips of pollen tubes. J Cell Biol 67: 488–492

JOHNSON GD, NOGUEIRA-ARANJO G, DE C (1981) A simple method of reducing the fading of immunofluorescence during microscopy. J Immunol Methods 43: 349–350

KAMIYA N (1962) Protoplasmic streaming. In: RUHLAND W (ed) Handbuch der Pflanzenphysiologie, vol 17/2. Springer, Berlin Göttingen Heidelberg, pp 979–1035

KITANISHI-YUMURA T, BLOOSE SH, FUKUI Y (1985) Role of the MT-MTOC complex in determination of the cellular locomotory unit in *Dictyostelium*. Protoplasma 127: 133–146

KURODA K, MANABE E (1983) Microtubule-associated cytoplasmic streaming in *Caulerpa*. Proc Japan Acad (Ser B) 59: 131–134

MAEKAWA T, TSUTSUI I, NAGAI R (1986) Light-regulated translocation of cytoplasm in green alga *Dichotomosiphon*. Plant Cell Physiol 27: 837–851

MANABE E, KURODA K (1983) Ultrastructural basis of the microtubule-associated cytoplasmic streaming in *Caulerpa*. Proc Japan Acad (Ser B) 60: 118–121

MENZEL D, SCHLIWA M (1986a) Motility in the siphonous green alga *Bryopsis*. I. Spatial organization of the cytoskeleton and organelle movements. Eur J Cell Biol 40: 275–285

— — (1986b) Motility in the siphonous green alga *Bryopsis*. II. Chloroplast movement requires organized arrays of both microtubules and actin filaments. Eur J Cell Biol 40: 286–295

MIZUKAMI M, WADA S (1981) Action spectrum for light-induced chloroplast accumulation in a marine coenocytic green alga, *Bryopsis plumosa*. Plant Cell Physiol 22: 1245–1255

MUROFUSHI H, KOTANI S, AIZAWA H, HISANAGA S, HIROKAWA N, SAKAI H (1986) Purification and characterization of a 190-kD microtubule associated protein from bovie adrenal cortex. J Cell Biol 103: 1911–1919

NUCCITELLI R (1983) Transcellular ion current: signals and effecters of cell polarity. Modern Cell Biol 2: 451–481

PASCHAL BM, VALLEE RB (1987) Retrograde transport by the microtubule-associated protein MAP1C. Nature 330: 181–183

— SHPETNER HS, VALLEE RB (1987) MAP1C is a microtubule-activated ATPase which translocates microtubules *in vitro* and has dynein-like properties. J Cell Biol 105: 1273–1282

PORTER ME, SCHOLEY JM, STEMPLE DL, VIGERS GPA, VALE RD, SHEETZ MP, MCINTOSH JR (1987) Characterization of the microtubule movement produced by sea urchin egg kinesin. J Biol Chem 262: 2794–2802

REISS H-D, HERTH W (1979) Calcium gradients in tip growing plant cells visualized by chlortetracycline fluorescence. Planta 146: 615–621

SPURR AR (1969) A low-viscosity epoxy resin embedding for electron microscopy. J Ultrastruct Res 26: 31–43

VALE RD, REESE TS, SHEETZ MP (1985) Identification of a novel force-generating protein, kinesin, involved in microtubule-based motility. Cell 42: 39–50

Protoplasma (1988) [Suppl 1]: 172–179

Chloroplast Orientation in *Mesotaenium*: the Phytochrome Effect Is Strongly Potentiated by Interaction with Blue Light

M. KRAML*, G. BÜTTNER, W. HAUPT, and H. HERRMANN

Institute of Botany, University of Erlangen-Nürnberg

Received January 5, 1988
Accepted April 19, 1988

Dedicated to Professor Dr. NOBURO KAMIYA on the occasion of his 75th birthday

Summary

The chloroplast of *Mesotaenium* is oriented by red light to the face position. This phytochrome effect can strongly be potentiated by blue light. Whereas a 10 min red irradiation is hardly effective, it induces a strong response if given together with 10 min blue (which by itself is also nearly ineffective). Even 1 min red can be rendered highly effective, if followed by 10 min blue. In this interaction, only red light is responsible for the directionality, whereas the direction from which blue light is given is not important for the response.

In order to potentiate the phytochrome response, blue light has been shown to have a dual and sequential effect. It has first to interact with P_{fr} for at least 1 min, after which time P_{fr} is no longer necessary for the orientation movement. Obviously, an internal signal (S) has been formed which stores the directional information. In a second phase, movement as controlled by S depends on blue light for its realization; *i.e.*, movement stops after blue light is switched off. Thus, the reorientation of the chloroplast is completed within the 10 min of blue irradiation, whereas the response in continuous red light may require more than 60 min.

Keywords: Action dichroism; Blue-red interaction; Chloroplast orientation; *Mesotaenium*; Movement of chloroplast; Phytochrome.

Abbreviations: P_r red-absorbing form of phytochrome; P_{fr} far-red-absorbing form of phytochrome; R red light; FR far-red light; B blue light.

1. Introduction

The unicellular alga *Mesotaenium* has strong structural similarities with the filamentous alga *Mougeotia* insofar, as the cell contains only one large ribbon-shaped chloroplast. This chloroplast orients, in both organisms, to the light direction by rotation. In red light the chloroplast rotates so as to expose its face towards the light source: this is a typical phytochrome response, and the chloroplast orients in a gradient of P_{fr} over the cell periphery (HAUPT 1983, KRAML 1986). However, there is an important difference between the two organisms concerning this response. In *Mougeotia*, on the one hand, a single pulse of red light, even if applied in the millisecond range or below (KRAML and HAUPT 1981, HAUPT and POLACCO 1979), can induce full chloroplast orientation during a subsequent dark period of half an hour or less. In *Mesotaenium*, on the other hand, orientation requires repetitive pulses of red light during the whole movement, with dark intervals not exceeding a few minutes (HAUPT and REIF 1979). This has been interpreted as being due to a rapid loss of physiological activity of P_{fr}; thus, the P_{fr} gradient established by a single red pulse cannot act during the whole time that is needed for the orientation to be completed in *Mesotaenium*. Indeed, HAUPT and REIF 1979 reported evidence for such a loss of P_{fr} activity. In contrast, the P_{fr} in *Mougeotia* proved to be remarkably stable, and its physiological activity for orienting the chloroplast has been demonstrated even 20 h after a red-light pulse (KRAML *et al.* 1987).

An additional difference between the two species, not explicitely referred to so far, becomes evident if the orientation movements in continuous red light are compared. In *Mougeotia* full face position in all cells can be attained after 15 min (SCHÖNBOHM 1969), whereas in *Mesotaenium* the orientation takes much longer time

* Correspondence and Reprints: Institut für Botanik und Pharmazeutische Biologie, Universität Erlangen-Nürnberg, Staudtstrasse 5, D-8520 Erlangen, Federal Republic of Germany.

(HAUPT and THIELE 1961; KRAML, unpublished results). This difference cannot be explained by a different duration of P_{fr} activity, as under continuous red light the "new" active P_{fr} is assumed to be continuously regenerated by $P_r \rightleftarrows P_{fr}$ cycling. For trying to explain this different behaviour of the two closely related algae in red light, an observation might be important, which has been made in preliminary experiments (unpublished): If, in *Mesotaenium*, red light is replaced by white light, movement is strongly accelerated in this continuous irradiation. This points to an additional photoreceptor pigment coacting or interacting with phytochrome, and this pigment should have its main absorption in the short-wavelength range of the light, *i.e.*, it is assumed to be a "cryptochrome" (*cf.*, SENGER and SCHMIDT 1986). It is the aim of this paper to demonstrate the contribution of the blue-light system to the phytochrome-mediated orientation movement in *Mesotaenium* (*i.e.*, to the so-called low-irradiance movement), and to investigate the coaction or interaction of these two photoreceptor systems.

2. Material and Methods

2.1. Plant Material

Mesotaenium caldariorum (strain 648-1) from the algal collection of the University of Göttingen (Federal Republic of Germany) was used for the experiments. The algae were cultured in test tubes on agar (1 g proteose pepton, Merck, and 10 g agar were dissolved in 1 l nutrient solution, which was described by GÄRTNER 1970). From these stock cultures, which were subcultured every 3 weeks, suspensions of algae in nutrient medium were made and poured out into one-way Petri dishes containing one layer of rectangular cover-glasses (15 × 15 mm) on the bottom. These liquid cultures were grown in a light : dark cycle of 12 : 12 h at a temperature of 16 ± 1 °C with 600 lx during the light period from fluorescent lamps (L40W/15 and L40W/32, Osram, Federal Republic of Germany). The algae settle on the cover-glasses and stick with their mucilage on the surface. After one week, a sufficient cell density for the experiments is obtained.

2.2. Preparations of *Mesotaenium* on Microscope Slides

Rectangular Plexiglas frames (25 × 25 mm; 0.8 mm high) were glued on Plexiglas slides (76 × 26 mm; 0.8 mm high) surrounding an area of 16 × 16 mm. A cover-glass from the liquid cultures was put into the areas and fastened by small ribbons of vaseline. After wetting the algae with a droplet of nutrient solution, each preparation was covered by a cover-glass (24 × 24 mm), which was adhered to the Plexiglas frame by vaseline. Under these conditions *Mesotaenium* remained alive and sensitive to inductions of chloroplast movement for several days. However, the experiments reported here were not extended beyond 24 h after preparing the slides. If the algae were irradiated from above, black diaphragms (23 × 23 mm; aperture 8 × 8 mm) were put upon the cover-glass in order to avoid light scattering, which might be caused by the edge of the nutrient solution or by the Plexiglas frame.

2.3. Light Pretreatments

As a starting position for all experiments the chloroplasts of the algae were oriented in profile position (as seen from above) by illuminating the preparations from the sides with white light (fluorescent tubes L40W/15; Osram, Federal Republic of Germany) for 20 h at 22 °C (*cf.*, HAUPT and REIF 1979). During this procedure the preparations were kept in a moist Petri dish and protected against light from above or below by sheets of black cardboard. After this preirradiation, unpolarized far-red light was applied for two minutes (250 W halogen lamps; colored glass filter RG 9, Schott and Gen., Mainz, Federal Republic of Germany; 3 mm thick) to saturate photoreversion of P_{fr} back to P_r.

2.4. Actinic Irradiations

Chloroplast orientation from profile to face position was induced by red light from below (250 W halogen lamps; interference filter AL 666, half-bandwidth 20 nm, Schott and Gen., Mainz, Federal Republic of Germany) and/or blue light from above (Xenon lamp XBO 450 ofr, Osram, Federal Republic of Germany; interference filter AL 448, half-bandwidth 21 nm, Schott and Gen.). In one experiment red light was applied from above and blue light from the sides. Polarized light was obtained with one sheet of Polarex P-W 44 (Käsemann, Oberaudorf, Federal Republic of Germany). The fluence rate of the blue irradiation was varied by neutral density filters (Schott and Gen.) and measured with a thermopile (Kipp and Zonen, Delft, Netherlands). During and after the irradiations the preparations were kept at a temperature of 22 ± 1 °C.

2.5. Quantification of Chloroplast Movement

Unless stated otherwise, 40 min after the beginning of actinic irradiations, the response was evaluated using a microscope with green safelight (glass filters BG 18 and GG 495; each 4 mm thick, Schott and Gen.). Cells with incomplete response were counted as follows:

$< 30\%$ face position = profile position (P)

30–70% face position = medium position (M)

$> 70\%$ face position = face position (F).

The positions of about 150 chloroplasts were determined in each preparation and the response calculated as follows:

Chloroplasts in face position (%) = numbers of

$$\frac{F + 0.5M}{P + M + F} \cdot 100$$

2.6. Statistical Analysis

Each data point in the figures is based on at least 5 preparations; but often more than 10 preparations were used. The standard error of the mean mostly is in the range of 1.5% with a maximum value of 2.5%.

3. Results

In a first experiment, *Mesotaenium* cells were irradiated for 10 min with red light, blue light, or red and blue light together, and the response was evaluated 40 min after beginning of the irradiation. Because of the action dichroism for the effect of red light (HAUPT and THIELE 1961), the latter was applied as linearly polarized light with the electrical vector vibrating perpendicular to the

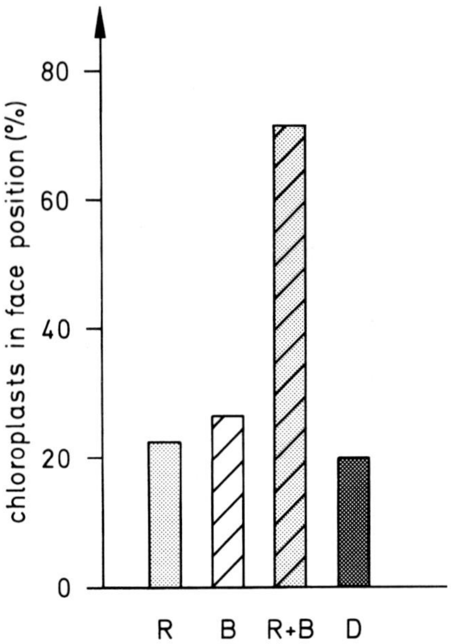

Fig. 1

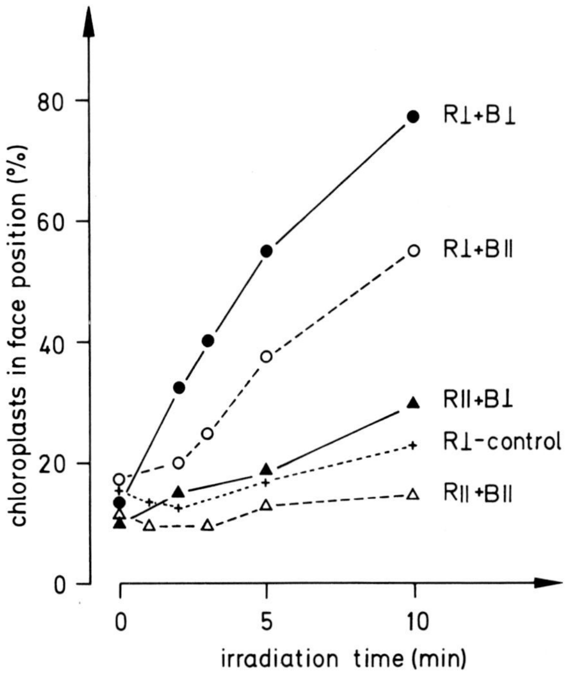

Fig. 2

Fig. 1. Chloroplast orientation as induced by 10 min irradiation with red (*R*, $8\,Wm^{-2}$), blue (*B*, $3\,Wm^{-2}$) or red and blue simultaneously. Blue irradiations from above, red from below, both linearly polarized with the electrical vector perpendicular to the cell axis. Ordinate: percent of cells with the chloroplast in face position 40 min after beginning of the irradiation. *D* Dark control. The standard error is in the range of 1.5% with a maximum value of 2.5%

Fig. 2. Chloroplast orientation as induced by simultaneous irradiations with red and blue (*R, B*) of various irradiation time. Polarized light with the electrical vector parallel (∥) or perpendicular (⊥) to the cell axis. Otherwise as in Fig. 1. Notice that the zero points of the curves are the respective dark controls

long cell axis. Thus, an optimal gradient of P_{fr} could be established. For technical reasons, blue light was given in the same way. The results are shown in Fig. 1. As expected, the effect of 10 min red light is minor, and the same is true for blue light: The response to either of these irradiations only slightly exceeds the dark control beyond the error range. Yet, the fluence rates used in these experiments were almost saturating: Doubling of red from 8 to $16\,Wm^{-2}$ or of blue from 3 to $6\,Wm^{-2}$ increases the response by only 2 to 3%, thus staying within twice the error range (data not shown). In contrast, simultaneous irradiation with red and blue results in a high percentage of orientation response. Thus, the two spectral ranges obviously act in a non-additive way: instead, an interaction is suggested, which will be studied in some of the following experiments.

Next, the experiment was extended to include shorter irradiation times, and the R/B combination was compared with R. In addition, a possible action dichroism for blue was tested by varying the directions of electrical vectors in all four combinations. The response was always evaluated 30 min after the end of the irradiation. The results are shown in Fig. 2. The curves confirm the strong action dichroism for red light; moreover, for blue light, a less spectacular but significant action dichroism is observed, with perpendicular vibrating light being more effective than parallel vibrating light. Thus, by far the strongest effect is found if both irradiations are given with the electrical vector vibrating perpendicular to the cell axis, and under these conditions a combined irradiation of as short as 2 min exhibits a significant response already. For the following experiments, therefore, both red and blue light are always given with the electrical vector vibrating perpendicular to the cell axis. The preceding experiment has shown that shortening of the irradiation time below 10 min strongly lowers the response. This might be due to the time being a critical factor for blue, for red or for both irradiations. However, for photoconversion of phytochrome with the fluence rate of $8\,Wm^{-2}$ red light, 10 min should be high in saturation. We therefore tested the effect of a shorter red pulse: The preceding experi-

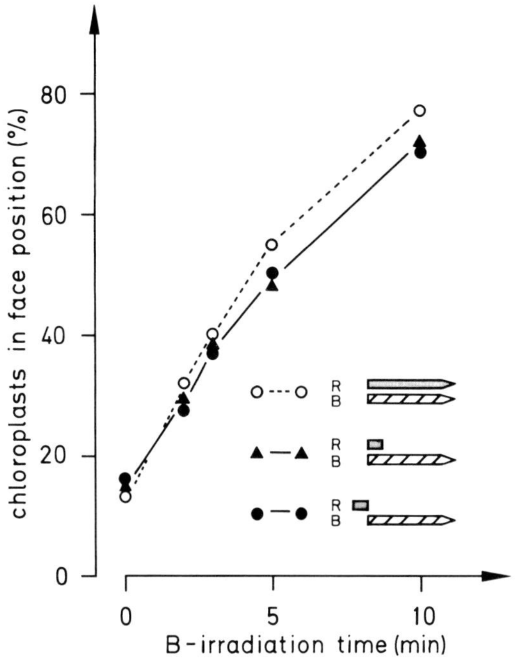

Fig. 3. Chloroplast orientation as induced by a red/blue (*R*, *B*) combination with the red irradiation shortened to 1 min. Two different protocols are shown in the inset, the data fit with a joint graph (solid line). For comparison, the dotted line shows the response to full-length irradiation with both red and blue (upper curve of Fig. 2). Otherwise as in Fig. 1

ment (Fig. 2, upper curve) was modified so as to shorten the red irradiation to 1 min, but to vary the irradiation time for blue from 0 to 10 min. As can be seen in Fig. 3, one minute red can fully replace, in this combination, the 10 min red irradiation, and this is equally effective whether blue irradiation is started together with red or after its termination. On the other hand, no interaction is found if red light is preceded by blue; these data will be presented and discussed in more detail in a forthcoming paper.

The results of Fig. 3 suggest that blue light interacts with P_{fr}. However, considering the rapid loss of P_{fr} activity in *Mesotaenium*, it might be doubted whether active P_{fr} is available during all the 10 min for blue light to interact with . Rather, it might be that interaction of blue light with P_{fr} is restricted to a shorter period and/or that in fact blue light interacts with an early product of P_{fr} action. This possibility is investigated next. The experimental approach is as follows (insets in Fig. 4): Red light is again given as a 1 min pulse, and this is followed by 10 min blue as in the previous experiment. This blue period, however, is split into two parts with a saturating far-red pulse of 2 min in between (solid curve); the duration of the first part of the blue

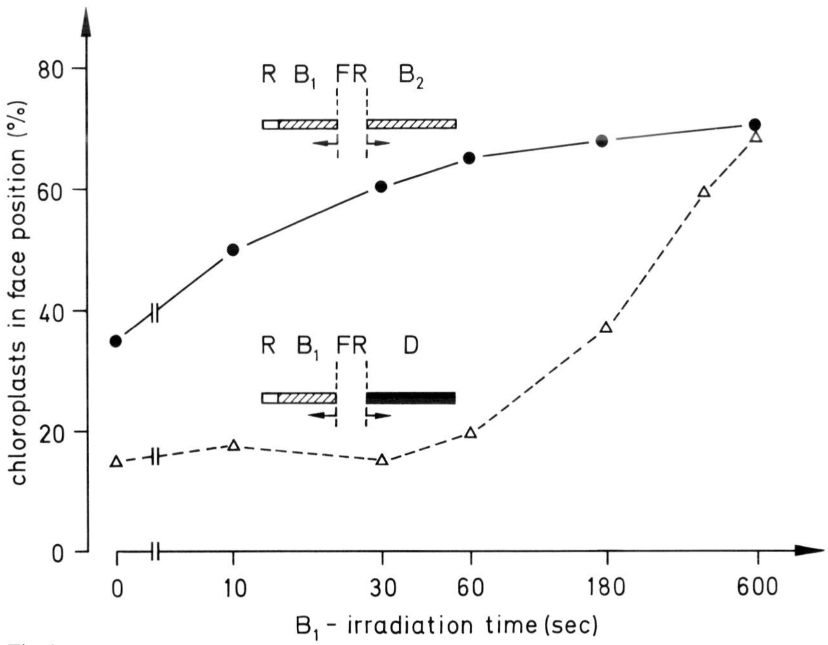

Fig. 4

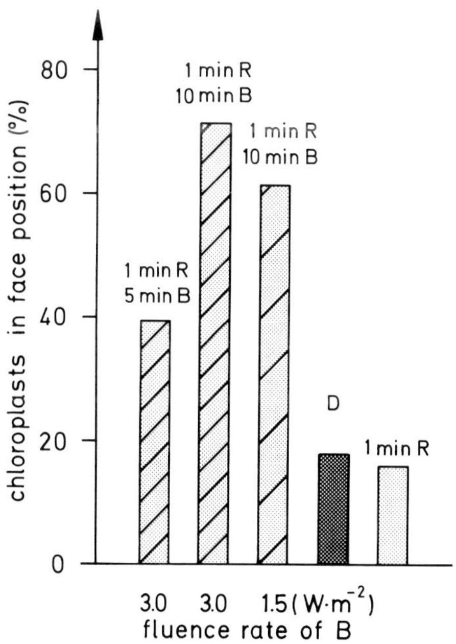

Fig. 5

Fig. 4. Chloroplast orientation as induced by sequence of 1 min red (*R*) and 10 min blue (*B*), but the latter divided into two parts (B_1, B_2), with 2 min unpolarized far-red (*FR*, 20 Wm^{-2}) in between. The abscissa shows the duration of the first blue period in log scale; the second blue period, then, amounts to $B_2 = 10\,min - B_1$. The dashed curve is a control without B_2. Otherwise as in Fig. 1. Notice that the zero points are equivalent to an uninterrupted blue period after *R-FR* and a dark control after *R-FR*, respectively

Fig. 5. Chloroplast orientation as induced by a sequence of 1 min red (*R*) followed by blue (*B*) with the fluence reduced either by the time or by the fluence rate. Otherwise as in Fig. 1

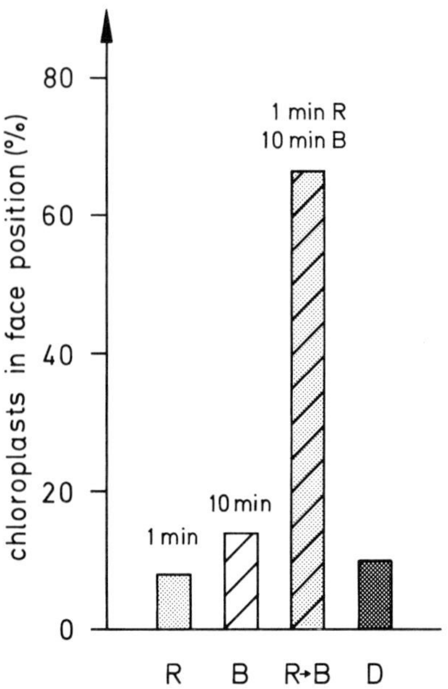

Fig. 6

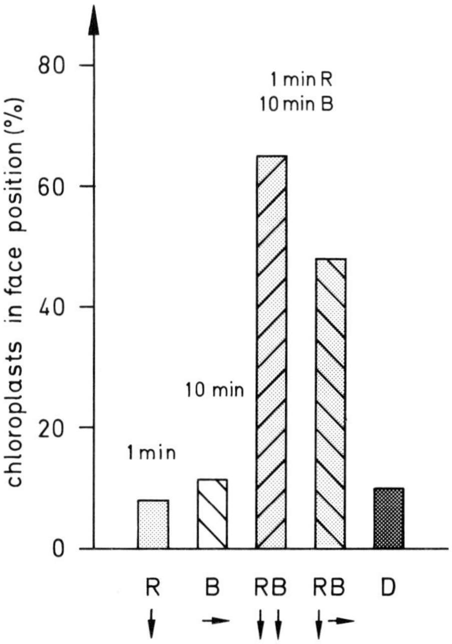

Fig. 7

Fig. 6. Chloroplast orientation as induced by a sequence of 1 min red (*R*) and 10 min blue (*B*), including the respective controls. The response was evaluated immediately after termination of *B* or 10 min after termination of *R*, respectively. Otherwise as in Fig. 1

Fig. 7. Chloroplast orientation as induced by a sequence of red (*R*) and blue (*B*) as in Fig. 6, but the direction of *B* varied, being applied either in the same direction as *R* (↓↓) or perpendicular to it (↓→). Evaluation of the response immediately after termination of *B* (or 10 min after termination of *R*, respectively). Otherwise as in Fig. 1

irradiation is given on the abscissa. As a control, the second part of the blue irradiation, *i.e.*, after far-red, is omitted (dashed curve).

Mainly three results have been obtained (Fig. 4):

i) The control curve with a varying uninterrupted blue period confirms that for a full effect blue light has to act approximately 10 min; thus, it must not be shortened to the same degree as the red-light pulse.

ii) If the blue period is interrupted by far-red, full response is obtained already, if P_{fr} was present only during the first of the ten minutes. Without the second blue period, however, this combined irradiation is not yet sufficient for a substantial response.

iii) Most remarkably, some response is even obtained if the total blue irradiation is applied after P_{fr} has been abolished by far-red (zero point of the upper curve).

These results allow for the conclusion that two stages of the induction process can be distinguished: In a first stage, blue has to coact/interact with P_{fr} or with a short-living result of P_{fr} action, and this reaction is nearly saturated after 1 min. In a second stage, blue has to coact/interact with a product of the first stage, and this reaction appears to be saturated after about 10 minutes.

From these results it may be assumed that for the blue-light effect the factor time is crucial. To test this assumption, part of the experiment of Fig. 3 is repeated with varying parameters of the blue irradiation. The result is given in Fig. 5:

Shortening of the blue irradiation to 5 min decreases the effect substantially. On the other hand, if the fluence rate is reduced by a factor of 2, but the irradiation time kept at 10 min, the response to the red-blue irradiation is only slightly diminished. Thus, reciprocity does not hold; with equal fluences, time is more important than fluence rate. This experiment confirms the crucial role of the irradiation time for blue.

So far, in all experiments the response was evaluated 40 min after the start of inducing irradiations. Thus, the 10 min of blue light, as usually applied, cover one forth of the total response time. It is interesting to know to what extent the response proceeds during that blue-light period, and what fraction of the response is continued as an aftereffect in the 30 min dark period. To answer this question, we irradiate again with a one-minute red pulse, followed by 10 min blue. But the response is evaluated immediately after the blue irradiation. As Fig. 6 shows, the response is almost completed; thus, the actual movement appears to be restricted to the blue-light period, and no aftereffect can be observed. This very fast movement is in strong contrast to that elicited by continuous red light; here only a weak response can be found, and even after 1 h continuous red no full response has been obtained in our experimental conditions (Kraml, in preparation).

This result is important also for methodical reasons. We now can evaluate the response immediately after termination of the blue-light irradiation.

The data obtained so far may suggest that the orientation movement is induced via phytochrome, as in *Mougeotia*, but that blue light is required to render this induction effective if red light is not given continuously or repeatedly. In this case, the directional information might be perceived exclusively by phytochrome, and the direction of blue light might not be crucial for its effect. This prediction is tested and the result shown in Fig. 7. In this experiment the response is evaluated immediately after termination of the blue irradiation (as recommended in the preceding paragraph).

The results strongly suggest that the direction of blue light is of minor importance for the orientation. Although in the 4th column blue light is acting for 10 min in the "wrong" direction (*i.e.*, expected to orient the chloroplast to a "zero" value as seen from above), it nevertheless strongly potentiates the orienting effect of the red preirradiation to face position (*cf.* 1st column). However, quantitative comparisons are not yet possible, as light from above or from the sides meets quite different optical conditions on its way to the cells. More detailed investigations are under way. Qualitatively, however, it is reasonable to conclude that the directional signal is perceived by phytochrome, and that blue light from any direction is required to render the orienting P_{fr} gradient effective.

4. Discussion

Whereas hitherto *Mesotaenium* required a long-term action of red light for orienting its chloroplast's face toward the light, this red-light irradiation could now drastically be shortened by the additional irradiation with blue light. Under these conditions, a response close to saturation can be induced by a combination of 1 min red of 8 Wm^{-2} and 10 min blue of 3 Wm^{-2}, both irradiations being almost ineffective if given alone. Moreover, under this combined irradiation protocol, movement proceeds much faster. On the basis of our experiments we can try to characterize and to interpret at least preliminary the blue-light effect.

4.1. *Characterization of the Blue-Light Effect*

As a photobiological characterization, it is important to note the failure of reciprocity: to be fully effective, blue light has to act a minimum time of 10 min. Shortening of this time cannot be compensated for by increasing the fluence rate. Moreover, the effect of blue light is independent of its direction. Thus, the directional signal is perceived exclusively by phytochrome, whereas the additional blue-light signal acts in a scalar way. A slight but significant action dichroism was shown for the blue effect, indicating a localization of the blue-light sensory pigment close to the cell membrane.

4.2. *The Interaction of Blue Light with Phytochrome*

Blue light is fully effective not only if applied simultaneously with red, but also if given afterwards. Thus, there is no interaction at the level of photoperception. Rather, blue light has to interact with the result of the red-light pulse, viz. with P_{fr}. Upon closer inspection, however, the situation is more complex: Blue light appears to act in a dual and sequential way. This has been concluded from the observation that the P_{fr} gradient has to be present only during the first minute of the blue-light period of 10 min. Thus, we can distinguish two phases.

i) In a first phase of 1 min, blue light interacts with P_{fr} in a way that after converting P_{fr} back to P_r a directional signal remains in the cell. This secondary signal will be called S and may be an unevenly distributed substance or structural change. By itself, S is hardly effective in sustaining the orientation movement.

ii) If, after formation of S, blue-light irradiation is continued, orientation movement occurs. The direction of movement, then, is controlled by the gradient of S, whereas blue light appears to have some activating function on S only. Remarkably, there is no aftereffect of blue, as movement stops as soon as blue light is switched off.

This distinction of two sequential blue light effects can be used to further analyze them. Some characteristics can be added and interpretations tried.

Concerning the first phase, the interaction of blue light with P_{fr} is saturated after 1 min already. Thus, we cannot explain the effect by assuming that blue light could intervene with the decay of P_{fr} activity and thus keeping P_{fr} active for controlling the movement. Rather, the production of a secondary signal (S), as proposed above, appears to be an inevitable assumption. Still more complicated, a significant effect of blue light—

although by far not saturating—has been found even if it is applied after P_{fr} has been abolished by far-red light. Thus, blue light might interact with an early product of P_{fr} action rather than with P_{fr} itself.

After the signal S has been formed, it appears to be relatively stable: it can orient the chloroplast during a subsequent 10 min period of blue, whereas the activity of P_{fr} is lost after a few minutes already (HAUPT and REIF 1979). Thus, this S appears also to have a longer life time than the blue-light-induced "product" that induces or controls orientation movements in *Mougeotia* and *Lemna* with a half life of about 2 min (ZURZYCKI *et al.* 1983, GABRYŚ *et al.* 1985). But quantitative investigations are still missing, and nothing can be speculated yet about the nature of S on a molecular level.

The second phase can be interpreted as an enormous speeding up of the movement as oriented by S. This might be due to an effect on the mechanical system for the response, as can be concluded from experiments in *Mougeotia*. In this alga, at temperatures slightly above the freezing point no orientation can be induced by red light, neither by a single pulse nor by continuous red. In continuous blue, however, chloroplast orientation proceeds. WEISENSEEL 1968 interpreted these findings as an effect of low temperature and of blue light on the "viscosity" of cytoplasm (whatever this means in molecular terms); *i.e.*, low temperature increases the viscosity so as to make intracellular movement impossible, but blue light antagonizes that effect. SCHÖNBOHM and HELLWIG 1979 supported this idea by centrifuging the chloroplast in longitudinal direction. At 2 °C in red light the chloroplast was anchored in the cytoplasmic layer much tighter than at 20 °C, but blue light caused a significant loosening of the chloroplast, thus facilitating centrifugation.

Stimulating as this hypothesis may be, there is not yet full consistency with these earlier results. In the experiments of SCHÖNBOHM and HELLWIG, some aftereffect of blue has to be postulated in *Mougeotia*, whereas in our *Mesotaenium* system no aftereffect has been found. However, the time resolution of the kinetics of movement as tested in our experiments do not allow to exclude a small aftereffect in the range of a few minutes, and therefore more detailed experiments on the kinetics of movement are in progress (KRAML, in preparation).

If we compare the blue-light effect on chloroplast orientation in *Mesotaenium* with that in *Mougeotia* and *Adiantum*, where phytochrome is the main photosensory system as well, we find a marked difference. As far as the so-called low-irradiance orientation is concerned, no blue-red interaction has been found in either *Mougeotia* or *Adiantum* (GABRYŚ *et al.* 1984, YATSUHASHI *et al.* 1985). Instead, blue light without red can induce and orient the movement as a directional signal, and at least for *Mougeotia* it has been proven that this blue-light effect is not mediated by phytochrome.

On the other hand, for the "high-irradiance orientation" of *Mougeotia* an interaction of phytochrome and cryptochrome is necessary. According to SCHÖNBOHM 1966 (see also SCHÖNBOHM 1980) the directional signal is perceived by phytochrome, whereas the cryptochrome-mediated blue-light effect is independent of its direction; its important function is to switch the transduction chain such that the chloroplast's edge no longer tries to avoid the highest levels of P_{fr}, but to approach them. Thus, there appears to be at least a basic similarity to our system: directional signal perception by phytochrome, and a scalar blue-light effect that ensures the realization of the response. This similarity, however, is superficial, as can be seen by comparing the effective sequences. In our *Mesotaenium* system blue has to act together with or after red to make it effective. In *Mougeotia*, on the contrary, the sequence blue-red is most effective, with a photoproduct of blue being assumed to interact with P_{fr} (GABRYŚ *et al.* 1985). Thus, extrapolation of blue-light effects between the various systems and response is not yet possible.

4.3. Concluding Remarks

Potentiation of phytochrome effects by blue light is not restricted to chloroplast orientation in *Mougeotia* and *Mesotaenium*. Rather, interactions of phytochrome and blue/UV photoreceptor pigments are well-known also in photomorphogenesis of higher plants, *e.g.*, hypocotyl elongation or anthocyanin synthesis (*cf.*, DRUMM and MOHR 1978, KIGEL and SCHWARTZ 1981, ATTRIDGE *et al.* 1984, GABA *et al.* 1984, MOHR *et al.* 1984). Again, P_{fr} is the effector of the morphogenetic response, and B/UV controls the responsiveness to P_{fr}, with different degrees of interaction depending on species (MOHR 1986, 1987, for review). Many of these responses are interpreted as regulation of gene expression by phytochrome, and relatively long-lasting actions of P_{fr} are mostly required. On the other hand, the responses in *Mesotaenium* and *Mougeotia* are assumed to be controlled by phytochrome effects on membrane properties, resulting *e.g.* in ion redistributions (WAGNER and GROLIG 1985). In these cases short-term effects are involved. Thus, it appears important to further analyze the phytochrome-cryptochrome interactions in the chloroplast-orientation systems and to compare them with those in higher plants.

Acknowledgements

This work was supported by the Deutsche Forschungsgemeinschaft (Ha 77/33-1; Kr 804/1-3).

References

ATTRIDGE TH, BLACK M, GABA V (1984) Photocontrol of hypocotyl elongation in light-grown *Cucumis sativus* L. A synergism between the blue-light photoreceptor and phytochrome. Planta 162: 422–426

DRUMM H, MOHR H (1978) The mode of interaction between blue (UV) light photoreceptor and phytochrome in anthocyanin formation of the *Sorghum* seedling. Photochem Photobiol 27: 241–248

GABA V, BLACK M, ATTRIDGE TH (1984) Photocontrol of hypocotyl elongation in de-etiolated *Cucumis sativus* L. Plant Physiol 74: 897–900

GABRYŚ H, WALCZAK T, HAUPT W (1984) Blue-light-induced chloroplast orientation in *Mougeotia*. Evidence for a separate sensor pigment besides phytochrome. Planta 160: 21–24

— — — (1985) Interaction between phytochrome and the blue light photoreceptor system in *Mougeotia*. Photochem Photobiol 42: 731–734

GÄRTNER R (1970) Die Bewegung des *Mesotaenium*-Chloroplasten im Starklichtbereich. II. Aktionsdichroismus und Wechselwirkungen des Photorezeptors mit Phytochrom. Z Pflanzenphysiol 63: 428–443

HAUPT W (1983) Movement of chloroplasts under the control of light. Prog Phycol Res 2: 227–281

— POLACCO E (1979) Phytochrome-mediated response in *Mougeotia* to very short laser flashes. Plant Sci Lett 17: 67–73

— REIF G (1979) "Ageing" of phytochrome P_{fr} in *Mesotaenium*. Z Pflanzenphysiol 92: 153–161

— THIELE R (1961) Chloroplastenbewegung bei *Mesotaenium*. Planta 56: 388–410

KIGEL J, SCHWARTZ A (1981) Cooperative effects of blue and red light in the inhibition of hypocotyl elongation of de-etiolated castor bean. Plant Sci Lett 21: 83–88

KRAML M (1986) Light direction and polarization. In: KENDRICK RE, KRONENBERG GHM (eds) Photomorphogenesis in plants. Martinus Nijhoff/Dr. Junk W, Dordrecht, pp 237–267

— HAUPT W (1981) Phytochrome-controlled chloroplast orientation in *Mougeotia*: Action dichroism in double-flash experiments. Plant Sci Lett 21: 145–150

— LEOPOLD K, WINKLER B (1987) Long-lasting activity of P_{fr} and P_{fr}-gradients in *Mougeotia* chloroplast movement? Acta Physiol Plant 9: 189–198

MOHR H (1986) Coaction between pigment systems. In: KENDRICK RE, KRONENBERG GHM (eds) Photomorphogenesis in plants. Martinus Nijhoff/Dr. Junk W, Dordrecht, pp 547–564

— (1987) Mode of coaction between blue/UV light and light absorbed by phytochrome in higher plants. In: SENGER H (ed) Phenomena, distribution and mechanisms of blue light effects. CRC Press, Boca Raton, pp 133–144

— DRUMM-HERREL H, OELMÜLLER R (1984) Coaction of phytochrome and blue/UV light photoreceptors. In: SENGER H (ed) Blue light effects in biological systems. Springer, Berlin Heidelberg New York Tokyo, pp 6–19

SENGER H, SCHMIDT W (1986) Diversity of photoreceptors. In: KENDRICK RE, KRONENBERG GHM (eds) Photomorphogenesis in plants. Martinus Nijhoff/Dr. Junk W, Dordrecht, pp 137–158

SCHÖNBOHM E (1966) Die Bedeutung des Phytochromsystems für die negative Phototaxis des *Mougeotia* Chloroplasten. Ber Dtsch Bot Ges 79: 131—138

— (1969) Untersuchungen über den Einfluß von Photosynthesehemmstoffen und Halogeniden auf die Starklicht- und Schwachlichtbewegung des Chloroplasten von *Mougeotia* spec. Z Pflanzenphysiol 60: 255—269

— (1980) Phytochrome and non-phytochrome dependent blue light effects on intracellular movements in freshwater algae. In: SENGER H (ed) The blue light syndrome. Springer, Berlin Heidelberg New York, pp 69–96

— HELLWIG H (1979) Zum Photorezeptor-Problem der Schwachlichtbewegung des *Mougeotia* Chloroplasten im Blau, bzw. Hellrot bei niederen Temperaturen. Ber Dtsch Bot Ges 92: 749–762

WAGNER G, GROLIG F (1985) Molecular mechanisms of photoinduced chloroplast movements. In: COLOMBETTI G, LENCI F, SONG P-S (eds) Sensory perception and transduction in aneural organisms. Plenum Press, New York, pp 281–298

WEISENSEEL M (1968) Vergleichende Untersuchungen zum Einfluß der Temperatur auf die lichtinduzierte Chloroplastenverlagerung. Z Pflanzenphysiol 59: 56—69

YATSUHASHI H, KADOTA A, WADA M (1985) Blue- and red-light action in photoorientation of chloroplasts in *Adiantum* protonemata. Planta 165: 43–50

ZURZYCKI J, WALCZAK T, GABRYŚ H, KAJFOSZ J (1983) Chloroplast translocations in *Lemna trisulca* L. induced by continuous irradiation and by light pulses. Kinetic analysis. Planta 157: 502–510

Protoplasma (1988) [Suppl 1]: 180–184

Differentially Adsorbed Vital Dyes Inhibit Chloroplast Movement in *Mougeotia scalaris*

U. Russ*, F. Grolig, and G. Wagner

Botanisches Institut 1, Justus-Liebig-Universität, Giessen

Received March 11, 1988
Accepted June 24, 1988

Dedicated to Professor Dr. Noburo Kamiya on the occasion of his 75th birthday

Summary

The vital dyes, neutral red and rhodamine B, specifically stain the calcium binding vesicles in *Mougeotia* (Grolig and Wagner 1987, Planta 171: 433–437). Increasing amounts of vesicularly bound neutral red or rhodamine B lead to increasing inhibition of chloroplast movement, probably due to dye-dependent effusion of vesicular calcium which overrides the cytoplasmic calcium homoeostasis. For the fast stain neutral red, the inhibitory effect mainly depends on the rate of dye uptake, less so on the absolute amount of dye bound to the vesicles. Locally restricted staining of a *Mougeotia* cell caused locally restricted inhibition of chloroplast movement, which could be overcome by the cell with time. Gathering of the vesicles, also observed in the treated area, *e.g.* upon penetration of a *Mougeotia* cell by a micropipette without dye, left the chloroplast movement unaffected.

Keywords: *Mougeotia*; Light-dependent chloroplast movement; Neutral red; Rhodamine B; Calcium homeostasis.

Abbreviations: IC_{50} half maximal inhibition; MES 2[N-morpholino]ethanesulfonic acid; NR neutral red; RB rhodamine B.

1. Introduction

Mougeotia, as well as *Mesotaenium*, and possibly other conjugatophycean green algae perform precise chloroplast reorientational movement with respect to light (Haupt and Wagner 1984). The mechanism of photosensory transduction often was studied here in a pharmacological approach in the living cell (Schönbohm and Schönbohm 1984, Serlin and Roux 1984, Wagner *et al.* 1984), confirmed by substructural and biochemical analysis *in situ* and *in vitro* (Grolig and Wagner 1987). Evidence resulting, *e.g.* from the use of cytochalasin B or chlorotetracycline *in vivo* was supplemented by actin identification after arrowhead decoration with rabbit heavy meromyosin or X-ray microanalysis of vesicular calcium *in situ* (Marchant 1976, Wagner and Klein 1981, Wagner and Rossbacher 1980). Similar phenomena of plant cell movement such as in *Euglena, Nitella* or the plasmodium of *Physarum* have elegantly been analyzed by Professor Kamiya, first during his stay with Professor Küster at the University of Giessen and followed ever since (Kamiya 1939, 1986). For *Mougeotia*, the pharmacological approach is extended here with a considerable effect of neutral red (NR) and rhodamine B (RB) on the chloroplast reorientational response.

* Correspondence and Reprints: Botanisches Institut 1, Justus-Liebig-Universität, Senckenbergstrasse 17—25, D-6300 Giessen, Federal Republic of Germany.

2. Materials and Methods

2.1. Plant Material

Mougeotia scalaris Hassel as determined via zygotes of the "Erlangen" strain (Haupt 1970), was grown on a large scale in an aerated synthetic medium with air plus 2% (v/v) CO_2 at 26 °C and at about 3000 lx white light (Stabenau 1978). After harvesting, the filaments were kept for 10 to 20 days in ENS-medium (Schönbohm 1963) before use in the experiments.

2.2. X-Ray Microanalysis

The calcium binding vesicles were isolated after vital staining with neutral red (NR) or rhodamine B (RB) as described by Grolig and Wagner (1987). For X-ray microanalysis *in vitro*, isolated unfixed vesicles were mounted on a discharged formvar/carbon coated copper grid and examined with a Kevex X-ray microanalyzer coupled to a Hitachi H-600 scanning electron microscope. Spot analysis of 0.2–0.5 μm in beam diameter was done within the energy range of 0–10.24 keV for 100 s at 100 kV acceleration voltage with beam current 10 μA. X-ray microanalysis of the calcium binding vesicles *in situ* was done by Wagner and Rossbacher (1980).

2.3. Cell Filament Experiments

Small amounts of algal filaments in petri dishes on a rotatory shaker (50 rpm) were incubated in MXS medium (Neuscheler-Wirth 1970) plus the vital dye under test. NR for 5 min staining was applied in concentrations of 20–80 μmol · l^{-1}; NR for 4 h staining was applied gradually by the addition of suitable amounts of dye each hour in constant steps to result in final concentrations of 8–28 μmol · l^{-1}. RB for 10 min staining was applied in concentrations of 50–2600 μmol · l^{-1}; RB for 4 h staining was applied gradually by the addition of suitable amounts of dye each hour in constant steps to result in final concentrations of 8–320 μmol · l^{-1}. High affinity of neutral red in particular to the calcium binding vesicles allowed the use of five times more diluted dye concentrations as used for vacuolar staining (Stadelmann and Kinzel 1972). After staining, the filaments were transfered onto a microscopic slide, covered by a cover glass with spacer, and rinsed with MXS medium buffered to pH 5 by 1 mmol · l^{-1} MES. The chloroplast response to "edge on" position (viewing through the microscope) was mediated by low irradiance white light (60 W/m^2) in horizontal direction for 30 min, and was evaluated, relative to the total number of cells, after additional 10 min in darkness. Screening effect by the photostable NR (λ_{max} of vesicle-adsorbed or free cationic NR = 540 nm (Grolig 1986); λ_{max} of free neutral NR = 460 nm) or by RB (λ_{max} = 555 nm), *e.g.* on the phytochrome-mediated low irradiance response in *Mougeotia* as studied here, appears improbable, namely under the light saturating conditions (Haupt 1970). Viability was tested via cell plasmolysis, induced by exchange of MXS-medium against MXS plus 0.6 mol · l^{-1} mannitol (pH 5). Finally, the amount of bound vital dye, relative to chlorophyll (chl), was determined as described by Grolig (1986).

2.4. Individual Cell Experiments

A single *Mougeotia* filament, bridging the gap between two mounted cover glasses, was fixed at the two ends with lanolin; the investigated cells were fully submersed in MXS medium. Before vital staining, the chloroplast was orientated in face on position by white light, irradiating from the microscope lamp (250 W/m^2) through the condenser lens. Partial staining of an individual cell was carried out with microcapillary(ies), inner tip diameter 5 μm, made with a vertical electrode-puller modified to a microforge (Getra, München, Federal Republic of Germany). The microcapillary(ies), filled with NR in aqua dest. (10^{-4} mol · l^{-1}) and connected to micromanipulator(s) (model DC 3, Märzhäuser, Wetzlar, Federal Republic of Germany), were positioned at one or at both flank(s) of the cell unter microscopic control. The drain of dye solution was regulated by hydrostatic pressure via communicating tubes. Staining was sufficient when the vesicles of the treated subcellular area appeared bright red (Grolig and Wagner 1987). Orientation of the chloroplast into "edge on" position (viewing through the microscope) was induced by continuous red light (OG 590, Schott, Mainz, Federal Republic of Germany, 1 W/m^2) through a light guide in horizontal direction (outer diameter 1 mm), ending in distance of about 2 mm next to the cell flank. Chloroplast movement was controlled every 5 min under dim green safe light up to 40 min. Recovery of reorientational capability of the inhibited part of the chloroplast was tested under identical conditions after an intermission of 40 min in white light.

For experiments on the effect of cytochalasin B or colchicine on the vesicular motility, the MXS-medium was exchanged against MXS with cytochalasin B (10^{-4} mol · l^{-1}) or colchicine (10^{-4} or 10^{-3} mol · l^{-1}), both dissolved in 1% (v/v) dimethylsulfoxid (DMSO). Inhibition proper of chloroplast movement by cytochalasin B was used to test the efficiency of the drug (Wagner and Klein 1981).

3. Results and Discussion

Staining of the calcium binding vesicles *in vivo* by the vital dyes neutral red and rhodamine B permits isolation of these vesicles (Grolig and Wagner 1987). After vesicle staining by neutral red, no calcium-indicating fluorescence by chlorotetracycline (Wagner and Rossbacher 1980) could be detected *in vivo*. In addition, X-ray microanalysis of vitally stained isolated vesicles shows them to lack calcium in contrast to the vesicles *in situ* (Fig. 1) or *in vivo* (Wagner and Rossbacher 1980). Presumably due to a dye-induced override of the cytoplasmic calcium homeostasis, staining of these vesicles inhibits chloroplast movement (Grolig 1986). The experiments here specify this supposed mechanism of inhibition in *Mougeotia*.

Upon staining with NR, half maximal inhibition (IC_{50}) of chloroplast reorientational response in continuous low irradiance white light was observed in *Mougeotia* close to 4.7 μmol/mg chl for 5 min staining, and 8,0 μmol/mg chl for sequential staining within 4 h, respectively (Fig. 2). The value for 5 min staining in continuous white light is consistent with the IC_{50}-value of 4 μmol/mg chl for a similar staining condition (Grolig 1986), with the chloroplast reorientational response, however, having been induced by a single pulse of red light. This consistency should exclude photodynamic side effects due to NR; also dye-mediated screening effects appear improbable (see Materials and Methods).

The inhibitory effect of NR on chloroplast response was seen independently whether vesicular staining proceeded rapidly with enhanced external concentration of NR over periods of 5 min or slowly with decreased external concentration of NR over periods of 4 h (Fig. 2); fast staining, however, resulted in enhanced inhibitory effect on the chloroplast response (Fig. 2 *a*). Viability, tested after evaluation of the chloroplast reorientational response by measure of cellular plasmolysis, was also affected by NR and appeared also sensitive to rapid staining (IC_{50} = 8.7 μmol/mg chl for 5 min staining, and 14.4 μmol/mg chl for staining within 4 h, respectively; Fig. 2 *b*).

In contrast, for RB the IC_{50} values were greatly independent of the applied procedure of staining, *i.e.* over

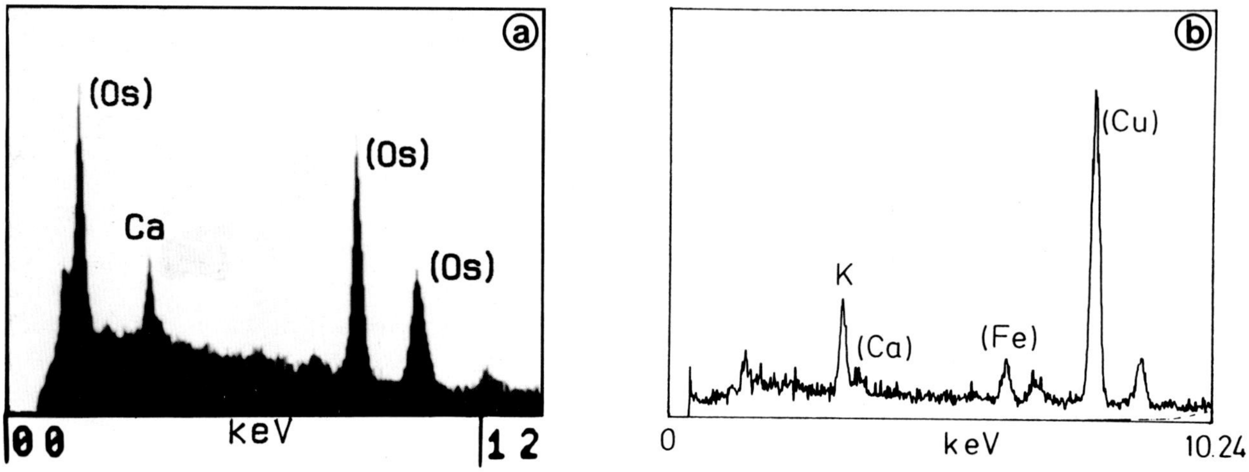

Fig. 1. Spectra of energy dispersive X-ray microanalysis in *Mougeotia*: *a* of a single calcium binding vesicle *in situ*, after freeze substitution and thin sectioning of the cell, *b* of an isolated calcium binding vesicle, after staining *in vivo* by rhodamine B. The vitally stained vesicle lacks calcium, but contains considerable amount of potassium, in contrast to the high amount of vesicular calcium seen *in situ*

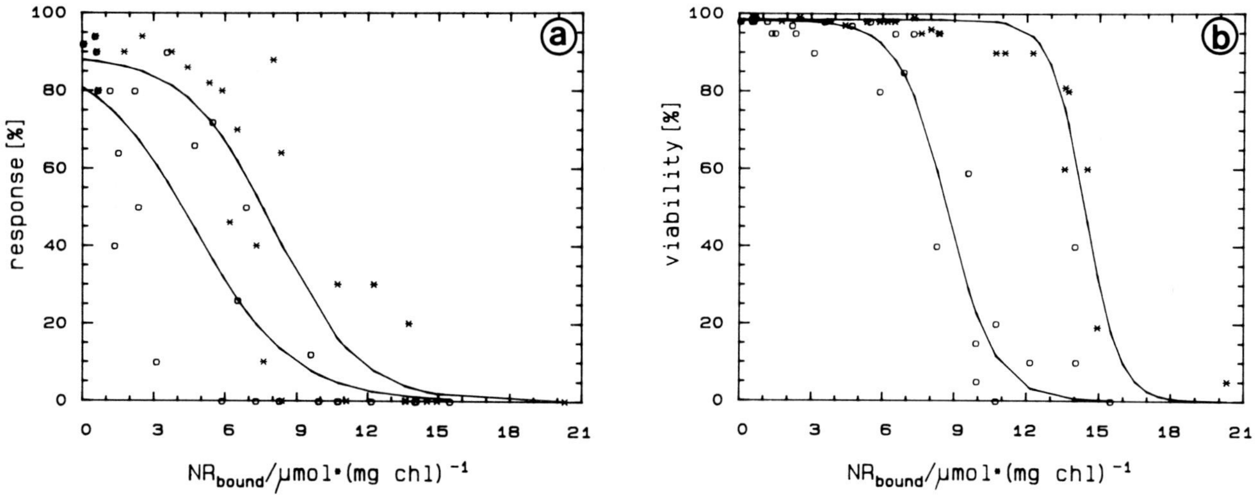

Fig. 2. Light induced chloroplast movement in *Mougeotia* (*a*) and viability (*b*) as influenced by the rate of uptake of neutral red to the *Mougeotia* calcium binding vesicles. Uptake was allowed to proceed for 5 min (○) or for 4 h (∗). Vesicularly bound neutral red (µmol), relative to the cellular content of chlorophyll (mg chl), was determined after monitoring chloroplast response in continuous low irradiance white light, followed by evaluation of cellular viability as tested by plasmolysis in 0.6 mol · l^{-1} mannitol (see Material and Methods)

10 min or 4 h: half maximal inhibition of chloroplast movement in both cases was close to 1.5 µmol/mg chl, with half maximal effect on viability close to 4 µmol/mg chl (Fig. 3, for 10 min staining).

Hence, RB turned out three- to sixfold more effective than NR in inhibiting chloroplast movement in the *Mougeotia* cell with a less than proportional increase in toxicity: for RB, the respective IC_{50}-ratio of cell viability versus chloroplast movement is 2.7, both, for fast and for slow staining; for NR, this ratio is 1.6, but also independent of the staining procedure which produces significantly different states of cell adaptation (see Fig. 2). Thus, dye-specific differences show up here, both in terms of inhibition of chloroplast movement and of toxicity, which may be expected from the difference in binding affinities of the vital dyes to the calcium binding vesicles in *Mougeotia*, with considerably higher affinities for neutral red compared to rhodamine B (Grolig and Wagner 1988 b).

Inhibition of chloroplast movement remained locally restricted upon local application of neutral red, *e.g.* draining from an externally applied microcapillary: the unstained parts of the *Mougeotia* cell performed chloroplast movement as normal (Fig. 4). Simultaneous gathering of the stained vesicles, also observed in the treated area, *e.g.* upon penetration of an unstained *Mougeotia* cell by a micropipette, did not affect chloroplast movement (Russ *et al.* 1988), and turned out unaffected by cytochalasin B up to concentrations of 10^{-4} mol/l^{-1} or colchicine up to 10^{-3} mol · l^{-1} (not

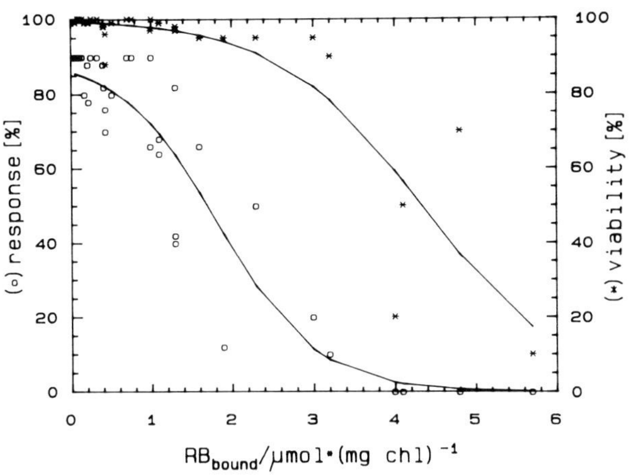

Fig. 3. Light mediated chloroplast movement in *Mougeotia* (○) and viability of the cells (∗), as influenced by rhodamine B, taken up for 10 min. Vesicularly bound rhodamine B, chloroplast response and viability were determined as in Fig. 2

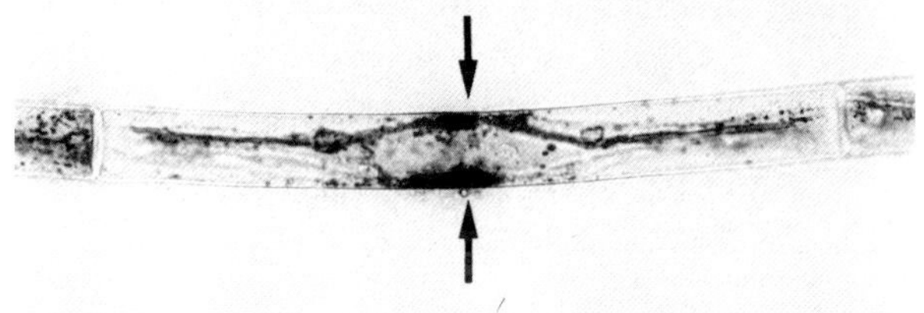

Fig. 4. Microview of an individual *Mougeotia* cell within the filament. The single chloroplast performed reorientational response into "edge on" position, as viewed through the microscope, at the left and right hand side of the cell, while the central part remained in "face on" position due to local application of neutral red at both flanks (see arrows); the result was reproduced more than ten times. Note also the gathering of vesicles where neutral red had been applied. Within an intermission of 40 min in white light (not shown), the inhibited part of the chloroplast recovered full capability for reorientational movement, inspite of the persisting aggregation of the stained vesicles

shown). In spite of the persisting aggregation of the stained vesicles, the inhibited part of the chloroplast recovered full capability for reorientational movement within 40 min in white light. This excludes inhibition of chloroplast movement by NR on its own, but indicates a NR-mediated impediment which can be overcome by the cell with time.

Local inhibition of the chloroplast movement as a consequence of locally restricted staining of the vesicles by NR appears fully consistent with the locally independent rotational behaviour of this organelle in *Mougeotia* (Grolig and Wagner 1988 a). This local inhibition possibly indicates local restriction of the probable, NR-induced effusion of vesicular calcium by binding to structures of hampered diffusion, such as components of the cytoskeleton (see review by Schnepf 1986).

Ackowledgements

We thank Dr. H. Felle for helpful advice and access to his micromanipulatory equipment. With financial support from the "Deutsche Forschungsgemeinschaft" (Wa 265/8-8).

References

Grolig F (1986) Calcium-Vesikel und lichtabhängige Chloroplastenorientierung bei *Mougeotia* spec. Ph.D. Thesis, Univ. Giessen, Federal Republic of Germany

Grolig F, Wagner G (1987) Vital staining permits isolation of calcium vesicles from the green alga *Mougeotia*. Planta 171: 433–437

— — (1988 a) Light-dependent chloroplast reorientation in *Mougeotia* and *Mesotaenium*: biased by pigment-regulated plasmalemma anchorage sites to actin filaments. Botanica Acta 101: 2–6

— — (1988 b) Characterization of the isolated calcium vesicles from the green alga *Mougeotia scalaris* and relevance for chloroplast movement. Planta, submitted

Haupt W (1970) Hellrot- und Dunkelrot-Wechselwirkungen bei der Chloroplastendrehung von *Mougeotia*. Wiss Z Ernst-Moritz-Arndt-Universität (Greifswald) 19: 47–54

— Wagner G (1984) Chloroplast movement. In: Colombetti G, Lenci F (eds) Membranes and sensory transduction. Plenum Press, New York London, pp 331–375

Kamiya N (1939) Die Rhythmik des metabolischen Formwechsels der Euglenen. Ber Deutsch Bot Ges 57: 231–240

— (1986) Cytoplasmic streaming in giant algal cells: a historical survey of experimental approaches. Bot Mag Tokyo 99: 441–467

Marchant HJ (1976) Actin in the green algae *Coleochaete* and *Mougeotia*. Planta 131: 119–120

Neuscheler-Wirth H (1970) Wachstumsgeschwindigkeit und Wachstumsrhythmik bei *Mougeotia*. Z Pflanzenphysiol 63: 352–369

Russ U, Altmüller D, Wagner G (1988) Calcium homeostasis, lithium and actin in *Mougeotia scalaris*. Eur J Cell Biol 46 [Suppl] 22: 58

Schnepf E (1986) Cellular polarity. Ann Rev Plant Physiol 38: 23–47

Schönbohm E (1963) Untersuchungen über die Starklichtbewegung des *Mougeotia*-Chloroplasten. Z Botanik 51: 233–276

— Schönbohm E (1984) Biophenole: Steuernde Faktoren bei der lichtorientierten Chloroplastenbewegung? Biochem Physiol Pflanzen 179: 498–505

Serlin BS, Roux SJ (1984) Modulation of chloroplast movement in the green alga *Mougeotia* by the Ca^{2+}-Ionophore A 23187, and by calmodulin antagonists. Proc Natl Acad Sci USA 81: 6368–6372

Stabenau H (1978) Wachstum von *Mougeotia* in der Durchlüftungskultur. Ber Deutsch Bot Ges 91: 251—255

Stadelmann EJ, Kinzel H (1972) Vital staining of plant cells. In: Prescot DM (ed) Methods in cell physiology, vol 15. Academic Press, New York, pp 325–372

Wagner G, Klein K (1981) Mechanism of chloroplast movement in *Mougeotia*. Protoplasma 109: 169–185

— Rossbacher R (1980) X-ray microanalysis and chlorotetracycline staining of calcium-vesicles in the green alga *Mougeotia*. Planta 149: 298–305

— Valentin P, Dieter P, Marmé D (1984) Identification of calmodulin in the green alga *Mougeotia* and its possible function in the chloroplast reorientational movement. Planta 162: 62–67

Protoplasma (1988) [Suppl 1]: 185–192

A Computerized Video System to Automatically Analyze Movements of Individual Cells and Its Application to the Study of Circadian Rhythms in Phototaxis and Motility in *Chlamydomonas reinhardtii***

T. Kondo[1], M. Kubota[1], Y. Aono[2], and M. Watanabe[1,]*

[1] National Institute for Basic Biology, Okazaki, Aichi, and [2] Yutaka Electronics Corporation, Hashima, Gifu

Received April 18, 1988
Accepted June 24, 1988

Dedicated to Professor Dr. Noburo Kamiya on the occasion of his 75th birthday

Summary

We have created a new computerized video method [the Tracker-Cell Movement Analyzer (TCMA)] for flexible and detailed time-related tracking and analyses of cell movements, especially behavioral responses to light stimuli. TCMA displays the swimming tracks of the individual cells in consecutive time segments of variable length using different colors to represent different time segments; it automatically performs statistical analyses of the swimming directions and velocities of the cells in each time segment. As an example of its many possible applications, an analysis of the circadian rhythms of phototactic orientation and motility in *Chlamydomonas* was performed. It is thus clearly demonstrated that the velocity of motile cells, the phototactic orientation and the percentage of motile cells, changed rhythmically in a circadian manner, all in the same phase, *i.e.* the maximum occurred at the "subjective dawn" and the minimum at the "subjective dusk". These facts can quantitatively account for the observed circadian change in the photoaccumulation activity.

Keywords: Tracker-Cell Movement Analyzer; Computerized video; Phototactic orientation; Motility; Photoaccumulation; Circadian rhythm; *Chlamydomonas reinhardtii*.

Abbreviations: CT circadian time (CT 0 is at the onset of the dim continuous light and 24 h in CT equals the period of the circadian rhythm); DD continuous darkness; LL continuous light; NTSC National Television System Committee; TCMA Tracker-Cell Movement Analyzer system; CMA Cell Movement Analyzer program.

* Correspondence and Reprints: National Institute for Basic Biology, Okazaki, Aichi 444, Japan.

** The Cell Movement Analyzer program developed in this study was written in Turbo Pascal (Boland International Inc., Scotts Valley, CA) with a short routine written in an assembler language for the V 30 (Nippon Electric Co., Ltd., Tokyo) microprocessor used in PC 9801 VM microcomputer (Nippon Electric Co., Ltd.). The source codes for the programs are available upon request. Graphics routines and binary codes must be changed depending on the individual application with respect to hardware used and the structure of the image data.

1. Introduction

In the study of tactic and/or phobic responses, including algal and bacterial photomovements (Häder 1979, Haupt 1986), the recording and analyses of the movements at the individual-cell level are important (Hand and Davenport 1970). For this purpose, photographic (Feinleib and Curry 1967, Watanabe and Furuya 1982), video- (Boscov and Feinleib 1979), computerized-video (Davenport *et al.* 1970) and computerized video-automatic analysis (Takahashi and Kobatake 1982, Häder and Lebert 1985) methods have been devised and used.

We have created a new computerized video-automatic analyses method [the Tracker-Cell Movement Analyzer (TCMA)] that enables a more flexible and detailed time-related tracking and analyses of cell movements, especially behavioral responses to light stimuli. The major unique points of this method are as follows: (1) it can display the swimming tracks of all the individual cells in the microscopic field in consecutive time segments of variable length using different colors to represent different time segments; (2) it can automatically measure the change in the swimming direction of each cell during any time segment as a function of the initial swimming direction; and (3) it can automatically perform statistical analyses of the swimming directions and velocities of the cells in each time segment. The TCMA system provides new possibilities for the experimental analyses of many cellular behavioral responses, including responses to various external stimuli. It is available to visiting users, as well as inside users, as a part of the Okazaki Large Spectrograph facility (WATANABE et al. 1982, WATANABE 1985), which is mainly dedicated to photobiological action spectroscopy.

The photoaccumulation activity of the unicellular alga, *Chlamydomonas*, is known to fluctuate in a diurnal fashion (BRUCE 1970). Since this rhythmicity persists for one week under continuous culture conditions including light conditions, it is controlled by an intracellular circadian clock. The photoaccumulation is a function of both the motility and the phototactic orientation of the cell. The TCMA system developed here was applied to the analysis of photoaccumulation in *Chlamydomonas* in order to determine the contribution of these two activities to the observed rhythmicity of photoaccumulation ability.

The details of the TRACKER-CELL MOVEMENT ANALYZER (TCMA) method and, as an example of its application, the analysis of the circadian rhythms of motility and phototactic orientation in *Chlamydomonas reinhardtii* are reported in this paper.

2. Materials and Methods

2.1. The TRACKER System—a Computerized Video-System to Digitally Record and Display the Swimming Tracks of the Moving Cells in Consecutive Time Segments

The major components of this system (Fig. 1) are: a 1-inch silicon vidicon camera (C 1000-02, Hamamatsu Photonics, Hamamatsu, Shizuoka) which has high sensitivity to both visible and infra-red light; circuits for binarization and interfacing which include a video slicer (M 1005, Hamamatsu Photonics) and an I/O buffer (M 998, Hamamatsu Photonics); an 8-bit microcomputer system (BIWAC DMS-808, Digi-Tek Laboratory Inc., Ootsu, Shiga) that includes a Z80A CPU at 4 MHz, two video processing boards, two 64 kbyte static RAM (= random-access memory) boards, combined with two floppy disk drives for 1-Mbyte floppy disks through a BIWAC Bus line.

Each of the video images produced by the video camera every $^1/_{60}$ sec (NTSC field) is binarized by the video slicer at an appropriate threshold level to make a black and white binarized image composed of 256 × 256 pixels of one bit (8 kbyte). This is performed by the combined action of the I/O buffer and the video slicer. The binarized fields are then summed up by the video processing boards to make a "superfield" in the static RAM boards. By this procedure, the images of each cell on the consecutive summed up video fields make up a swimming track for the time segment between the first and the last video fields. In many cases where not every field is necessary, a certain number a ($0 \leqslant a \leqslant 255$) of fields defined by the user can be thinned out between every addition. Therefore, if the number of field

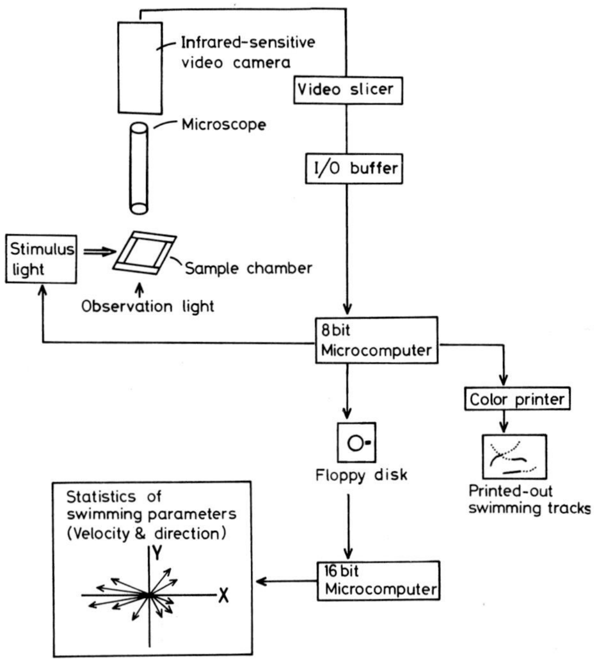

Fig. 1. Block diagram of the TRACKER system

additions per "superfield" is b ($1 \leqslant b \leqslant 65536$) and the number of "superfields" is c ($1 \leqslant c \leqslant 16$), then each of the consecutive "superfields" shows swimming tracks of the cells for a time segment duration of $(^1/_{60}) \times (a+1) \times b$ seconds. Therefore, the total duration for which the swimming track is recorded is equal to c times the above time segment duration. These consecutive "superfields" rapidly created in the static RAM boards (frame buffer) at the real-time speed, are then transferred onto the floppy disk memory, which has a greater capacity than the static RAM boards. These "superfield" data can be read out from the floppy disk memory and printed out by color printers (G 500, Mitsubishi Electric Corp., Tokyo; JP-80, Seiko Epson Corp., Shiojiri, Nagano) with different colors (in any sequences defined by the user of magenta, yellow, cyan, red, green, violet and black) to represent different "superfields" (Fig. 2), or the data on the floppy disk can be automatically analyzed by a 16-bit microcomputer with the CELL MOVEMENT ANALYZER program described below to give a statistical report of the cell movement. In addition, 4 different channels of electric circuits can be independently and automatically switched on or off at the beginning of each time segment represented by the "superfield". This function is used to control the timing and duration of different light stimuli and other signals for experimental procedures and recordings.

2.2. The CELL MOVEMENT ANALYZER Program—a Program that Automatically Analyzes the Data Provided by the TRACKER

First, the data are processed by the thinning algorithm of HILDITCH (1969). This algorithm scans each pixel and deletes any pixel that does not contribute eight-neighbouring connections. The thinning procedure is written in TURBO Pascal (Boland International Inc.) linked with a V 30 (Nippon Electric Co., Ltd.) assembler to address each pixel. It takes 1–2 min to complete the thinning of a "superfield" that has an average density of pixels (ca. 10%). After the thinning procedure is completed, a tracing procedure searches for the tracks which appear in at least one "superfield". The following criteria are used to decide whether or not the candidate track is suitable for further analyses: (1) Both ends of the movement should be continuous with the tracks in the previous and the next frame; this condition means the cell did not go out of focus during the "superfield" period. (2) If a track crosses another track, the direction showing the smaller change in angle is selected. The tracing procedure traces every track in the "superfield" and reads the x- and y-coordinates of the initial and the final points of each track and the length along each track. (3) The selected tracks are then analyzed statistically with a subprogram written in TURBO Pascal. Before the statistical analyses, a numerical "filter" can be placed to select tracks having velocities within a limited range defined by the user. In this study, a range of 20 to 100 $\mu m\,s^{-1}$ was used in order to analyze the directions of movement of vigorously swimming cells. The program performs a statistical analysis of the cell velocities (lengths along the tracks) and the directions of the tracks in each "superfield". The results of these analyses are graphically displayed and/or printed out (Fig. 3). In this study, the number of the tracks in each "superfield" was 20–100. Consequently, the number of the cells used in the statistical analyses of the movements during a 6 s period (6 "superfields") ranged from 120 to 600 (Figs. 4 and 6).

2.3. *Chlamydomonas Cells*

2.3.1. Algal Material

A clone (from Dr. C. Sato, Kagawa Medical School) of a wall-less mutant of *Chlamydomonas reinhardtii*, CW 15 mt+, was used for this study. In our culture conditions for the assay of circadian rhythms, this clone swims more vigorously than the clone (CC 400) of CW 15 mt+ that was directly obtained from the Culture Center at Duke University.

2.3.2. Incubation

Experimental cultures were started by innoculating 10^5 cells into 100 ml of Sager's medium (SAGER and GRANICK 1953) supplemented with 1% sodium acetate. The culture was illuminated continuously with 2 $W\,m^{-2}$ of white light from fluorescent tubes under continuous reciprocal shaking at a speed of 50 strokes per min. To avoid the effect of cell division on the motility, the culture was incubated for 5 days, when the exponential growth phase was over (cell density was 3×10^6 cells ml^{-1}). Then, the culture was subjected to 12 h of darkness to reset the clock. After the dark period, 3 ml aliquots of the culture were pipetted into plastic Petri dishes (Falcon 1008; 42 mm in diameter) and set into a photoaccumulation assay system (described below) under dim continuous light (LL) of 0.6 $W\,m^{-2}$. The time of the onset of a 12 h dark period was taken as CT 12 ("subjective dusk"). The culture temperature was 25 °C.

2.3.3. Photoaccumulation Assay

The 3 ml aliquots of *Chlamydomonas* in Petri dishes were kept under dim continuous light (background light, 0.6 $W\,m^{-2}$) with continuous shaking. At minute 0 of each hour, the background light was turned off and the shaking was stopped. Subsequently, a white light beam from a small tungsten lamp was passed from beneath the bottom of each dish through a heat-absorbing filter (HA 30, Hoya Corporation, Tokyo). A phototransistor was placed just above the lid of each dish to monitor the light transmittance through the cell suspension. The measurement was conducted every 5 s for 11 min (128 measurements) and the difference in transmittance between the initial level and the lowest level was calculated and stored in an Apple II+ computer. After the measurements, the background light was turned on and shaking resumed until the beginning of the next measurement period.

2.3.4. Measurement of Velocity and Phototactic Orientation

At various phases of the photoaccumulation rhythm, a 100 µl aliquot of the cell suspension was transferred into a chamber (300 µm in depth) on a microscopic slide glass. STRALEY and BRUCE (1976) reported that stickiness of the *Chlamydomonas* cells to the glass changed in a circadian fashion. To avoid such sticking of the cells, the glassware (the chamber and pipettes) was soaked in a conditioned medium (used medium without cells; KAMIYA R, personal communication) before use, and the measurement was performed within 1 min after putting the cell suspension into the chamber. As the field of focus was the central portion of the chamber, at a halfway from the top, the effects of the cells that were stuck on the surface of the glass chamber could be minimized by these means. The movements of the cells under a microscope (BHC, Olympus Optical Co., Ltd.) were recorded by the TRACKER system using a 20 × objective lens and a 2.5 × camer lens. The illuminating light used for monitoring the cell movement was filtered by a band-pass filter transmitting light of 660–700 nm which did not affect the phototactic movement (data not shown). A stabilized DC (direct current) power supply was used for illumination. During the first 7 s, movement in the absence of the stimulus light was recorded. Subsequently, the cells were irradiated horizontally with a white stimulus light (18 $W\,m^{-2}$ as measured with a UDT radiometric sensor; United Detector Technology Inc., Santa Monica, CA) for 8 s during which time the orientation of the cell toward the light was recorded. The light was provided by a microscopic illuminator (Model LSD, Olympus Optical Co., Ltd.) equipped with a tungsten lamp (6 V 30 W) operated at 6 V. The light was filtered through a heat-absorbing filter (HA 50, Hoya Corporation).

3. Results

3.1. *Movements of the Cells in the Absence or Presence of the Stimulus Light*

Figure 2 shows the swimming tracks of the *Chlamydomonas* cells as recorded and displayed by the TCMA system. In the absence of the stimulus light (Fig. 2 *a*), the directions of the movements are random. By contrast, in the presence of the stimulus light (Fig. 2 *c*), it is evident that most of the cells swim toward the stimulus light. The cells changed their swimming direction with a time lag of approximately 1–2 s (data not shown) after the onset of the stimulus light. Figure 2 *b* and *d* shows the tracks after the thinning procedure. The tracks have been converted to the lines of unit thickness (one pixel) without altering any other features.

Examples of the data analyses performed by the CELL MOVEMENT ANALYZER program are shown in Fig. 3. The random movements in the absence of the stimulus light and the increasingly oriented movements in the presence of the stimulus light are presented as vector diagrams (Fig. 3 *a–d*). These vector diagrams present the complex data in an easily comprehensible manner.

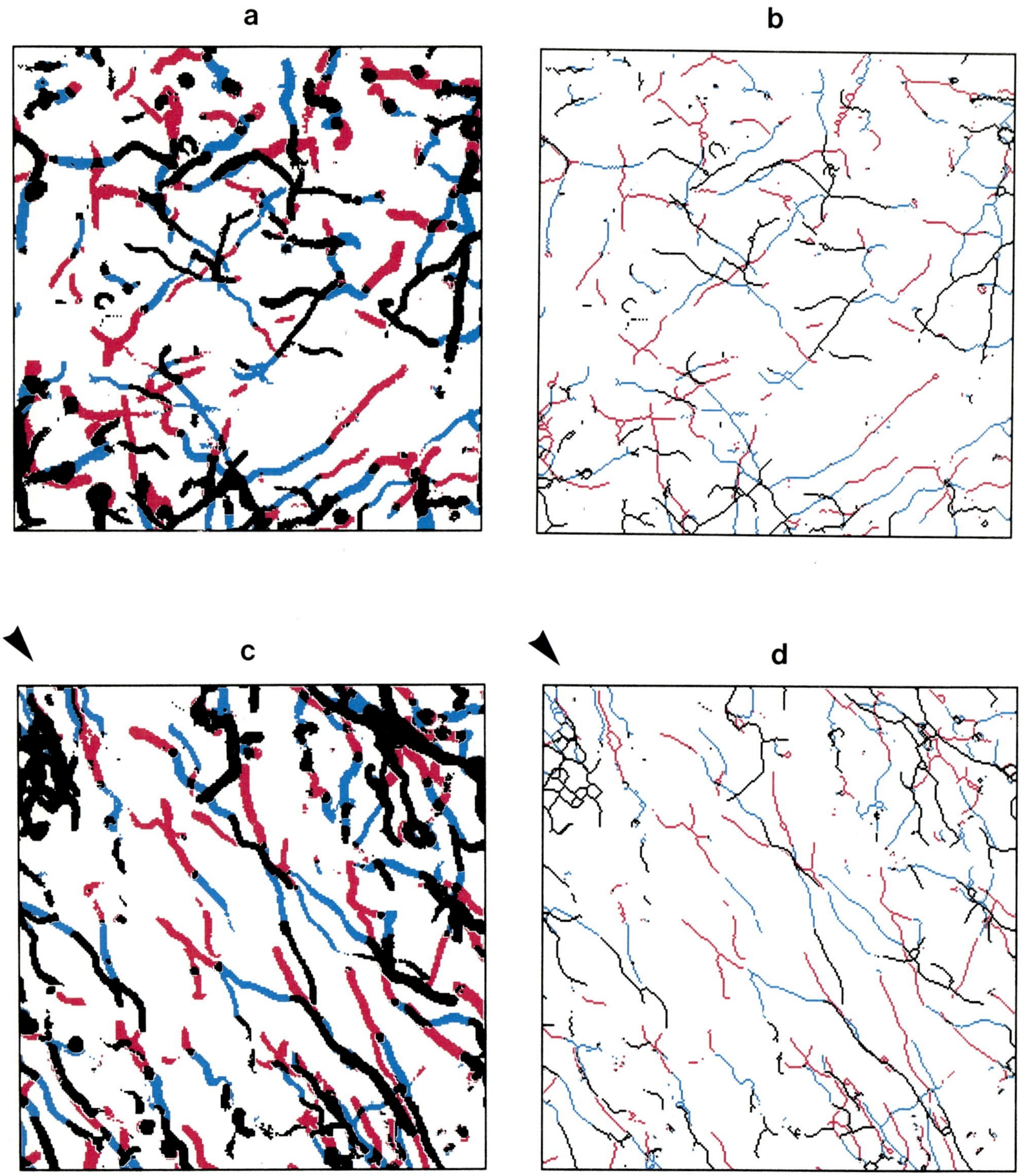

Fig. 2. TRACKER charts showing the cell movements in the presence and absence of the stimulus light. Recorded tracks of the cell movements of *Chlamydomonas* are shown. Each plane is a superimposed sum of three consecutive "superfields" of 1 s duration (in the order of black, cyan and magenta). *a* In the absence of the stimulus light; *c* in the presence of the stimulus light (18 W m^{-2}), the first "superfield" (black) is for the 5th s after the onset of the light. *b* and *d* After the thinning procedure, of *a* and *c*, respectively. The large arrowhead shows the incident direction of the stimulus light. The width and depth of the frame of each panel (*a–d*) correspond to 560 μm in the microscopic field

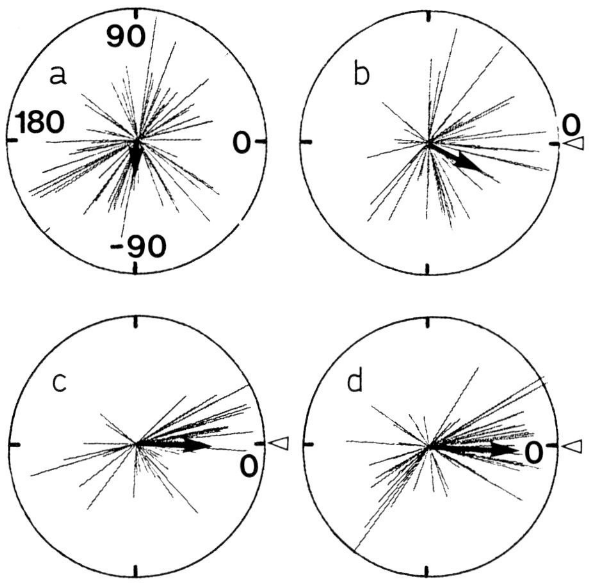

Fig. 3. Vector diagrams of cell movement as analyzed and presented by the CELL MOVEMENT ANALYZER. The movements of the *Chlamydomonas* cells at the high photoaccumulation phase are shown. The vectors of all the tracks in two or three consecutive 1 s "superfields" are shown. The open arrowhead indicates the direction of the stimulus light, *a* Before the onset of the stimulus light, *b* 0–3 s, *c* 3–5 s, and *d* 5–8 s after the onset of the stimulus light. The mean vector is shown in each vector diagram by an arrow (the length of which is enlarged twice to facilitate viewing). The radius of the enclosing circle is 60 μm s^{-1}

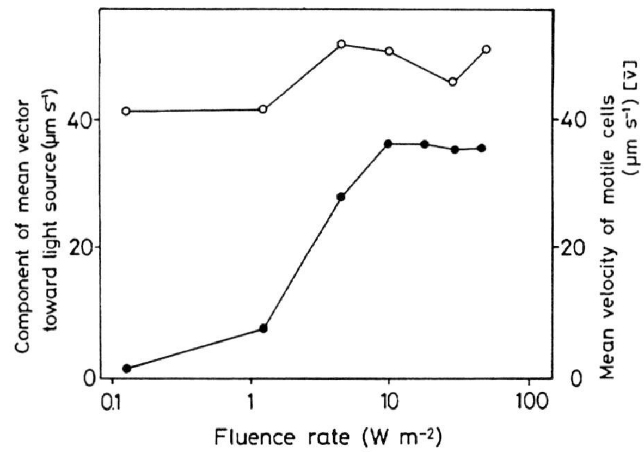

Fig. 4. Fluence rate-response relationship of phototactic orientation and swimming velocity. The component of the mean vector toward the stimulus light direction (●-●) and the mean velocity of motile cells (○-○) during a 6 s period after the onset of stimulus light are plotted against the fluence rate of the stimulus light

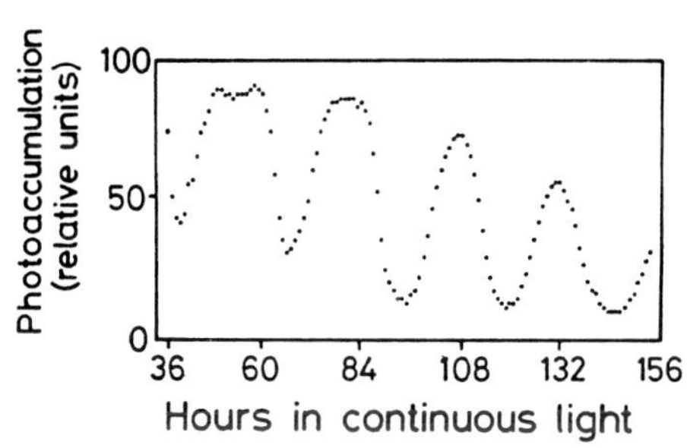

Fig. 5. Circadian rhythm of photoaccumulation ability. A dark 12 h pulse was given to the *Chlamydomonas* culture in Sager's medium. After the dark pulse, the photoaccumulation was examined in the first 10 min of every hour and plotted against the time in dim continuous light. The photoaccumulation is shown as the decrease in the transmittance of the stimulus and monitoring light through the cell suspension

In addition, the results of the statistical analyses, namely the mean vectors (Fig. 3 *a–d*, arrows) provide quantitative indices of the orientation of the cells.

3.2. Phototactic Orientation Toward the Stimulus Light of Various Fluence Rates

The fluence-rate relationship of the phototactic orientation of the cells is shown in Fig. 4. The component of the mean vectors toward the stimulus light is plotted as a function of the fluence rate of the stimulus light. In the presence of fluence rates of 4 W m^{-2} and greater, the direction of the movement focus to the direction of the stimulus light as shown by an increase in the component of the mean vector toward the stimulus light. As the response of the cells saturated at about 10 W m^{-2}, the fluence rate of 18 W m^{-2} was chosen for the subsequent examination of phototactic orientation. In addition, the mean velocity of the motile cells is somewhat increased by elevating the fluence rate of the stimulus light at 4 W m^{-2} and greater.

3.3. Circadian Rhythm in Photoaccumulation Activity

Figure 5 shows the changes in the photoaccumulation activity of *Chlamydomonas* cultures under the dim LL (continuous light). The photoaccumulation activity oscillated for five days or longer under dim LL with a period of 26 h. The maximum photoaccumulation was observed at the "subjective dawn" (hours 53, 79, 105 ... of the dim LL) and the minimum at the "subjective dusk" (hours 40, 66, 92 ... of the dim LL).

3.4. Circadian Change in Velocity and Phototactic Orientation

The velocity and the phototactic orientation were measured at various times during day 3 and 4 of the dim LL. The mean velocity (Fig. 6 *b*) changed in parallel

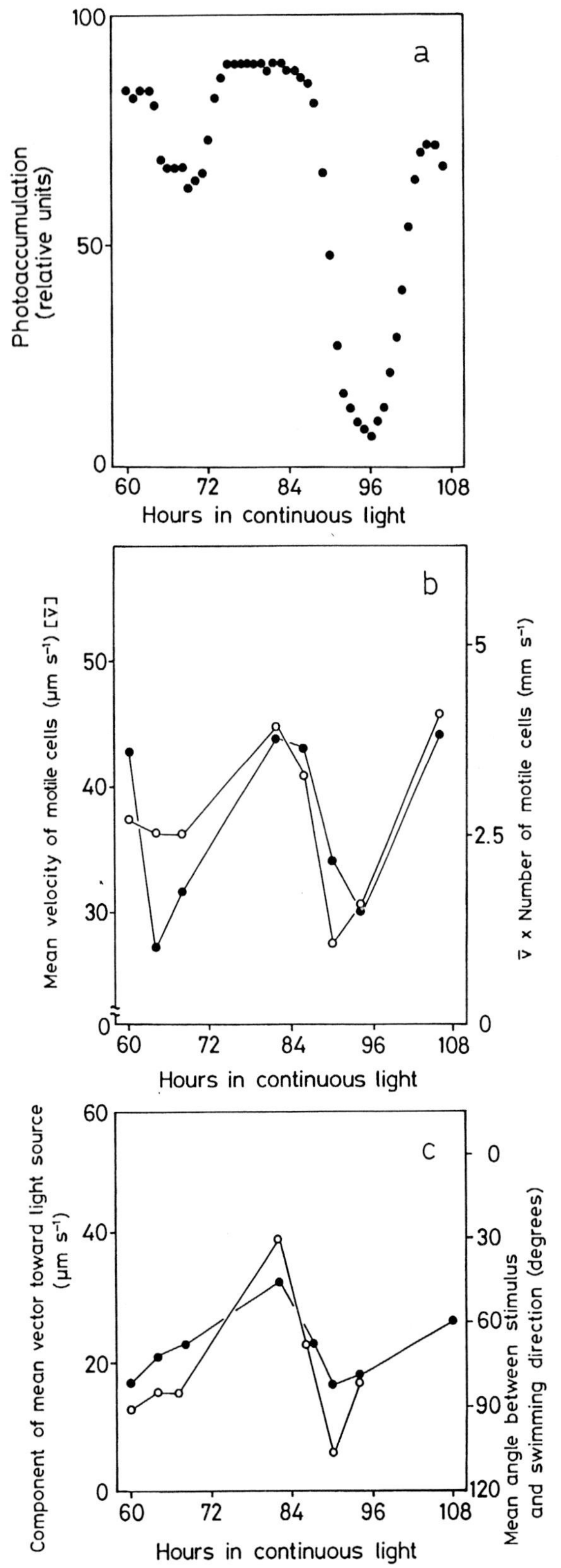

Fig. 6. Circadian changes in the photoaccumulation ability, motility and phototactic orientation. During hours 60 to 108 of the continuous dim light (LL), a batch of cells showing a circadian rhythm in photoaccumulation was intermittently sampled and examined for motility and phototactic orientation using the Tracker-Cell Movement Analyzer system. *a* Photoaccumulation plotted in the similar fashion as in Fig. 5. *b* Mean velocity of the motile cells (○-○) and the total motility (mean velocity times the number of the motile cells) (●-●) during a 6 s period as recorded and analysed using the TCMA system are plotted against time in the dim LL. *c* The component of the mean vector of the cells swimming at 20–100 μm s^{-1}, toward the stimulus light direction (○-○). The cell paths for a 6 s period were analysed by the TCMA system and plotted against time in dim continuous light. The mean of the angles (in absolute value) between the stimulus light and the swimming directions of the individual cells (●-●) are also plotted. Data shown are typical examples from three equivalent experiments

with the photoaccumulation rhythm (Fig. 6 *a*); the maximum was observed at the "subjective dawn" and the minimum at the "subjective dusk". However, the span of the change in velocity was only 20% of the maximum velocity while that of the photoaccumulation was 90%. Taking into account that the number of swimming cells is smaller in the low photoaccumulation phase (immotile cells settled down at the bottom of the chamber), the total motility of the cultures was estimated by multiplying the mean velocity by the number of motile cells. These value of the total motility changed with a span of 60–70% of the maximum value (Fig. 6 *b*). Figure 6 *c* shows that the gross orientations of the cell population (the components of the mean vectors toward the stimulus light) also changed in parallel with the photoaccumulation rhtythm. It should be noted that this change in the gross orientation of the cell population is not only due to the change in total motility as shown in Fig. 6 *b* but also due to the change in the ability of the cells to change the swimming directions, because the mean angles between the stimulus light and the swimming directions (in absolute values) also change in phase with the photoaccumulation rhythm (50 and 90 degrees at the maximum and minimum accumulation phase, respectively). Zero, 90, and 180 degress of these indices respectively mean that (1) all the cells swim directly toward the stimulus light; (2) the cells swim at random or normal to the direction at the light; and (3) the cells swim directly away from the stimulus light. In addition, this change in the mean angle was found in the group swimming at low velocities as well as in the group swimming at high velocities.

4. Discussion

4.1. Methodology

4.1.1. Advantages of the Tracker-Cell Movement Analyzer System

The time resolution, as expressed by the minimum time segment duration of each "superfield", of the Tracker system is 16.7 ms (= $^{1}/_{60}$ s, see Methods section). The unique display method using different colors to represent consecutive superfields greatly facilitates time-related and flexible analyses, especially of complex and unpredictable behavior as well as simple ones.

The results of the statistical analysis of the movement data (Figs. 3, 4, and 6) indicate that the CELL MOVEMENT ANALYSER program developed for this study succeeds in extracting the biological responses of the cell to the stimulus light. Seeing that it is very time-consuming to manually perform statistical analyses of the movements of even 50 cells for example, the use of this system for automatic analyses is advantageous. The CELL MOVEMENT ANALYZER program greatly facilitates quick, accurate and flexible analyses of a large number of movement data needed for physiological studies including action spectroscopy. We are continuing the improvement and enrichment of the automatic analysis program, which will greatly contribute to the analysis of various types of cell behaviors.

4.1.2. Further Applications of the TRACKER-CELL MOVEMENT ANALYZER System

TCMA can and will be used for time-related analyses, including action spectroscopy, of step-up and step-down photophobic responses as well as phototactic orientation in many kinds of algal flagellates, in which the involvement of different photoreceptor systems has been suggested (HÄDER 1979, FOSTER *et al.* 1984, HAUPT 1986). TCMA is also suited for the analysis of protoplasmic streaming in giant algal cells (KAMIYA 1986). Such analyses at the individual cell level will provide important bases for further studies at the subcellular level, either biochemical or organellar.

4.1.3. The Choice of Cell Density and Magnification Affects the Thinning and Tracing Efficiencies

The thinning and tracing subprograms select for analysis the tracks of the cell movement on the basis of two conditions as described in the Methods section. In the case of the *Chlamydomonas* experiments described here, the subprograms with these criteria select more than 90% of the tracks in each "superfield" if the number of tracks is less than 100 per "superfield". When the track density was higher than 100, the probability of a correct selection of the tracks decreased, because the thinning arlgorithm failed to separate overlapping tracks or yielded abnormal branching patterns (PAVLIDIS 1982). This limiting cell density is also influenced by the size of the cell in the microscopic field and the motility, both of which determine the length of the tracks. Therefore, the cell density and the magnification of the microscope should be chosen carefully. The magnification should be chosen so as to obtain optimum lengths of the tracks which are 10–50 pixels. When the data includes 16 "superfields" in which 100 cells of *Chlamydomonas* are swimming, the subprograms complete the processing in 40 min using a NEC 9801 VM microcomputer. The program is flexible and the conditions used in the tracing subprogram may be replaced by other conditions depending on the requirement of the individual project.

4.1.4. Focusing of the Microscope and the Depth of Field

The aperture size of the condensor lens and the level of the threshold of binarization greatly influence the binarized images of the cells. Therefore, these parameters and the depth of the chamber on the slide glass should be carefully chosen depending on the nature of the sample organism and the purpose of the experiments. In this study, these parameters were fixed for a series of observations: The condensor aperture was selected to give a depth of field of approximately 100 µm. The objective lens was focused at a depth one halfway from the top of the chamber. Consequently, the density of the cells appearing in the "superfield" was influenced by the percentage of immotile cells although the total number of cells put into the sample chamber was constant.

4.2. Circadian Rhythm

4.2.1. Circadian Clock Controls Both Motility and Phototactic Orientation

Theoretically, the change in the photoaccumulation ability of the cells can result from the changes in motility, phototactic orientation and photophobic responses. The studies in this paper clearly demonstrate that both the motility and phototactic orientation change in parallel with the photoaccumulation ability rhythm, indicating that the circadian clock controls both activities. Recently, KAMIYA and HASEGAWA (1987) developed a method to measure the beat frequency of the flagella of *Chlamydomonas* by the fast Fourier analysis of the fluctuation in the light fluence in microscope images of swimming cells. Examination of the beat frequency would be of interest to know a more precise mechanism of the circadian change in motility.

4.2.2. Further Analyis on the Linkage Between the Pacemaker and the Overt Rhythm

The experiments in this study show that both the motility and the phototactic orientation changed in a circadian fashion. The change in the former activity has a larger contribution than the latter to the photoaccumulation rhythm. But the change in the orientation is not a secondary effect of the change in the motility, since the change in orientation was observed in both the vigorously swimming and sluggish cells at all the phases of the rhythm. In the dinoflagellate, *Gonyaulax*, McMurry and Hastings (1972) demonstrated that four activities of the cell (flash- and glow-bioluminescences, cell division and photosynthesis) are independently controlled by a single circadian clock. That is, these four rhythms have the same period and inhibition of any of four rhythms did not affect the other rhythms. Such a relation was also observed in various ion uptake rhythms in *Lemna* (Kondo 1982). Similarly, the motility and phototactic orientation in *Chlamydomonas* might well be controlled independently by a single clock. Further analysis of the two parameters under various conditions are required to know whether or not a linkage of the two rhythms exists, which is an essential question for the understanding of the structure of the control pathways from the circadian clock to the various overt rhythms.

Acknowledgements

Part of this study was carried out under the Cooperative Research Program for the use of the Okazaki Large Spectrograph (Project Nos. 86-529, 87-524 and 87-530) of the National Institute for Basic Biology, Okazaki, Japan. We thank Mr. Shigeru Matsumura (Digi-Tek Laboratory Inc.) and Mr. Jun-Ichi Hattori (Yutaka Electronics Corporation) for their skilled technical contributions. We thank Dr. Randy Wayne (Cornell University) for carefully reading the manuscript, and Mrs. Junko Watanabe for assistance in preparing the manuscript.

References

Boscov JS, Feinleib ME (1979) Phototactic response of *Chlamydomonas* to flashes of light—II. Response of individual cells. Photochem Photobiol 30: 499–505

Bruce VG (1970) The biological clock in *Chlamydomonas reinhardtii*. J Protozool 17: 328–334

Davenport D, Culler GJ, Greaves JOB, Forward RB, Hand WG (1970) The investigation of the behavior of microorganisms by computerized television. IEEE Trans Biomed Engineer 17: 230–237

Feinleib MEH, Curry GM (1967) Methods for measuring phototaxis of cell populations and individual cells. Physiol Plant 20: 1083–1095

Foster KW, Saranik J, Patel N, Zarilli G, Okabe M, Kline T, Nakanishi K (1984) A rhodopsin is the functional photoreceptor for phototaxis in the unicellular eukaryote *Chlamydomonas*. Nature 311: 756–759

Häder D-P (1979) Photomovement. In: Haupt W, Feinleib ME (eds) Encyclopedia of plant physiology, new series, vol 7. Springer, Berlin Heidelberg New York, pp 268–309

— Lebert M (1985) Real time computer-controlled tracking of motile microorganisms. Photochem Photobiol 42: 509–514

Hand WG, Davenport D (1970) The experimental analysis of phototaxis and photokinesis in flagellates. In: Halldal P (ed) Photobiology of microorganisms. Wiley, London, pp 253–282

Haupt W (1986) Photomovement. In: Kendrick RE, Kronenberg GHM (eds) Photomorphogenesis in plants. Martinus Nijhoff, Dordrecht, pp 415–441

Hilditch CJ (1969) Linear skeletons from square cupboards. Machine Intelligence 4: 403–420

Kamiya N (1986) Cytoplasmic streaming in giant algal cells: a historical survey of experimental approaches. Bot Mag (Tokyo) 99: 441–467

Kamiya R, Hasegawa E (1987) Intrinsic difference in beat frequency between two flagella of *Chlamydomonas reinhardtii*. Exp Cell Res 173: 299–304

Kondo T (1982) Correlation between potassium uptake rhythm and nitrate uptake rhythm in *Lemna gibba* G 3. Plant Cell Physiol 23: 909–915

McMurry L, Hastings JW (1972) No desynchronization among four circadian rhythms in the unicellular alga, *Gonyaulax polyedra*. Science 175: 1137–1139

Pavlidis T (1982) Algorithms for graphics and image processing. Computer Science Press, Rockville, pp 195–214

Pfau J, Nultsch W, Ruffer U (1983) A fully automated and computerized system for simultaneous measurements of motility and phototaxis in *Chlamydomonas*. Arch Microbiol 135: 259–264

Sager R, Granick S (1953) Nutritional studies with *Chlamydomonas reinhardtii*. Ann NY Acad Sci 56: 831–838

Straley SC, Bruce VG (1976) Stickiness to glass. Circadian changes in the cell surface of *Chlamydomonas reinhardtii*. Plant Physiol 63: 1175–1181

Takahashi T, Kobatake Y (1982) Computer-linked automated method for measurement of the reversal frequency in phototaxis of *Halobacterium halobium*. Cell Struct Funct 7: 183–192

Watanabe M, Furuya M (1982) Phototactic behavior of individual cells of *Cryptomonas* sp. in response to continuous and intermittent light stimuli. Photochem Photobiol 35: 559–563

— Furuya M, Miyoshi Y, Inoue Y, Iwahashi I, Matsumoto K (1982) Design and performance of the Okazaki Large Spectrograph for photobiological research. Photochem Photobiol 36: 491–498

Watanabe M (1985) The Okazaki Large Spectrograph and its application to action spectroscopy. In: Longworth JW, Jagger J, Shropshire W Jr (eds) Photobiology 1984. Praeger, New York, pp 37–44

Index of Keywords